T0135273

Lecture Notes in Electrical Engineering

Volume 464

** Indexing: The books of this series are submitted to ISI Proceedings, EI-Compendex, SCOPUS, MetaPress, Springerlink **

Lecture Notes in Electrical Engineering (LNEE) is a book series which reports the latest research and developments in Electrical Engineering, namely:

- Communication, Networks, and Information Theory
- Computer Engineering
- Signal, Image, Speech and Information Processing
- Circuits and Systems
- Bioengineering
- Engineering

The audience for the books in LNEE consists of advanced level students, researchers, and industry professionals working at the forefront of their fields. Much like Springer's other Lecture Notes series, LNEE will be distributed through Springer's print and electronic publishing channels.

For general information about this series, comments or suggestions, please use the contact address under "service for this series".

To submit a proposal or request further information, please contact the appropriate Springer Publishing Editors:

Asia:

China, *Jessie Guo, Assistant Editor* (jessie.guo@springer.com) (Engineering)

India, *Swati Meherishi, Senior Editor* (swati.meherishi@springer.com) (Engineering)

Japan, *Takeyuki Yonezawa, Editorial Director* (takeyuki.yonezawa@springer.com) (Physical Sciences & Engineering)

South Korea, *Smith (Ahram) Chae, Associate Editor* (smith.chae@springer.com) (Physical Sciences & Engineering)

Southeast Asia, *Ramesh Premnath, Editor* (ramesh.premnath@springer.com) (Electrical Engineering)

South Asia, *Aninda Bose, Editor* (aninda.bose@springer.com) (Electrical Engineering)

Europe:

Leontina Di Cecco, Editor (Leontina.dicecco@springer.com)
(Applied Sciences and Engineering; Bio-Inspired Robotics, Medical Robotics, Bioengineering; Computational Methods & Models in Science, Medicine and Technology; Soft Computing; Philosophy of Modern Science and Technologies; Mechanical Engineering; Ocean and Naval Engineering; Water Management & Technology)

(christoph.baumann@springer.com)
(Heat and Mass Transfer, Signal Processing and Telecommunications, and Solid and Fluid Mechanics, and Engineering Materials)

North America:

Michael Luby, Editor (michael.luby@springer.com) (Mechanics; Materials)

More information about this series at http://www.springer.com/series/7818

Jason C. Hung · Neil Y. Yen
Lin Hui
Editors

Frontier Computing

Theory, Technologies and Applications
(FC 2017)

 Springer

Editors
Jason C. Hung
Department of Information Technology
Overseas Chinese University
Taichung
Taiwan

Lin Hui
Department of Innovative Information
  and Technology
Tamkang University
Yilan County
Taiwan

Neil Y. Yen
School of Computer Science and
  Engineering
University of Aizu
Aizu-Wakamatsu
Japan

ISSN 1876-1100   ISSN 1876-1119   (electronic)
Lecture Notes in Electrical Engineering
ISBN 978-981-13-3947-9   ISBN 978-981-10-7398-4   (eBook)
https://doi.org/10.1007/978-981-10-7398-4

Printed on acid-free paper

This Springer imprint is published by the registered company Springer Nature Singapore Pte Ltd.
part of Springer Nature
The registered company address is: 152 Beach Road, #21-01/04 Gateway East, Singapore 189721,
Singapore

The original version of the book was revised:
Second editor affiliation has been included.
The erratum to the book is available at
https://doi.org/10.1007/978-981-10-7398-4_51

Contents

Removal of Impulse Noise Using Gain Factors Adapted by Noise-Free Pixel Number and Pixel Variation

Ching-Ta Lu[1,2], Jun-Hong Shen[1,2(✉)], Mu-Yen Chen[3],
Ling-Ling Wang[1], and Chih-Chan Hsu[1]

[1] Department of Information Communication, Asia University,
Taichung 413, Taiwan
lucasl@ms26.hinet.net, {shenjh,ling}@asia.edu.tw,
criss.tiger1991@gmail.com
[2] Department of Medical Research, China Medical University Hospital,
China Medical University, Taichung 404, Taiwan
[3] Department of Information Management, National Taichung University
of Science and Technology, Taichung 404, Taiwan
mychen@gm.nutc.edu.tw

Abstract. Impulse noise impacts an image, causing the quality of image to be deteriorated in image transmission or capture. In this paper, we propose a gain factor for the removal of the impulse noise. A 3×3 fixed-size local window is employed to analyze each extreme pixel (0 or 255 for an 8-bit gray-level image). All non-extreme pixels are sorted in an ascending order and are grouped according to the variation of pixel levels. If the pixel level between adjacent two sorted pixels varies seriously, a new group is created. Hence, the ratio and median value of each group are computed to determine the values of the gain factors. They are multiplied with the median value of each group to obtain the weighted value which is employed to replace the center pixel with an extreme value, enabling noise-corrupted pixels to be restored. Experimental results show that the proposed method can effectively remove salt-and-pepper noise from a corrupted image for various noise corruption densities (from 10% to 90%); meanwhile, the denoised image is freed from the blurred effect.

Keywords: Image denoising · Salt-and-Pepper noise · Gain factor
Majority and variation adaptation

1 Introduction

An image would be interfered by impulse noise. This interference noise may be caused by malfunctioning pixels in camera sensors or transmission in a noisy channel, bit errors in transmission, and fault memory locations in hardware et al. Salt-and-pepper noise is a major type of impulse noise which can seriously impact an image. This noise corrupts pixels, causing the gray levels of the interfered pixels to be either the minimum or maximum values. This noise deteriorates the quality of an image significantly.

© Springer Nature Singapore Pte Ltd. 2018
J. C. Hung et al. (Eds.): FC 2017, LNEE 464, pp. 1–11, 2018.
https://doi.org/10.1007/978-981-10-7398-4_1

How to remove the salt-and-pepper noise effectively is an important research task for image processing.

Recently, many novel methods have been proposed for the reconstruction of the images corrupted by salt-and-pepper noise [1–15]. An adaptive median filter proposed by Hwang and Hadded [1] selects the median value in an adaptive window for each pixel, enabling the noise-corrupted pixels to be removed. This method performs well at low noise densities. Wang et al. [2] proposed an adaptive fuzzy-switching-weighted-mean filter to cope with salt-and-peeper noise. They computed the maximum absolute luminance difference of processed pixels next to possible noise pixels to classify them into three categories: uncorrupted pixels, lightly corrupted pixels, and heavily corrupted pixels. The gray level of a lightly corrupted pixel is replaced by the weighted average value of the weighted mean and its own value. A heavily corrupted pixel is modified by the weighted mean. In [3], a modified decision based un-symmetric trimmed median filter was proposed. This method replaces noise-corrupted pixels by trimmed median value when other pixel values, 0's and 255's are present in the selected window and when all the pixel values are 0's and 255's then the noise pixel is replaced by mean value of all the elements present in the selected window. Deivalakshmi and Palanisamy [4] proposed a tolerance based adaptive masking selective arithmetic mean filter which is combined with wavelet thresholding. This method can cope with a heavily noise-corrupted image. Lu et al. [5] proposed a three-values-weighted approach for image denoising. Non-extreme pixels in an analysis window are classified and placed into the minimum, middle, or maximum group. The distribution ratios of the three groups were utilized to weight the neighbor noise-free pixels, enabling a noise-interfered pixel to be reconstructed. An iterative adaptive fuzzy filter using an alpha-trimmed mean was proposed by Ahmed and Das [6], which employed an adaptive, iterative, fuzzy filter for denoising images corrupted by the impulse noise. The detection of the noise-corrupted pixels had an adaptive fuzzy detector followed by denoising process, using a weighted mean filter on the noise-free neighboring pixels. Experimental results showed that this method could operate in heavy noise densities. The proposed algorithm replaces the noise-corrupted pixel by the trimmed median value when other pixel values, 0's and 255's are present in the selected window and when all the pixel values are 0's and 255's then the noise-corrupted pixel is replaced by the mean value of all the elements present in the selected window. In [7], a directional-weighted-median (DWM) filter was proposed to remove random-valued noise and salt-and-pepper noise. This method selects the variation direction of neighbor pixels in four directions. The noise-corrupted center pixel was modified by the weighted median on the selected direction. In [8], a modified DWM (MDWM) filter was proposed. This method utilized twelve directions to find a better pixel variation direction for noise removal. Experimental results showed that this method can improve the performance of the DWM filter [7] significantly.

Based on the above findings, how to develop an effective approach for the removal of salt-and-pepper noise is an important research task, in particular at heavy-noise corruption conditions (the noise density exceeding 70%). In this paper, we propose a novel approach to remove salt-and-pepper noise by using a gain factor. In order to prevent blur effect in the denoised image, a small size window (window size equaling 3 × 3) is employed to analyze each extreme pixel (0 or 255 for an 8-bit gray-level image). Initially, non-extreme pixels are utilized to roughly estimate the noise density which is employed to determine the type of gain factor. This gain factor is adapted by the distribution ratio of noise-free pixels and the pixel-variation level of a group. In the conditions of slight noise density (noise density is less than 20%), the number of noise-free pixels is sufficient. A multiplication-type gain factor is utilized. Conversely, a weighting-type gain factor is used for the middle and high levels of noise densities (noise density is greater than 20%). In a local window, all non-extreme pixels are sorted in an ascending order and are grouped according to the pixel-variation level. If the pixel level between adjacent two sorted pixels varies seriously, a new group is created. Hence, the ratio of noise-free pixels and the median value of each group are computed for the determination of the gain factors. The gain factor are multiplied with the median values of the classified groups to obtain the weighted value which is employed to replace the center pixel with an extreme value, enabling noise-corrupted pixels to be restored. Experimental results show that the proposed method can effectively remove salt-and-pepper noise from a corrupted image for different noise corruption densities (from 10% to 90%); meanwhile, the denoised image is free from the blurred effect.

The rest of this paper is organized as follows. Section 2 introduces the proposed gain factors for the removal of salt-and-pepper noise. Section 3 demonstrates the experimental results. Conclusions are finally drawn in Sect. 4.

2 Proposed Gain Factor

A local window that slides from the left to right and from the top to bottom in an image is utilized to analyze the neighbor properties for each extreme pixel. Non-extreme pixels in the local window are employed to restore the noise-corrupted center pixel. The local window $W(i,j)$ can be expressed as

$$W(i,j) = \{X(i + \Delta i, j + \Delta j)| \text{ where } \Delta i, \Delta i \in (-1 \sim 1)\} \qquad (1)$$

where $X(i, j)$ denotes the input pixel at the location i^{th} row and j^{th} column of the local window. The window size is equal to 3 × 3.

This gain factor is determine according to the distribution ratio of noise-free pixels and the pixel-variation level of a group. In the conditions of slight noise density (noise density is less than 20%), A multiplication-type gain factor is utilized. Conversely, a weighting-type gain factor is used for the conditions of noise densities being greater than 20%. The number of the non-extreme pixels is employed to determine the weighting factor. A non-extreme flag $F^{non-extreme}(i,j)$ is determined to denote whether a pixel is non-extreme, given as

$$F^{non-extreme}(i,j) = \begin{cases} 1, & \text{if } X(i,j) \in \text{non - extreme pixels} \\ 0, & \text{otherwise} \end{cases} \qquad (2)$$

If the value of the pixel $X(i, j)$ is non-extreme, then the non-extreme flag $F^{non-extreme}(i,j)$ is set to unity. This represents that the pixel is noise-free. It will be employed to adapt the gain factor for the modification of the center pixel with an extreme value. Conversely, the non-extreme flag is set to zero when the pixel value of $X(i, j)$ is extreme. This represents the pixel being regarded as a noise candidate. It will not be employed for the reconstruction of a noise-corrupted pixel. In addition, we can employ $F^{non-extreme}(i,j)$ to roughly estimate the noise density $\hat{\eta}$ in a local window, given as

$$\hat{\eta} = 1 - \frac{\sum_{i=0}^{M-1} \sum_{j=0}^{N-1} F^{non-extreme}(i,j)}{M \cdot N} \tag{3}$$

The estimate of noise density $\hat{\eta}$ in (3) is employed to determine the computation method of the gain factor $G_k(i,j)$, given as

$$G_k(i,j) = \begin{cases} \dfrac{p_k(i,j) \cdot \lambda_k(i,j)}{\sum_{k=0}^{K-1} p_k(i,j) \lambda_k(i,j)}, & \text{if } \hat{\eta} \leq T^L \\ \beta \cdot p_k(i,j) + (1-\beta) \cdot \dfrac{\lambda_k(i,j)}{\sum_{k=0}^{K-1} \lambda_k(i,j)}, & \text{otherwise} \end{cases} \tag{4}$$

where T^L denotes the threshold of noise density for using the multiplication-type gain factor which is empirically chosen to be 20%. $p_k(i,j)$ and $\lambda_k(i,j)$ represent the ratio of noise-free pixels and the weighting of pixel-variation level in a local window, respectively. β is the weighting factor between $p_k(i,j)$ and the pixel-variation level. The subscript k denotes the index of classified group of each local window.

The restored pixel can be obtained by multiplied the gain factors with the median values of classified groups, given as

$$\hat{X}(i,j) = \sum_{k=0}^{K-1} G_k(i,j) \cdot \mu_k(i,j) \tag{5}$$

where $\mu_k(i,j)$ denotes the median value of the kth group. It can be obtained by

$$\mu_k(i,j) = median\{\tilde{X}(i+\Delta i, j+\Delta j), \text{ where } \tilde{X}(i+\Delta i, j+\Delta j) \in k^{th} \text{ group}\} \tag{6}$$

where $\tilde{X}(i,j)$ denotes the non-extreme pixels.

In (6), the non-extreme pixels $\tilde{X}(i,j)$ are obtained from the noise-corrupted pixels where the pixels with extreme values 0 or 255 are excluded, given as

$$\tilde{X}(i,j) = \{X(i,j), X(i,j) \neq 0 \text{ and } X(i,j) \neq 255\} \tag{7}$$

The non-extreme pixels are employed to estimate the ratio of noise-free pixels and pixel-variation level. In (4), the pixel-variation level is small when the pixels are in a smooth region or in the same edge position of an image. The value of the gain factor $G_k(i,j)$ is increased when the number of noise-free pixels increases. Thus the gain factor $G_k(i,j)$ varies proportional to the ratio of noise-free pixels $p_k(i,j)$. The non-extreme pixels $\tilde{X}(i,j)$ are classified to several groups. The ratio of non-extreme pixels and the pixel-variation of each group are employed to define the value of the gain factor.

2.1 Group Classification

Non-extreme pixels are classified to several groups. The gain factor is computed for each group. Initially, the non-extreme pixels are sorted in an ascending order, we obtain

$$\tilde{X}_n^s(i,j) = \{\tilde{X}_1^s(i,j), \tilde{X}_2^s(i,j), \ldots, \tilde{X}_L^s(i,j)\} \qquad (8)$$

where L denotes the number of non-extreme pixels in a local window.

The number of groups is defined according to pixel distance between the sorted pixels $\tilde{X}_n^s$. If the distance between two adjacent sorted pixels exceeds a threshold, the pixels may lie in different regions. A new group is created.

In (4), the ratio of non-extreme pixels $p_k(i,j)$ of the group k can be obtained by

$$p_k(i,j) = \frac{I_{k+1}(i,j) - I_k(i,j)}{L(i,j)} \qquad (9)$$

where $I_k(i,j)$ denotes the first index of group k in the sorted pixels $\tilde{X}_n^s$. $L(i,j)$ represents the number of non-extreme pixels in a local window.

The weighting of pixel-variation level of a group $\lambda_k(i,j)$ in (4) can be computed by

$$\lambda_k(i,j) = \frac{1}{v_k(i,j) + \varepsilon} \qquad (10)$$

where ε is a constant to prevent the zero value of the denominator. $v_k(i,j)$ represents the pixel-variation of a group. It is computed by

$$v_k(i,j) = \frac{\displaystyle\sum_{\tilde{X}_i^s(i,j) \in group\, k} \tilde{X}_{i+1}^s(i,j) - \tilde{X}_i^s(i,j)}{N_k} \qquad (11)$$

where N_k represents the number of non-extreme pixels in group k.

3 Experimental Results

Standard gray-level test images were employed to measure the performance of a denoising algorithm in the experiments. These test images include "Lena" and "boat" images, each with size 512×512. The test images were corrupted by salt-and-pepper impulse noise with different noise densities (from 10% to 90%). The switch median filter (SM) [15], the directional-weighted-median (DWM) filter [7], the modified decision based unsymmetric trimmed median (MDBUTM) [3] were implemented for comparisons. Restoration results were quantitatively measured by the peak signal-to-noise ratio (PSNR).

A restored image can be quantitatively measured by the PSNR as an objective measure. The PSNR can be expressed as

$$\text{PSNR(dB)} = 10 \cdot \log_{10}(\frac{\text{MAX}}{\text{MSE}}) \tag{12}$$

where MAX represents the largest value of the energy of gray level, it is 255^2 for an 8-bit gray level image. The MSE denotes the mean-square-error between the original and the reconstructed images. It is computed by

$$\text{MSE} = \frac{1}{M \cdot N} \sum_{i=0}^{M-1} \sum_{j=0}^{N-1} |S(i,j) - \hat{X}(i,j)|^2 \tag{13}$$

where $S(i,j)$ represents the original pixels. M and N are the sizes of an image for the width and the height, both of which are 512.

Table 1. Comparisons of restoration results in PSNR (dB) for Lena image with resolution 512×512.

Noise density	Denoising method			
	SM	DWM	MDBUTM	Proposed
10%	36.12	40.78	37.91	43.12
20%	33.42	37.02	34.78	39.47
30%	31.36	34.63	32.29	37.04
40%	29.88	32.51	30.32	35.24
50%	28.54	30.23	28.18	33.82
60%	26.76	27.69	26.43	32.07
70%	24.47	25.23	24.30	30.32
80%	19.52	21.00	21.70	28.10
90%	8.80	15.45	18.40	24.86

Table 2. Comparisons of restoration results in PSNR (dB) for boat image with resolution 512 × 512.

Noise density	Denoising method			
	SM	DWM	MDBUTM	Proposed
10%	32.84	36.19	40.18	40.26
20%	30.05	32.81	36.46	36.51
30%	28.15	30.44	33.85	34.03
40%	26.76	28.61	32.08	32.41
50%	25.08	26.70	30.34	30.83
60%	23.16	24.25	28.13	29.12
70%	20.54	22.10	25.31	27.32
80%	13.25	18.36	21.43	25.07
90%	7.83	14.42	16.74	21.94

Tables 1 and 2 present the performance comparisons for the various image denoising filters in terms of the PSNR for the Lena and boat images, respectively. According to (12), the larger the value of the PSNR, the better is the quality of the restored image. It can be found that the proposed method outperforms the other methods for both images in all noise density conditions.

In order to explore the visual quality, we show the reconstructed images for various denoising methods with the noise density equaling 80% for Lena and boat images in Figs. 1 and 2, respectively. Figure 1 shows Lena image, which is heavily corrupted by salt-and-pepper noise with 80% noise density (Fig. 1(b)). It can be found that the SM (Fig. 1(c)) and DWM (Fig. 1(d)) filters fail to restore the noise-corrupted images. The MDBUTM filter (Fig. 1(e)) and the proposed method (Fig. 1(f)) can restore the noise corrupted image well. However, the restored image of the MDBUTM filter suffers from plenty of residual noise. The quality of denoised image is not satisfied. On the contrary, the proposed method not only can effectively remove interference noise, but also can reconstruct the image well.

Figure 2 shows a boat image, which is heavily corrupted by salt-and-pepper noise with 80% noise density (Fig. 2(b)). By observing Figs. 2(c) and 2(d), the SM (Fig. 2(c)) and DWM (Fig. 2(d)) filters also fail to restore the noise-corrupted image. Although the MDBUTM filter (Fig. 2(e)) can reconstruct the image well, the restored image suffers from a quantity of residual noise. These results are similar to that shown in Fig. 1. The proposed method (Fig. 2(f)) can effectively remove background noise; meanwhile the details of the towrope on the mast and the words on the back hull of the boat can be reconstructed under such heavy noise corruption. Accordingly, the proposed method outperforms the compared filters.

Fig. 1. Restored images of various denoising filters for Lena image with 80% noise density. (a) Original image; (b) noise-corrupted image; (c) restored image using the SM filter; (d) restored image using the DWM filter; (e) restored image using the MDBUTM filter; (f) restored image using the proposed method.

Fig. 2. Restored images of various denoising filters for boat image with 80% noise density. (a) Original image; (b) noise-corrupted image; (c) restored image using the SM filter; (d) restored image using the DWM filter; (e) restored image using the MDBUTM filter; (f) restored image using the proposed method.

4 Conclusions

This study proposes using distribution ratio of noise-free pixels and the pixel-variation level of a group to define the value of the gain factor. A multiplication-type gain factor is utilized in the conditions of slight noise density (noise density is less than 20%). A weighting-type gain factor is used for the middle and high levels of noise densities (noise density is greater than 20%). Experimental results show that the proposed method can well restore a noise-corrupted image without blurred effect; meanwhile the salt-and-pepper noise can be effectively removed for various corruption densities (ranging from 10% to 90%).

Acknowledgments. This research was supported by the Ministry of Science and Technology, Taiwan, under contract numbers MOST 105-2410-H-468-012 and MOST 104-2221-E-468-007.

References

1. Hwang, H., Hadded, R.: Adaptive median filter: new algorithms and results. IEEE Trans. Image Process. **4**, 499–502 (1995)
2. Wang, Y., Wang, J., Song, X., Han, L.: An efficient adaptive fuzzy switching weighted mean filter for salt-and-pepper noise removal. IEEE Signal Process. Lett. **23**, 1582–1586 (2016)
3. Esakkirajan, S., Veerakumar, T., Subramanyam, A.N., PremChand, C.H.: Removal of high density salt and pepper noise through modified decision based unsymmetric trimmed median filter. IEEE Signal Process. Lett. **18**, 287–290 (2011)
4. Deivalakshmi, S., Palanisamy, P.: Removal of high density salt and pepper noise through improved tolerance based selective arithmetic mean filtering with wavelet thresholding. Int. J. Electron. Commun. (AEU) **70**, 757–776 (2016)
5. Lu, C.-T., Chen, Y.-Y., Wang, L.-L., Chang, C.-F.: Removal of salt-and-pepper noise in corrupted image using three-values-weighted approach with variable-size window. Pattern Recog. Lett. **80**, 188–199 (2016)
6. Ahmed, F., Das, S.: Removal of high density salt-and-pepper noise in images with an iterative adaptive fuzzy filter using alpha-trimmed mean. IEEE Trans. Fuzzy Syst. **22**, 1352–1358 (2014)
7. Dong, Y.Q., Xu, S.F.: A new directional weighted median filter for removal of random-valued impulse noise. IEEE Signal Process. Lett. **14**, 31–34 (2007)
8. Lu, C.-T., Chou, T.-C.: Denoising of salt-and-pepper noise corrupted image using modified directional-weighted-median filter. Pattern Recog. Lett. **33**, 1287–1295 (2012)
9. Xiao, L., Li, C., Wu, Z., Wang, T.: An enhancement method for X-ray image via fuzzy noise removal and homomorphic filtering. Neurocomput. **195**, 56–64 (2016)
10. Deng, X., Ma, Y., Dong, M.: A new adaptive filtering method for removing salt and pepper noise based on multilayered PCNN. Pattern Recog. Lett. **79**, 8–17 (2016)
11. Wang, X., Shen, S., Shi, G., Xu, Y., Zhang, P.: Iterative non-local means filter for salt and pepper noise removal. J. Visual Commun. Image Represent. **38**, 440–450 (2016)
12. Li, Z., Liu, G., Xu, Y., Cheng, Y.: Modified directional weighted filter for removal of salt & pepper noise. Pattern Recog. Lett. **40**, 113–120 (2014)

13. Li, Z., Cheng, Y., Tang, K., Xu, Y., Zhang, D.: A salt & pepper noise filter based on local and global image information. Neurocomput. **159**, 172–185 (2015)
14. Liu, L., Chen, C.L.P., Zhou, Y., You, X.: A new weighted mean filter with a two-phase detector for removing impulse noise. Inform. Sci. **315**, 1–16 (2015)
15. Zhang, S., Karim, M.A.: A new impulse detector for switching median filters. IEEE Signal Process. Lett. **9**, 360–363 (2002)

Clustering of Freight Vehicle Driving Behavior Based on Vehicle Networking Data Mining

Liangbin Yang[✉] and Xiao Wang

School of Information Science and Technology,
University of International Relations, Beijing 100091, China
ylb@uir.cn

Abstract. [Purpose/meaning] The massive information of car under the network environment is of special significance and value to analyze the characteristic of the freight vehicle driving. Through mining vehicle speed, acceleration and other driving data is advantageous to help research vehicle drivers driving behavior and to standard the driver's driving behavior, and realize the intelligent management of the vehicle. [Methods/processes] In this paper, the part of freight vehicles operating within the bounds of Hebei province as the research object to obtain the characteristic parameters of vehicle driver's driving behavior, and using data mining method based on factor analysis to convert the parameters as indicators of the K-Means clustering method to analyze driving behavior. [Results/conclusions] According to the analysis results, dangerous driving behavior in the process of freight vehicles on the road exists and there is a certain effect for road transport safety, but they are all very few. By studying the characteristic parameters of velocity and acceleration of the vehicle can better response freight vehicles in operation in the process of driving behavior, and it contributes to the intelligent vehicle management.

Keywords: Connected vehicle · Data mining · Factor analysis
Cluster analysis · K-Means clustering

1 Introduction

With the rapid development of the Internet and its related technologies, people's lives become more intelligent. One of the important manifestations is the vehicle networking which based on the generation and rise of the Internet of things networking technology. As the name suggests, vehicle networking is a huge interactive network which consisted by the vehicle location, speed, route and other vehicle information [1]. The network is based on the electronic identification of the vehicle installation, and uses the radio frequency and other electronic technologies to realize display of the attribute information and static or dynamic information of all the vehicles on the information

Supported by "the Fundamental Research Funds for the Central Universities", Project No. 32620 17T26, Topic: Research on the construction and key technologies of information dissemination model in the big data environment of social network. Project Leader: Liangbin Yang.

© Springer Nature Singapore Pte Ltd. 2018
J. C. Hung et al. (Eds.): FC 2017, LNEE 464, pp. 12–23, 2018.
https://doi.org/10.1007/978-981-10-7398-4_2

network platform. The information is extracted and used effectively, so as to effectively monitor the operation status of all vehicles according to different functional requirements and provide comprehensive services. The core of the vehicle networking technology is to use the vehicle's sensor terminal to collect the vehicle information uploaded to the traffic information network control platform in order to achieve the effective management of the vehicle, and further provide the corresponding integrated services [2].

The development of vehicle networking technology has a positive effect. First of all, the vehicle networking can enhance the scientific of road transport management, and improve the efficiency of the use of the road. The application of vehicle networking technology can reduce the traffic congestion by about 60%, improve the short-distance transport efficiency by about 70% and increase the existing road network capacity by 2 to 3 times [3]. If a car runs in the vehicle networking technology structure of the intelligent road network, the number of parking can be reduced by 30%, driving time reduced by 13% to 45% and the vehicle's efficiency can be increased by 50%. Secondly, the development of vehicle networking technology is of great significance for automotive transportation enterprises to achieve intelligent management, and also contributes to the improvement of operational efficiency and operational safety. According to the relevant research, that the application of vehicle networking technology can make 81% of traffic accidents to be avoided (no person injuries). In view of this, the development of vehicle networking has important significance and value.

The massive vehicle network data which show the driving behavior of vehicle drivers have important research value. There is a close contact between car driving behavior and road transport safety [4]. According to the relevant statistics, in all traffic accidents, due to the driver does not regulate the operation led to the occurrence of accidents is one of the main incentives for traffic accidents. Therefore, it is of importance to the protection of road transport safety by constraining the vehicle operator's driving behavior. At the same time, along with the maturity of the vehicle networking technology and the extensive application in road traffic safety management, the vehicle realizes real-time monitoring and produces a lot of data. For analyzing the driving behavior characteristics of vehicle drivers and the intelligent management of road traffic safety, it is of great value and practical significance to analyzing the general characteristics of these data and extracting the valuable characteristic parameters.

This article mainly from the data mining point of view to analyze the driver's driving behavior and the relationship between road safety. This paper takes the massive real-time vehicle data generated by the network environment as the research object and uses the data mining technology to cluster the driving characteristics, and studies the distribution pattern of the typical driving characteristics so that the relationship between the driver's driving behavior and the road transportation safety can be obtained. In view of the current research in the relevant field, the data produced by most of the freight vehicles in real time are only kept as historical data in the database and not fully exploited and utilized. And foreign research scholars such as Herrera [5] and so on are only to build a GPS-based system platform which get traffic data by mobile phone positioning, but did not achieve associating with the associated driving behavior. Although Greaves [6] used GPS to obtain driving data to compensate the lack of data accuracy, but only to study the speeding behavior.

2 Data Mining

2.1   A Brief Introduction to the Present Situation of Data Mining Technology

Data mining technology, as the name suggests, is a process of extracting hidden and valuable data in the massive, mixed, random data which based on the different need [7]. For today's explosive growth in the amount of data, it is of great significance to applicate the data mining technology so that making a large number of cumbersome data re-applied value. Therefore, application software which based on the data mining technology has been a leap-forward progress.

The following table are data mining tools which widely used in recent years (Table 1).

Table 1. Data mining tools widely used in recent years

Company	Data mining system name
IBM	Intelligent Miner
SGI	SetMiner
SPSS	Clementine
SAS	Enterprise Miner
Sybase	Warehouse Studio
RuleQuest Research	See5

Through the above table we can see that data mining technology has achieved some results after years of research and has been widely used in the application areas, especially in Europe and North America. Many well-known companies are studying application tools which based on data mining technology. At the same time, China's data mining research also achieved gratifying results and has played an important role in various fields and industry.

2.2   Introduction to Data Mining Technology

Data mining technology has made great progress with the relevant research continues. Related technology has also been various. Under normal circumstances, data mining technology is divided into two categories [8]:

(1) Statistical type. Including clustering analysis, correlation, probability statistics and other data mining technology [9];
(2) Machine learning type. Mainly in the field of artificial intelligence it has a wide range of applications. Based on a large number of samples, it can obtain the required model or related parameters through training and learning to meet the relevant needs.

The following table lists some classic data mining algorithms used in the field.

Table 2. Classic data mining algorithm

Algorithm name	Introduction	Advantages and disadvantages
C4.5	A classification decision tree algorithm of machine learning algorithm, the core is ID3 algorithm	Advantages: understand easily high accuracy Disadvantages: efficiency is relatively low
K-Means	A clustering algorithm invented in 1956	Advantages: running fast Disadvantages: inappropriate parameter k may lead to a error in the results
SVM	It is a supervised learning algorithm, mainly in the statistical classification and regression analysis has a wide range of applications	Advantages: to solve the problem of high dimensions Disadvantages: sensitive to missing data
Apriori	An Algorithm for Mining Frequent Itemsets of Boolean Association Rules. The core is based on the two-stage frequency set of the first recursive algorithm	Disadvantages: easy to produce a large number of candidate sets or repeat the scan database, resulting in lower efficiency
Expectation Maximization Algorithm	Mainly used in machine learning and data aggregation areas of computer vision	Disadvantages: easier to fall into the local optimal
CART	Classification and Regression tree algorithm	Advantages: the calculation rules are easy to understand Disadvantages: the field of continuity is more difficult to predict
NBM	Originated in the classical mathematical theory, which has a solid mathematical theory and a stable classification efficiency	Advantages: efficiency is stable, less parameters Disadvantages: the data is really insensitive, easily lead to errors

It can be seen from Table 2 that various algorithms have their own advantages and disadvantages, and are widely used in many fields. Such as: Yonghua Qu, Jinti Wang [10] and so on who created a mixed inversion model for vegetation surface parameter estimation which based on Bayesian network theory to; Hui Dong [11] researched the prediction of the geotechnical nonlinear deformation behavior which based on support vector machine algorithm.

The following focuses on the data mining techniques used by the article - factor analysis and clustering analysis.

Data mining technology based on factor analysis and clustering analysis is one of the most efficient data mining techniques which can deal with multivariable and high dimensional data among modern disciplines.

Factor analysis is a statistical technology which research variable group and extract a common factor from it [14]. This data mining technology inventor is the British psychologist Spielman. Factor analysis as a statistical data mining technology, has more than 10 methods to analyze the data. Such as the least squares method, the center of gravity method, the impact analysis method and so on. The basis of these methods is the correlation coefficient matrix and the difference is that the correlation coefficient matrix diagonal value is different. In the analysis using factor analysis, the core concerns are mainly the following aspects: the construction of factor variables and how to name and explain the variables. The specific steps of the factor analysis are also based on these two core issues for specific work. The specific steps of the factor analysis are as follows:

(1) For the initial sample to be studied to confirm that they satisfy the requirements of using factor analysis;
(2) Construct factor;
(3) Rotate the factor variables so that the factor variables have a better descriptibility;
(4) Calculate Factor Variable Score.

The advantages of factor analysis are the simplification of the data, and the factoricity of the variables becomes better by factor rotation which leads to the clarity of naming. The main disadvantage is that the method which used in the factorial score calculation is the least squares method. The method under the influence of certain conditions may fail, resulting in analysis of large errors and even lead to failure analysis.

Clustering analysis, as the name suggests, refers to the characteristics of the study object to classify them in order to reduce the number of research objects. The purpose is to collect and analyze data which based on similarity. Cluster analysis as an exploratory data analysis method, there is no need to develop standards in the analysis process and classify the sample data according to the similarity.

In the cluster analysis, the clustering method is mainly used in the following two categories:

Hierarchical clustering. Mainly including the merger method, decomposition method and other methods.

Non-hierarchical clustering. Mainly including the division of clustering and spectral clustering.

The application of factor analysis and cluster analysis is of great significance to increasing the accuracy of data analysis. At present, factor analysis and cluster analysis are mainly used in economic, sociology, archeology, medicine, geology and many other areas of research work.

In this paper, the principal component analysis method is used to reduce the characteristic parameters of driving behavior and simplify it into a few comprehensive variables [15], which clearly shows the information about the driving behavior characteristics represented by most of the data before processing. Rotation factor is obtained by factor rotation, which leads to more clear driving behavior information. Then, the score of the factor of the freight vehicle is taken as the clustering index, and the clustering method is used to realize the scientific clustering of the driving behavior of the freight vehicle.

3   Extraction of Driving Characteristics of Freight Vehicles

3.1   Data Attributes

The data used in this paper is the vehicle operation data collected by the vehicle terminal which mainly using the car networking technology to realize the intelligent management of the freight vehicle. Data accuracy is consistent with data analysis requirements. These data mainly include the location information of the vehicle, the driving road information and the vehicle speed information and so on.

3.1.1   Extraction of Driving Behavior Characteristic Parameters

(1) vehicle average speed V_a, speed standard deviation V_s
     According to the relevant research, the average speed of vehicles is higher, the greater the probability of accidents in the process of driving, especially freight vehicles. And the speed standard deviation reflects the speed of the vehicle's degree of dispersion, the speed distribution of the more discrete, much easier lead to accidents. It can be seen that the average speed and velocity standard deviation during vehicle travel is one of the important parameters that embody the driving behavior. The formula for calculating the mean speed and velocity standard deviation are as follows:

$$V_a = \frac{1}{n} \sum_{k=1}^{n} V_k \tag{1}$$

$$V_s = \sqrt{\frac{\sum_{k=1}^{n} (V_k - V_a)^2}{n}} \tag{2}$$

In the above formula, v_k represents the kth speed value of the freight vehicle in the data; n is the number of samples of the same vehicle in the data; v_s represents the standard deviation of the speed of the freight vehicle; v_a represents the same vehicle speed average.

(2) average acceleration a, forward average acceleration a_a^+, reverse average acceleration a_a^-, acceleration standard deviation a_s.

The vehicle driving behavior extracted in this paper also includes changes in vehicle acceleration, such as average acceleration a, forward average acceleration a_a, reverse mean acceleration a_a, and acceleration standard deviation a_s. The acceleration, as one of the characteristic parameters of the driver's driving behavior, can clearly reflect the vehicle driver's manipulation of the accelerator pedal and the brake pedal during vehicle travel. Forward acceleration speed reflects the driver's control of the accelerator pedal, which relates with the road traffic environment. The main reflection of the reverse acceleration is the driver's control of the brake pedal which influence the freight vehicle driver's physical feel and the security of the freight vehicles and cargo. Therefore, it is of great significance to making the average acceleration, forward average acceleration, and reverse average acceleration as the parameters of vehicle driving behavior for research.

$$a_s = \sqrt{\frac{\sum\limits_{k}^{n}(a_k - \bar{a})}{n}} \tag{3}$$

$$a_a^+ = \frac{1}{n}\sum_k^n \frac{V_k - V_{k-1}}{t} \tag{4}$$

$$a_a^- = \frac{1}{n}\sum_k^n \frac{V_k - V_{k-1}}{t} \tag{5}$$

$$\bar{a} = \frac{1}{n}\sum_k^n a_k \tag{6}$$

a_k represents the acceleration value of the vehicle acquired at the kth time; a represents the acceleration average in the course of vehicle travel; a_s represents the acceleration standard deviation in the course of vehicle travel; a_a represents the acceleration of the vehicle during travel Positive acceleration average; a_a represents the vehicle's reverse acceleration average. The extraction of the above data as a characteristic parameter of vehicle driving behavior has important research value for studying the driving characteristics of freight vehicles.

4   Data Mining Analysis of Freight Vehicle Driving Behavior

4.1   Data Processing

The data used in this paper is the data of the vehicle networking which is the data of the vehicle terminal of 16 freight vehicles traveling on a national highway in Hebei Province from October 10 to 19, 2016. The total number of data is 20 million. From the practical point of view, in the data processing process has abandon the value of speed is 0. The following table shows the results of the data processing (Table 3).

Table 3. Data processing results

No.	v_a	v_s	$\bar{a}$	a_s	a_a^+	a_a^-
1	39.07407407	16.4372064	0.0312561	0.00097362	0.07526852	0.08006162
2	39.45185185	22.3501030	0.0680887	0.00463720	0.080907969	0.08024391
3	36.14021164	6.38760325	0.0121656	0.00015169	0.053987348	0.055600127
4	36.69879063	14.0322975	0.0288903	0.00083791	0.04364673	0.042978326
5	32.17032164	17.0608701	0.0202406	0.00042334	0.045543502	0.04688192
6	35.08119658	6.79511637	0.0379753	0.00147510	0.047369301	0.051393968
7	40.76888163	8.25695032	0.1297381	0.01685605	0.091619831	0.065180994
8	40.37234043	16.8962979	0.2065940	0.04163738	0.201829546	0.144393234
9	35.12304527	13.4159075	0.0544900	0.00298145	0.132530661	0.148506883
10	31.26564592	13.1913135	0.0860399	0.00740593	0.171276706	0.188343803
11	33.47408591	13.2476561	0.0735031	0.00540198	0.150310888	0.16828775
12	35.32175926	13.2877933	0.0829702	0.00688784	0.060606796	0.048531994
13	37.18376068	12.940691	0.0127224	0.00016248	0.025648779	0.025124543
14	39.09074074	12.9260717	0.0240660	0.00058876	0.089068407	0.098087894
15	35.99067599	15.0795909	0.2495871	0.06229243	0.30841735	0.228294208
16	32.95051195	14.0761410	0.1892549	0.03581581	0.218269768	0.168812762

4.2 Data Analysis

4.2.1 Driving Behavior Characteristic Parameter Analysis

The tool used in the data analysis of the driving behavior parameters is SPSS19.0 statistical software. The factor extraction method adopts the principal component analysis method. The factor rotation is the maximum variance method. Through the factor analysis of the driving behavior characteristic parameters, the KMO measure between variables is 0.61, which indicates that there is a certain correlation between the characteristic parameters. At the same time, the result of spherical hypothesis test is 0.000, that is the significant level is 0.000, which rejects the null hypothesis and indicates that the experimental data used in the sample sufficient. The factor analysis results of the driving behavior characteristic parameters are shown in the following Table 4.

Table 4. Variance

Explain the total variance									
Ingredients	Initial eigenvalue			Extract squares and load			Rotate squares and load		
	Total	Variance %	Accumulated %	Total	Variance %	Accumulated %	Total	Variance %	Accumulated %
1	3.608	60.134	60.134	3.608	60.134	60.134	3.587	59.779	59.779
2	1.186	19.765	79.899	1.186	19.765	79.899	1.207	20.120	79.899

As it can be seen from the above table, when the cumulative variance reaches 79.899%, the analysis of driving characteristics of the parameters obtained two main factors and the two main factors contain the original data of 79.899% of the information so that it meets the factors Analysis of the requirements. Before and after the rotation of the cumulative contribution rate has not changed. In order to discover the meaning of the factor, the factor scalar is rotated. The purpose of factor rotation is to assign the variance ratio of each factor by changing the position of the coordinate axis, thus reducing the complexity of the factor structure and making it easier to explain. Two rotation factors are obtained by rotation, and the variance contribution rate of rotation factor is 59.7% and 20.12%. It shows that the two factors have obvious and great contribution to driving behavior (Table 5).

Table 5. Composition matrix

Component matrix[a]

	Ingredients	
	1	2
Forward average acceleration	.987	−.099
Average acceleration	.942	.156
Acceleration standard deviation	.932	.166
Reverse average acceleration	.889	−.276
Average speed	−.092	.921
Speed standard deviation	.280	.447

Extraction method: principal component

[a]Two ingredients have been extracted.

It can be seen from the above table that the load of the characteristic parameters related with acceleration (average acceleration, acceleration standard deviation, forward average acceleration, reverse average acceleration) of factor is 0.987, 0.942, 0.932, 0.889, and it is significantly higher than that of other variables, which Indicating that the factor 1 has a high degree of correlation with the characteristic parameters associated with the acceleration (Table 6).

Table 6. Rotate the composition matrix

Rotation component matrix[a]

	Ingredients	
	1	2
Forward average acceleration	.992	−.006
Average acceleration	.923	.244
Acceleration standard deviation	.913	.253
Reverse average acceleration	.910	−.191
Average speed	−.178	.908
Speed standard deviation	.237	.471

Extraction method: principal component
Rotation method: Orthogonal rotation with
Kaiser normalization

[a]The rotation converges after 3 iterations.

From the above table, it can be found that the correlation between the rotation factor 1 and the acceleration (average acceleration, acceleration standard deviation, forward average acceleration, reverse average acceleration) is 0.992, 0.923, 0.913, 0.910 after the rotation of the main factor 1 and the main factor 2. The correlation between the rotation factor 1 and the acceleration correlation is higher than the average speed and the standard deviation. And the correlation degree of the factor related with velocity (average speed, velocity standard deviation) and the rotation factor 2 is 0.908 and 0.471. Compared with other characteristic variables, the correlation with the average speed is higher.

4.2.2   Analysis Based on K-Means Clustering

It can be seen from the correlation between the main factor, the rotation factor and the driving behavior characteristic variables that there is a strong correlation between the main factor 1 and the acceleration-dependent variables. Acceleration characteristics variable mainly reflects the changement of vehicle acceleration in the process of driving [16]. To a certain extent, it reflects that the dramatic changes in vehicle acceleration may bring some risks to the vehicle and the goods under the influence of the vehicle drivers in a special road driving environment (such as night environment, complex road conditions, etc.) and other subjective driving behavior (such as robbery, gat, etc.) Thus, the main factor 1 can be used as a criterion for K-Means clustering. At the same time, the main factor 2 can be clustered as a clustering criterion for speed-related driving behavior.

The clustering results of the main factor 1 as the standard for K-Means clustering are shown in the following Table 7.

Table 7. Clustering results

The number of cases in each cluster		
Clustering	1	13.000
	2	3.000
Effective		16.000
Missing		.000

It can be seen from the above table that the characteristic parameters of the driving behavior of the freight vehicle cluster two class is more appropriate according to the characteristics of the main factor 1. Freight transport drivers have different degrees of speed in the process of running a truck, but the number of drastic changes is relatively small [17]. It is shown that the operation of the vehicle driver causes the vehicle to have different degree of speed change due to the influence of various factors during the running of the freight vehicle. The speed change behavior reflects the driver's use of the accelerator pedal and the brake pedal. Frequent acceleration and deceleration will not only increase the fuel consumption, but also lead to the loss of the brake. But because of the driver's better driving behavior and safety awareness, the speed of the vehicle has a better control and avoid the emergence of severe acceleration and deceleration situation.

The clustering results of the main factor 2 as a standard for K-Means clustering are shown in the following Table 8.

Table 8. Clustering results

The number of cases in each cluster		
Clustering	1	1.000
	2	2.000
	3	6.000
	4	7.000
Effective		16.000
Missing		.000

As it can be seen from the above table, the main factor 2 used as a standard to cluster the characteristics of the driving behavior of freight vehicles, and clustering for the four categories is more appropriate. The vast majority freight vehicle drivers did not appear speeding behavior in the process of driving freight vehicles, only a very small number of vehicles in the process of running have a different degree of speeding behavior. The increase of speeding behavior in the course of operation will have some dangerous effects, but because the driver has good driving habits and safe driving awareness, there is no large number of speeding behavior and protect the freight vehicles and drivers of security.

5 Conclusion

In this paper, the vehicle data collected from the vehicle terminal installed on the freight vehicle are taken as the research object, and the characteristic parameters of the driving behavior of the freight vehicle are extracted and further refined. Processing the mass of the vehicle networking data obtained within the scope of Hebei Province so that it can be more clearly reflect the freight vehicles driving behavior characteristics in the driving process. In the process of analysis, factor analysis and cluster analysis are used to analyze the characteristics of driving behavior characteristics of freight vehicles so that the correlation between the characteristic parameters is well reflected and the distribution of the characteristics behavior of the freight vehicle during the running process is clearly reflected. Through the above experimental study, it has a clearer understanding about the important parameters which may affect the safety in the process of driving how to influence the safety of driving. Through the combination of factor analysis and cluster analysis, the relationship between massive data, driving behavior and road transport safety is formed. Based on the research, realizing the intelligent management of freight vehicle operation is of great significance to the safety of vehicle and road transportation.

References

1. Tang, H.: Research on the architecture and key technologies of IOV. Netw. Secur. Technol. Appl. **9**, 42–43 (2013)
2. Yang, S.: Research on Nantong City Transport Vehicle Safety equipment supervision system. Chang'an University, Xi'an (2014)
3. Yang, X., Chen, Y.: Research on the development of vehicle network. China Radio **2**, 45–47 (2013)
4. Pei, Y., Zhang, X.: Analysis on the characteristics of bad car driving. Traffic Inf. Secur. **3**, 81–84 (2009)
5. Herrra, J.C., Work, D.B., Herring, R., et al.: Evaluation of traffic data obtained via GPS-enabled mobile phones: the mobile century field experiment. Transp. Res. Part C Emerg. Technol. **18**(4), 568–583 (2010)
6. Greavses, S.P., Ellison, A.B.: Personality, risk aversion and speeding: an empirical investigation. Accid. Anal. Prev. **43**(5), 1828–1836 (2011)
7. Zhang, Z.: Research on Methods of Scientific and Technical Monitoring Information Visualization Based on Scientific and Technological Literature. China Science and Technology Information Institute, Beijing (2007)
8. Yang, L.: Research on the Methods of System Identification and Pattern Classification Based on SVM. Tongji University, Shanghai (2010)
9. Ren, C.: The Uncertainty Modeling and Simulation of Motor Vehicle Driver Behavior. Hefei University of Technology, Hefei (2015)
10. Qu, Y., wang, J., Liu, S., Wan, H., Zhou, H., Lin, H.: Research on mixed inversion model of surface parameters supported by Bayesian network. J. Remote Sens. **10**(1), 6–14 (2006)
11. Dong, H.: Geotechnical Nonlinear Deformation Behavior Prediction Based on SVM. Central South University, Changsha (2007)
12. Zhang, D.Q., Chen, S.C.: A comment on "Alternative c-means clustering algorithms". Patt. Recogn. **37**(2), 173–174 (2004)
13. Provost, F.J., Domingos, P.: Tree induction for probability-based ranking. Mach. Learn. **52**(3), 199–215 (2003)
14. Ge, L.: Research on the Evaluation and Promotion of Entrepreneurship Education Ability in Colleges and Universities Based on CIPP. Dalian University of Technology, Dalian (2014)
15. Xu, C.: Application of Cluster Analysis in Data Mining. Shanxi Finance University, Taiyuan (2004)
16. Pang, M.: Research on Vehicle Scheduling Problem Based on Dynamic Environment. Hebei University of Technology, Tianjin (2009)
17. Su, J.: Research on Algorithm of Critical Vehicle Distance in Vehicle Active Safety. Hunan University, Changshi (2011)

Performing Iris Segmentation by Using Geodesic Active Contour (GAC)

Yuan-Tsung Chang, Chih-Wen Ou, Timothy K. Shih[✉], and Yung-Hui Li

Department of Computer Science and Information Engineering,
National Central University, Taoyuan, Taiwan
harry.500net@gmail.com, chihwen.frankou@gmail.com,
timothykshih@gmail.com, yunghui@csie.ncu.edu.tw

Abstract. A novel iris segmentation technique based on active contour is proposed in this paper. Our approach includes two important issues, pupil segmentation and iris circle calculation. If the correct center position and radius of pupil can be found in the tested image, we can precisely segment the iris. The accuracy of our proposed method for ICE dataset is around 92%, and also reached high accuracy level of 79% for UBIRIS. Our results demonstrate that the proposed iris segmentation method can perform well with high accuracy for Iris segmentation in images.

Keywords: Iris segmentation · Active contour · Biometrics

1 Introduction

Biometrics matching has been widely adopted as a secure way for identification and verification purpose. Iris recognition is an important biometric modality which can be used to identify individuals with high accuracy. Large-scale biometric systems such as face and iris recognition are now used in identity cards and mobile secure payment methods in some countries [1]. Big data is now widely available for biometric matching related applications like iris matching, due to the increasing availability of data from multiple sources such as social media, online transactions, network sensors, or mobile devices.

Ratha et al. in [1] presents how the large scale biometric systems such as iris-based recognition becoming very famous in critical security systems. They identified four challenges that involve the effective managing of the complex life cycle and operations of identity information as, the enrollment database size (volume), transaction response-time (velocity), fraudulent (veracity), and multiple (variety) biometric identifiers. They further believe that by virtue of dealing with some of the most critical entities, namely identity and entitlement, biometric systems are likely to emerge as among the most critical of the Big Data systems. Liu et al. in [2] researched on the computation demand associated with running biometric matching algorithms on a big data set with different underlying hardware platforms. Fernándeza et al. in [3] reviewed Latin American

© Springer Nature Singapore Pte Ltd. 2018
J. C. Hung et al. (Eds.): FC 2017, LNEE 464, pp. 24–35, 2018.
https://doi.org/10.1007/978-981-10-7398-4_3

contributions in pattern recognition and related fields on—but not restricted to—applications in the fields of computer vision and image analysis with large scale characteristics.

In this work, we proposed a new iris segmentation algorithm based on geodesic active contour (GAC) and tested it on UBIRIS [15] and ICE [16] dataset. Our contributions include the novel iris segmentation algorithm which enhances the accuracy of the iris segmentation results. In the beginning of the study, we predict that the major bottleneck would be the implementation of GAC. However, we soon discovered that the most difficulties can be traced back to the pupil segmentation and iris boundary localization. In other words, if we can precisely find out the correct center position and radius of pupil in the test images, then we will be able to segment the iris precisely. The factors whether we can find the correct pupil depends on the process we perform among the input images. The quality of images also plays an important role in this problem.

There are two datasets used for the experiment: ICE and UBIRIS. The ICE dataset is composed of grayscale images and contains roughly three thousand images. The other dataset, UBIRIS, is consisted of about 1200 color images. Both datasets show diversities among image quality and characteristics. For example, there are images with dense eyelash. Some of images are blurry. Some eyes are even not well opened. Irises are partially occluded in most of the images. Furthermore, the pupils are often not identifiable in the images. Such datasets are similar to the real cases if we apply this technology in the real world. Therefore, if our algorithm can successfully work on these datasets, it is highly likely to be practically useful in real life scenario.

The rest of the paper is organized as follows. Section 2 provides a brief of the related work on iris segmentation carried out in recent past. Section 3 explains how the iris segmentation is implemented. The implementation includes pupil segmentation, GAC, and iris localization. The experimental results and discussions are provided in Sect. 4. Conclusion and future works are stated in the Sect. 5.

2 Related Work

According to [4], we can perform iris segmentation based on the result of pupil segmentation. The result of the pupil segmentation is an estimated circle parameter well describing the boundary of the pupil. We can run the GAC algorithm from the pupil boundary outward to the iris boundary according to the gradient of the grayscale iris, then, reach the boundary between the iris and sclera. Several iris segmentation algorithms have been proposed based on similar ideas.

Ma et al. [5] used the Hough transform to detect the inner and outer boundaries of the iris. Local sharp variation points, which denote the appearing or vanishing of an important image structure, were utilized to represent the characteristics of the iris. They constructed a set of one-dimensional intensity signals to effectively characterize the most important information of the original two-dimensional image, and a position sequence of local sharp variation points are recorded as features used in a particular class of wavelets. Xu et al. in [6] presented an algorithm to estimate pupil center and pupil radius.

They used pupil parameters estimation to create a weight ring mask, and then, a weighted Hough transform is used to extract pupil features.

Zuo et al. in [7] presented a segmentation methodology utilizing shape, intensity, and location information that is intrinsic to the pupil/iris. They reliably segment non-ideal imagery affected with such factors as specular reflection, blur, lighting variation, and off-angle images. Most commercial iris recognition systems use patented algorithms developed by Daugman, and these algorithms are able to produce perfect recognition rates [8]. Especially it focuses on image segmentation and feature extraction by using Log Gabor wavelet and Laplacian of Gaussian filter for iris recognition process. For iris segmentation, most of the researchers assume that the iris is circular or elliptical [8, 9]. Zuo et al. [7] used Randomized Elliptical Hough Transform Weighted Integro-differential operator.

Proença and Alexandre [10], Arvacheh and Tizhoosh et al. [11] used near circular active contour model (snakes), interpolation process to improve performance. Jarjes et al. [12] used angular integral projection function (AIPF) for pupil boundary localization and for the localization of limbus boundary. The AIPF is applied again within two rectangles on both iris sides. Proenca presented a segmentation method that can handle degraded images acquired in less constrained conditions. They considered the sclera as the most easily distinguishable part of the eye in degraded images [13].

3   Proposed Method

3.1   Pupil Segmentation

The purpose of pupil segmentation is to recover the pupillary boundary, which can be approximated using a circle. The first process is to preprocess the input image. We first apply a Gaussian filter with sigma equal to 0.9 to the image. Figure 1 shows the result after applying a Gaussian filter to the input image.

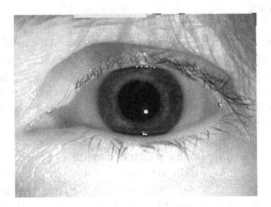

Fig. 1. An eye image filtered by Gaussian.

After applying the Gaussian filter, the second step is to convert the gray-scale images to binary images with a pixel value threshold of 0.18. The third step is to use a Gaussian filter to process the binary image again with sigma set to 2, as shown in Fig. 2.

Fig. 2. The binarized image for the purpose of pupil segmentation.

It is obvious that the black area is the pupil. If we can obtain the boundary of this black area, we can find the potential circle on this boundary. Therefore, we first apply Canny edge detection to recover the boundary in the binary image, as shown in Fig. 3.

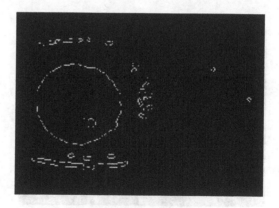

Fig. 3. Result after applying edge detection to Fig. 2.

And then, we use Hough transform to find whether there is any circle inside the edge map. The search parameters include a sensitivity value and a range for the radius of a circle. Several sensitivities are used in our implementation as a sensitivity enumeration. The candidate values of sensitivity value are 0.65, 0.7, 0.75, 0.8, 0.85, 0.9, 0.95. The range of radius starts from 20 to 80. The circle search is implemented as a loop procedure based on the iteration among sensitivities. An example of circular Hough transform is shown as Fig. 4.

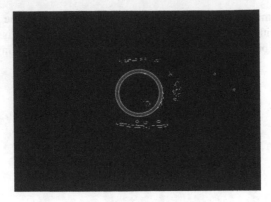

Fig. 4. The recovered pupillary boundary.

3.2 Geodesic Active Contour

Before applying GAC to the eye image, we convolved the image with a Gaussian filter with $\sigma = 3.5$, and binarized the image with pixel threshold 0.5. The initial position of the GAC is set on the pupillary boundary. An example of the initial position of GAC is shown in Fig. 5.

Fig. 5. The initial contour and the complement image for running GAC

After examining the GAC result, specified as blue edges in Fig. 6, there is a potential circle, specified as a green circle, along a part of the final contour. This green circle is not mentioned in the study [4]. However, we will use it to improve our result in the next stage.

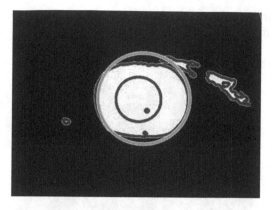

Fig. 6. The final contour (blue) and the potential circle (green) (Color figure online)

3.3 Iris Boundary Estimation

Since we have obtained the final contour, the goal of the last stage is to discover the iris circle in the final contour. The iris circle should ideally match the real boundary between the iris and the sclera. The study [4] proposes a method based on points sampled from the final contour edges. These points are sampled according to the six angles [−30, 0, 30, 150, 180, 210] from the pupillary center to the contour edges. These six points are grouped as two sets based on their position. The grouping criterion is a vertical line crossing the pupillary center. Those points located at the same side from the vertical line are grouped in a set. The study [4] draws a circle for the iris boundary based on three points of a set having larger area between the vertical line and final contour. However, it does not correctly discover an iris circle for some cases, as shown in Fig. 7.

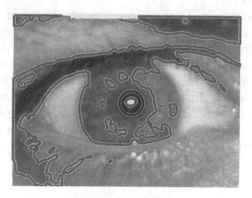

Fig. 7. The case which cannot be correctly segmented by the method in previous study [4]

It is obvious that the right part of final contour is much larger to generate a reasonable iris circle according to the method proposed by the study [4], because the real iris boundary does not fit to the final contour. We discover that there still exists a potential circle, so that we apply the potential circle discovered in the GAC result here and defined it as the estimated iris circle (EIC). In this stage, the iris circle comes from the estimated iris circle will be specified as the green circle, as shown in Fig. 8.

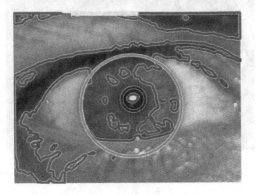

Fig. 8. The segmentation result of the image in Fig. 7, which demonstrate that with our method, it correctly estimates the iris zoundary for the image. (Color figure online)

For detection of the limbic boundary of the iris, method based on a level sets representation of the Geodesic active contour (GAC) model is used. This approach is based on the relation between active contours and the computation of geodesics (minimal length curves). The GAC scheme is contour evolution scheme based on image content and curve regularity. The technique is to evolve the contour from inside the iris under the influence of geometric measures of the iris image. GACs combine the energy minimization approach of the classical "snakes" and the geometric active contours based on curve evolution. Evolution terminates when contour encounters the "boundary" of the object and able to detect boundary even if gaps exist in the boundary. GAC can split and merge at local minima [14].

We soon test this idea and get a positive result. And then, a classification is defined. If there is at least one estimated iris circle in the final contour, we will directly apply it as the iris circle only if its center locates near the pupillary center. The distance between pupillary and iris circles is the key factor in our iris circle calculation. If there are multiple estimated iris circles, the one having largest radius will be selected as the iris circle. For rest of the samples with no any estimate iris circle, we simplified the method proposed by the study [4]. Only two angles [0, 180] are used in the method. The criteria for determine the iris circle is different. And from [2] we can get the formula as (1) and (2):

$$C_t = K(c + \varepsilon\kappa)\vec{N} - (\nabla K \cdot \vec{N})\vec{N} \tag{1}$$

$$K(x, y) = \frac{1}{1 + \left(\dfrac{\|\nabla(G(x, y) * I(x, y))\|}{\kappa}\right)^{\alpha}} \tag{2}$$

We directly find the potential circle which may have radius same as the distances from the pupillary center to the final contour based on these two angles. If there is still nothing found, the mean of these two distances will be defined as the radius of the final iris circle. The proposed algorithm is shown in Fig. 9.

```
1: load img
2: binaryObject=rgb2gray and binaryForPupil(img)
3: edgeObj=edge detector(binaryObject)
4: [centers,radii,~]=find the circle(edgeObj)
5: if centers !=0
6:    pcenter=centers(ind,:);  pradii=val;
7:    [a,b,c]=buildEquation(pcenter,pradii);
8:    p=[1,a,(y^2+b*y+c)];
9:    imgGAC=initImgForGAC(p)
10:   [acResult, gacCenter,gacRadius]=doGAC(imgGAC,acmask,pcenter,pradii)
11:   irisBoundary=edge(acResult)
12:   if gacCenter == 0
13:           [irisCenter,irisRadius]=
              calculateRadiusFromBoundaryImg(irisBoundary,pcenter);
14:   elseif d(1) > 1
15:           irisCenter = gacCenter(ind,:);  irisRadius = mr;
16:   else
17:           irisCenter = gacCenter;  irisRadius = gacRadius;
18:   end if
19: end if
20: Return viscircles(irisCenter, irisRadius)
```

Fig. 9. Pseudo code for the proposed algorithm for iris boundary localization.

We decide to obtain the estimated iris circle (EIC) from the final contour due to the case studies that the Iris boundary is able to be located by EIC. From this point of view, the EIC will be taken as Iris Circle (IC) if it can be found in final contour.

But however, there is an issue to be considered. The EIC cannot be taken as IC if the selected EIC cannot form a concentric circle with PC. Hence, the distance check for the concentric circle is performed in our method. In general, the unique IC will be determined based on the length of radius if multiple ICs are obtained.

4   Results and Discussion

To further enhance the efficiency of the proposed algorithm, we modified code file, which is used for running automatic image segmentation for entire dataset. The Fig. 10 shows an example of running segmentation for a certain ICE image. The Fig. 11 shows an example of running segmentation for a certain UBIRIS image.

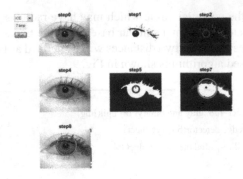

Fig. 10. An example of the segmentation result of an ICE image

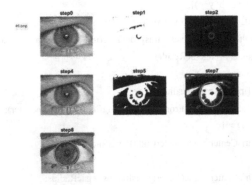

Fig. 11. An example of the segmentation result of an UBIRIS image

Figure 12 shows the elapsed time of iris segmentation using the proposed algorithm on each image.

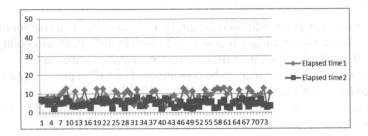

Fig. 12. The elapsed time of iris segmentation using the proposed algorithm on each image.

The configuration of computational environment for running the experiment is listed in Table 1. When we perform iris segmentation using Algorithm 1, the average time needed for each image segmentation was about 5 s. After we enhance the segmentation accuracy using Algorithm 2, the average time needed for each image segmentation increases to about 6 s. The runtime performance is listed in Table 2.

Table 1. Computational environment for the experiment

Item	Description
CPU	Intel i5-4210u 1.70 GHZ
RAM	4 GB RAM
OS	Windows7 Pro 64-bit

Table 2. The average CPU and memory usage, as well as the average processing time for the proposed algorithms.

Item	Mean time	CPU usage	RAM usage
Algorithm 1	5 sec/image	33%	1.1 GB
Algorithm 2	6.3 sec/image	28%	1.4 GB

Our experiment is randomly pick N photos from the sample database, and to find the position of pupils in a photo, it's accuracy is:

$$Accuracy = \frac{correct\ number\ of\ samples}{sampling\ numbers}$$

If we repeat this process M times, it's average accuracy is:

$$Average\ accuracy = \frac{\sum accuracy}{M}$$

The verification of the segmentation accuracy for all images is another important topic in this study due to the large size of the dataset. We propose to use a self-defined criteria to judge whether two circular boundaries (the ground-truth and the hypothesis) matched to each other. We set the criteria for a correct match to be having the overlapping region of two circles larger than 95%.

Two circles may intersect in two points, a single degenerate point, or does not intersect with each other. The intersections of two circles determine a line known as the radical line. If three circles mutually intersect on a single point, their point of intersection is the intersection of their pairwise radical lines, known as the radical center. Figures 13 and 14 illustrate this idea.

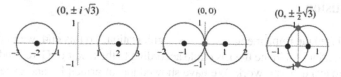

Fig. 13. Three cases of the relation between two circles. From left to right: no intersection, tangent to each other, overlapped.

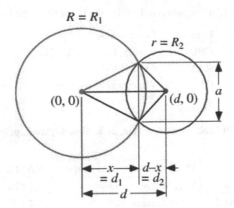

Fig. 14. The calculation of the intersection of two circles.

The running time is increased as the accuracy improved. The final accuracies for ICE and UBIRIS are about 92% and 79% respectively, as shown in Table 3. Note that when we ran this experiment before the code was optimized, the accuracy for ICE dataset was only about 55%. The accuracy for UBIRIS was just about 10%. Therefore, the final result shows significant improvement of accuracy.

Table 3. Result of ICE and UBIRIS datasets

Dataset	# Success	# Failure	NA	Total	Accuracy
ICE	2708	242	3	2953	91.7%
UBIRIS	951	231	21	1203	79.1%

How to fine-tune the parameters in our algorithm such that it can be applicable to any iris database is a challenge in this work. In the beginning, we fine-tune the parameters for only ICE dataset. Later, we discovered that such parameters are only good for ICE, but not UBIRIS. It takes us a lot of time and effort to try to fine-tune a good set of parameters which can be used for both datasets. It requires iterative processes of trials. We believe that there must be some efficient solutions to deal with this issue. Therefore, the future work is to pursue a new parameter optimization method which can work for all kinds of database.

5   Conclusion

We have finished the implementation of iris segmentation. To this end, we implement the pupil segmentation, calculate the iris circle according to the result of GAC, and design the experiment to evaluate our work. We have showed lots of principles and experiments for applying machine learning based solution if we encounter some requirements in our future project. Finally, we have shown proposed algorithm and good result with experiments.

Acknowledgment. Thanks for Prof. Li and Prof. Shin for their valuable suggestion and advice.

References

1. Ratha, N.K., Connell, J.H., Pankanti, S.: Big data approach to biometric-based identity analytics. IBM J. Res. Dev. **59**(2/3), 4:1–4:11 (2015)
2. Liu, C., Petroski, B., Cordone, G., Torres, G., Schuckers, S.: Iris matching algorithm on many-core platforms. In: 2015 IEEE International Symposium on Technologies for Homeland Security (HST), pp. 1–6 (2015)
3. Fernández, A., Gómez, Á., Lecumberry, F., Pardo, Á., Ramírez, I.: Pattern recognition in latin America in the 'Big Data' Era. Patt. Recognit. **48**(4), 1185–1196 (2015)
4. Shah, S., Ross, A.: Iris segmentation using geodesic active contours. IEEE Trans. Inf. Forensics Secur. **4**(4), 824–836 (2009)
5. Ma, L., Tan, T., Wang, Y., Zhang, D.: Efficient iris recognition by characterizing key local variations. IEEE Trans. Image Process. **13**(6), 739–750 (2004)
6. Xu, Z., Shi, P.: A robust and accurate method for pupil features extra. In: 2006 18th International Conference on Pattern Recognition, ICPR 2006, vol. 1, pp. 437–440 (2006)
7. Zuo, J., Kalka, N.D., Schmid, N.A.: A robust iris segmentation procedure for unconstrained subject presentation. In: 2006 Biometrics Symposium: Special Session on Research at the Biometric Consortium Conference, pp. 1–6 (2006)
8. Chouhan, B., Shukla, S.: Comparative analysis of robust iris recognition system using log gabor wavelet and laplacian of gaussian filter. Int. J. Comput. Sci. Commun. IJCSC **2**(1), 239–242 (2011)
9. Ross, A., Shah, S.: Segmenting non-ideal irises using geodesic active contours. In: 2006 Biometrics Symposium: Special Session on Research at the Biometric Consortium Conference, pp. 1–6 (2006)
10. Proença, H., Alexandre, L.A.: Iris segmentation methodology for non-cooperative recognition. In: IEE Proceedings of Vision, Image and Signal Processing, vol. 153, pp. 199–205 (2006)
11. Arvacheh, E.M.: A study of segmentation and normalization for iris recognition systems (2006)
12. Jarjes, A.A., Wang, K., Mohammed, G.J.: Iris localization: detecting accurate pupil contour and localizing limbus boundary. In: 2010 2nd International Asia Conference on Informatics in Control, Automation and Robotics (CAR), vol. 1, pp. 349–352 (2010)
13. Proenca, H.: Iris recognition: on the segmentation of degraded images acquired in the visible wavelength. IEEE Trans. Pattern Anal. Mach. Intell. **32**(8), 1502–1516 (2010)
14. Pawar, M.K., Lokhande, S.S., Bapat, V.N.: Iris segmentation using geodesic active contour for improved texture extraction in recognition. Int. J. Comput. Appl. **47**(16) (2012)
15. Proena, H., Alexandre, L.A.: UBIRIS: a noisy iris image database. In: Proceedings of International Conference on Image Analysis and Processing (ICIAP), vol. 1, pp. 970–977 (2005). http://iris.di.ubi.pt
16. The ICE Iris Image Database. https://www.nist.gov/programs-projects/iris-challenge-evaluation-ice

Building Emotion Recognition Control System Using Raspberry Pi

Hung-Te Lee[1], Rung-Ching Chen[1(✉)], and David Wei[2]

[1] Department of Information Management, Chaoyang University of Technology,
Taichung, Taiwan
crching@cyut.edu.tw
[2] Department of Computer and Information Sciences, Fordham University,
New York, USA
wei@dsm.fordham.edu

Abstract. Facial expression analysis for human-computer interaction, the driver's state monitoring, or user emotion state monitoring has always been a very important issue in emotion recognition. People have a few emotions, and in different emotions, their facial expressions will have different characteristics. For example, if a person is happy, he/she may be with smiling face or smiling eyes, and if a person is angry or sad, he/she may frown. Once the system identified the user's facial expression, there will be a corresponding action. This paper presents a framework for the use of raspberry functions to develop emotion recognition systems. We use Raspberry Pi's camera module to detect the user's facial expressions and use the Microsoft emotion recognition API to identify the user's emotion. If the recognition result is angry or sad emotion, the system will broadcast gentle music and tune the light to be soft to smooth the person's mood. Experiments results indicated the system can interact with users' emotions.

Keywords: Human-computer interaction · Microsoft Emotion API
Raspberry Pi · Emotion recognition

1 Introduction

IoT is becoming a popular research topic due to the rapid advances of the network technologies and the increasing scope of IT applications. Meanwhile, the demand for high performance computer systems is getting higher and higher, hoping that the computer can bring us a good quality of life. We are thus increasingly paying attention to human-computer interaction. Humans tend to convey their emotions through their behaviors in which not only show human emotions, but also entrain a lot of messages. Through these messages, one can understand the state of the person at that time. For example, in the image care system, by observing a patient's facial expressions and behavior to confirm whether the patient's condition is stable, or in the drive monitoring system, when a driver is fatigue, the driver's face will show signs of fatigue and his/her facial expressions will change. Then the monitoring system will take appropriate action by reminding the driver to have a rest.

J. C. Hung et al. (Eds.): FC 2017, LNEE 464, pp. 36–45, 2018.
https://doi.org/10.1007/978-981-10-7398-4_4

Computer vision plays an important role in the development of human-computer interaction. If the computer vision can help identify a person's facial expressions so that the computer can provide appropriate treatment, then the relationship between human and computer becomes closer. Therefore, facial expression recognition has become a popular research topic in recent years. Based on the captured information of facial features to quantify and classify, a computer will be able to learn to identify a person's emotional state and can provide better services to human. For example, when a person's emotion is recognized as joy or in good mood, the system will broadcast brisk rhythm music, or the light will be tuned into a more vivid color.

There are two categories of methods in facial image feature extraction. One method extracts facial expression by using geometric features [4–7] which is based on Facial Action Coding System (FACS) [10]. FACS can describe the displacement situation of specific region in face area. The other method extracts facial expression using facial appearance features to process texture features [8, 9]. Zhang et al. [12] compare the two methods for facial expression. According to the result, the method of using regional texture features has better performance than the method of using geometric features. P. Viola and M. Jones presented Viola-Jones object detection framework [1, 11] that becomes one of most popular classifier in facial expression recognition. In addition, Viola-Jones object detection framework has great performance in computation speed.

In this paper, we propose a method to apply the research of expression recognition to real life. The raspberry pi is used as a tool to collect human facial images, and the images are sent to the emotional recognition API for identification through the R language. The result is passed back to the raspberry pi. Then, based on the received result, the Raspberry sends corresponding order to control devices.

The rest of this paper is organized as follows: Sect. 2 reviews the related image recognition technology. Section 3 presents our research methods, and Sect. 4 discusses the experimental results. Finally, we give a conclusion in Sect. 5.

2 Background

2.1 Image Processing on Raspberry Pi

Raspberry pi is a kind of micro-computer that has a variety of expansion modules. It can use the camera module to collect image information. Raspberry pi supports Python which supports cross-platform computer vision library (OpenCV). Thus, the Raspberry pi can be used to develop applications of image recognition.

2.2 Viola-Jones Target Detection Framework

Viola-Jones target detection framework is a classifier [1, 11], which has three characteristics. First, the classifier can use each pixel difference to find the rectangle features, and then uses these rectangular features to define the framework of the human face. Figure 1 is an example that uses rectangular features to find out the face frame. This method has a very good performance in terms of the speed of calculation. The second feature is that this method is with real-time characteristics, and its cross-platform computer vision library also provides functions for system developers to use. The third feature is that the Viola-Jones target detection framework is an adaptive enhancement classifier (AdaBoost). The feature of the adaptive enhancement classifier is that the samples that have been misclassified will be used to train the next classifier. Although at the beginning, it is a Weak Classifier, after training, it can also turn to a Strong Classifier.

Fig. 1. Using rectangular features to find out the face frame

2.3 Convolutional Neural Network

Convolutional Neural Network (CNN) [2] is a kind of feedforward neural network. Feedforward neural network is the simplest and the earliest invention in neural network. A convolutional neural network consists of a fully connected layer and one or more convolutions. It also contains the associated weights and pooling layers. Convolution neural network has good performance in image and speech recognition.

2.3.1 Convolutional Layer

Each convolution layer [2] consists of a number of convolution unit. The goal of convolution operations is to obtain the different features of the input. The first convolution layer may only be able to obtain the primary features such as corners, edges, or lines. More advanced features can be obtained from the primary features through iterations.

2.3.2 Rectified Linear Units Layer

Rectified Linear Units layer (Re LU layer) [2] uses linear rectification as the activation function of the neural layer.

$$f(x) = \max(0, x) \tag{1}$$

It optimizes the nonlinearity of the decision function and the entire neural network, and the linear rectifier does not affect the convolution layer. In addition, other functions can also be used to improve the nonlinear characteristics of the network, such as hyperbolic tangent function,

$$f(x) = \tanh(x), f(x) = |\tanh(x)| \tag{2}$$

The advantage of the Re LU function is that it increases the training speed of the neural network several times but would not affect the generalization accuracy of the whole model.

2.3.3 Pooling Layer

Pooling [2] is an important part in convolutional neural networks, where pooling is a down-sampling of a pattern. There are many models for nonlinear pooling functions, and "Max pooling" is one of the most commonly used function. Pool maximization divides input image into multiple rectangular area, and then for each rectangular area output maximum value.

This mechanism can is efficient because after the discovery of a feature, the exact position of the feature is far less important than the relationship between the feature and the relative position of other features. Pooling layer will continue to reduce the size of the data space so that the number of parameters and calculation will also decline, and it thus can control over fitting. The pooled layers are periodically interposed between the CNN convolutional layers.

Pooled layers typically operate on each imported feature and reduce the size of the feature. The more commonly used form of the pool layer is the one that for every 2 elements there will be a block 2 * 2 separated from the image, and then the maximum value among the 4 values of the block will be used. This way it can reduce the amount of data by 75%. Figure 2 shows the concept of pooling.

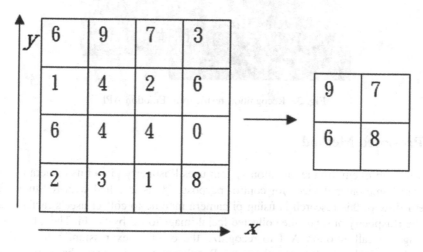

Fig. 2. Concept of pooling

Because pooled layers shrink the size of data too quickly, the current trend is to use pooled filters instead of pooled layers.

2.3.4 Loss Layer

Loss layer [2] is used to judge the training process, compare the predicted results with the real results, and then decide how to adjust the next training. Loss layer usually is the final layer in CNN.

2.4 Emotion API

Microsoft Emotion API [3] processes images uploaded by users. It gives every facial set a score. Emotion API also uses Face API that can detect the facial location.

There are eight kinds of emotion in the feedback result, including anger, contempt, disgust, fear, happiness, neutral, sadness, and surprise. These emotions are with different facial expression features. Facial expressions are also important part in the communications between people. Figure 3 is an example of the result of uploading image to Emotion API.

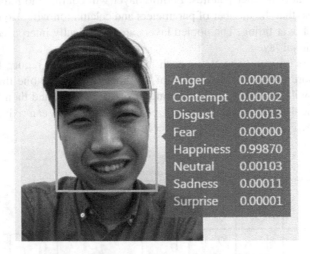

Fig. 3. Recognition result from Emotion API

3   Proposed Method

Our research of emotion recognition system uses Raspberry pi and its camera module, PC, and some other devices for control purpose. PC is installed with R Studio. The system flow of this research is using pi camera module to collect user's facial image, and the Raspberry pi sends the collected facial image to PC by Wi-Fi. Then, PC uses R language to call emotion API to recognize the emotion expression. When PC gets result, it sends the result to Raspberry pi. Raspberry pi will control the corresponding devices based on the recognized result. Figure 4 shows the architecture of our system.

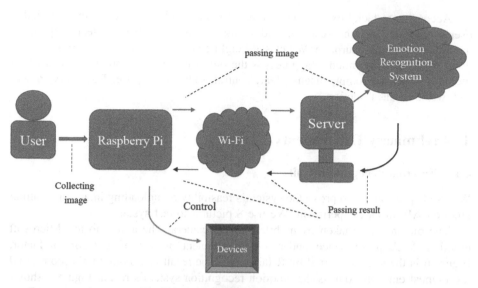

Fig. 4. Structure of the system

We may get 27 landmark points from Emotion API [1] (See Fig. 5). These 27 landmark points include Nose Root Left, Eyebrow Left Inner, Eyebrow Left Outer, Eye Left Top, Pupil Left, Eye left Outer, Eye Left Bottom, Eye Left Inner, Nose Left Tip, Nose Left Alar Top, Nose Left Alar Out Tip, Mouth Left, Nose Root Right, Eyebrow Right Inner, Eyebrow Right Outer, Eye Right Top, Pupil Right, Eye Right Outer, Eye Right Bottom, Eye Right Inner, Nose Right Tip, Nose Right Alar Top, Nose Right Alar Out Tip, Mouth Right, Upper Lip Top, Upper Lip Bottom, Under Lip Top, Under Lip Bottom.

Fig. 5. The landmark points of emotion API

According to facial expression features, emotion API can process image to judge user's emotion, and the result includes anger, contempt, disgust, fear, happiness, neutral, sadness, and surprise. We classify eight emotions into two categories, namely good mood and bad mood. Happiness is the only case of good mood, and bad mood includes anger, contempt, disgust, fear, neutral, sadness, surprise. Raspberry Pi will follow the results to action.

4 Preliminary Experiments

4.1 Emotion Recognition Unit

We use R programing to prove the system's feasibility by uploading image and calling emotion API to start experiment. We use 8 pictures to test system.

The pictures were taken by mobile phone's camera. The aim is to test Microsoft emotion API's performance under the environment of free of any control factor. Figure 6 is the result exported by R language. The result has 8 emotion's scores, and the highest emotion point is the emotion recognition system's result. Figure 6 shows that happiness's point has 0.997591 that is highest point in 8 emotion's scores. Thus, the emotion recognition system's result is happiness.

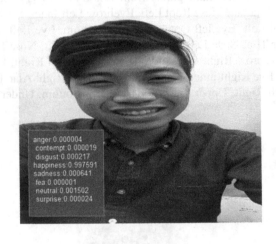

0:maxXY

Fig. 6. Result exported by R language

We upload 8 kinds of emotion pictures to system, and we obtain 8 results shown in Table 1. 8 kinds emotion pictures includes AN (anger), CO (contempt), DI (disgust), FE (fear), HA (happiness), NE (neutral), SA (sadness), SU (surprise), and table fields are divided into image input and result. We can see that the system does not perform good on processing anger, contempt, disgust, fear, sadness, surprise, but works well on processing happiness and neutral.

Table 1. The result of testing each emotion in emotion recognition system

AN image input	Result	CO image input	Result	HA image input	Result	NE image input	Result
	AN: 0.0026 CO: 0.0004 DI: 0.0025 FE: 0.0020 HA: 0.0004 **NE: 0.9110** SA: 0.0550 SU: 0.0215		AN: 0.0009 CO: 0.0154 DI: 0.0010 FE: 0.0021 HA: 0.0015 **NE: 0.7750** SA: 0.1940 SU: 0.0100		AN: 0.0056 CO: 0.0001 DI: 0.05320 FE: 0.0001 **HA: 0.9280** NE: 0.0133 SA: 0.0033 SU: 0.0017		AN: 0.0015 CO: 0.0244 DI: 0.0042 FE: 0.0002 HA: 0.0002 NE: 0.8770 **SA: 0.0876** SU: 0.0053

DI image input	Result	FE image input	Result	SA image input	Result	SU image input	Result
	AN: 0.0015 CO: 0.3350 DI: 0.0424 FE: 0.0066 HA: 0.0004 **NE: 0.4460** SA: 0.1310 SU: 0.0372		AN: 0.0010 CO: 0.0023 DI: 0.0026 FE: 0.0051 HA: 2.94e-05 **NE: 0.7810** SA: 0.0408 SU: 0.1670		AN: 0.0155 CO: 0.0283 DI: 0.0025 FE: 0.0051 HA: 0.0002 **NE: 0.5190** SA: 0.4200 SU: 0.0104		AN: 0.0001 CO: 0.0002 DI: 0.0030 FE: 0.0011 HA: 4282e-06 NE: 0.0928 SA: 0.0041 **SU: 0.8990**

4.2 Control Unit

The system sends result to Raspberry pi after it obtained the result from emotion recognition system. Raspberry pi will judge the result, and send instruction to control Bluetooth color ball via Bluetooth. Figure 7 shows the images of various colors of Bluetooth color ball that are controlled by different instructions of Bluetooth.

Fig. 7. Bluetooth color ball

Table 2. Every color represents different emotions

Emotion	Color	Emotion	Color	Emotion	Color	Emotion	Color
Anger	Red	Happiness	Orange	Disgust	Green	Sadness	Blue
Contempt	Lake green	Neutral	Yellow	Fear	Purple	Surprise	Pink

We set eight emotions to eight different colors. These eight colors include red for anger, lake green for contempt, green for disgust, purple for fear, orange for happiness, yellow for neutral, blue for sadness, and pink for surprise. Table 2 shows color information of Bluetooth color ball, and we can see user's emotion by its corresponding color.

5 Conclusions

The experiment results presented in this paper are just preliminary research results. There is still room to improve in this research. So far, emotion API has good enough performance on Happy, but not the case when processing other types of emotion. Our future research will focus on processing the feedback of emotion API, or joining image preprocessing action. We will also work on improving emotion recognition system's performance. We also expect the system can control Audio-visual equipments or temperature equipments in future.

Acknowledgements. This paper is supported by Ministry of Science and Technology, Taiwan, (Grant Nos. MOST-104-2221-E-324-019-MY2 and MOST- 103-2632-E-324-001-MY3).

References

1. https://dataholic.wordpress.com/2016/10/29. Accessed 29 Oct 2016
2. https://en.wikipedia.org/wiki/Convolutional_neural_network
3. Microsoft Emotion API. https://www.microsoft.com/cognitive-services/en-us/emotion-api
4. Kotsia, I., Pitas, I.: Using geometric deformation features and support vector machines. IEEE Trans. Image Process. **16**(1), 172–187 (2007)

5. Pantic, M., Patras, I.: Dynamics of facial expression: recognition of facial actions and their temporal segments from face profile image sequences. IEEE Trans. Syst. Man Cybern. Part B Cybern. **36**(2), 433–449 (2006)
6. Pantic, M., Rothkrantz, L.J.M.: Facial action recognition for facial expression analysis from static face images. IEEE Trans. Syst. Man Cybern. Part B Cybern. **34**(3), 1449–1461 (2004)
7. Pantic, M., Rothkrantz, L.: Expert system for automatic analysis of facial expression. Image Vis. Comput. **18**(11), 881–905 (2000)
8. Bartlett, M.S., Littlewort, G., Frank, M., Lainscsek, C., Fasel, I., Movellan, J.: Recognizing facial expression: machine learning and application to spontaneous behavior. In: Proceedings of IEEE Conference on Computer Vision and Pattern Recognition, vol. 2, pp. 568–573 (2005)
9. Bartlett, M.S., Movellan, J.R., Sejnowski, T.J.: Face recognition by independent component analysis. IEEE Trans. Neural Netw. **13**(6), 1450–1464 (2002)
10. Ekman, P., Friesen, W.V.: Facial Action Coding System. Consulting Psychologists Press, Palo Alto (1978)
11. Viola, P., Jones, M.: Robust real time object detection. In: IEEE ICCV Workshop on Statistical and Computational Theories of Vision (2001)
12. Zhang, Z., Lyons, M.J., Schuster, M., Akamatsu, S.: Comparison between geometry-based and Gabor-wavelets-based facial expression recognition using multi-layer perceptron. In: Proceedings of IEEE International Conference on Automatic Face and Gesture Recognition, pp. 454–459 (1998)

A Study on the Binding Ability of Truncated Aptamers for the Prostate Specific Antigen Using Both Computational and Experimental Approaches

Hui-Ting Lin[1], Wei Yang[2], Wen-Yu Su[3], Chun-Ju Chan[3], Wen-Yih Chen[4], Jeffrey J. P. Tsai[3], and Wen-Pin Hu[3,5(✉)]

[1] Department of Physical Therapy, I-Shou University, Kaohsiung City 82445, Taiwan
huitinglin@isu.edu.tw
[2] Institute of Chemical Engineering, National Taipei University of Technology, Taipei City 10608, Taiwan
wei38@ntut.edu.tw
[3] Department of Bioinformatics and Medical Engineering, Asia University, Taichung City 41354, Taiwan
{wenyusu,president,wenpinhu}@asia.edu.tw, y5180927wk@gmail.com
[4] Department of Chemical and Materials Engineering, National Central University, Jhong-Li 32001, Taiwan
wychen@ncu.edu.tw
[5] Department of Medical Laboratory Science and Biotechnology, China Medical University, Taichung City 40402, Taiwan

Abstract. Prostate-specific antigen (PSA) test is a commonly used clinical examination to evaluate the risk of prostate cancer, with the antibodies used normally as the recognition molecules for measuring PSA levels in serum. Alternatively, aptamers that are able to bind target molecules with high affinity and specificity similar to antibodies could be generated much easier and cheaper than the production of antibodies. In this study, we used computaional and experimental approaches to select truncated PSA-binding aptamers generated from the sequence information of PSA-binding aptamers previously reported in a literature. Genetic algorithm, the analysis of secondary structure, and molecular simulation were utilized in the *in silico* analysis. The top 4 ranked sequecnes *in silico* analysis were evaluated through their PSA-binding ability on the quartz crystal microbalance (QCM) biosensor. Finally, We identified a truncated aptamer obtained from the selection showing a nearly 3.5-fold higher measured signal than the response produced by the best known DNA sequence in the QCM measurement.

Keywords: Prostate specific antigen (PSA) · Aptamer
Quartz crystal microbalance (QCM) · Molecular simulation

1 Introduction

Prostate-specific antigen (PSA) is an important and commonly used biomarker to diagnose prostate cancer. PSA is secreted by prostate epithelial cells to the circulatory system

© Springer Nature Singapore Pte Ltd. 2018
J. C. Hung et al. (Eds.): FC 2017, LNEE 464, pp. 46–55, 2018.
https://doi.org/10.1007/978-981-10-7398-4_5

of patient with prostate cancer of which the PSA concentration is much higher when comparing to the PSA level in health people. The cut-off value of serum PSA which can be used positively to make the diagnosis of prostate cancer is 4.0 ng/ml [1, 2]. In patients with PSA levels between 4 and 10 ng/ml, they have a high risk of prostate cancer. For the PSA level of patient is above 10 ng/ml, risk of prostate cancer is much higher. The PSA test commonly use antibodies to recognize the circulation PSA in the serum of patient.

Most aptamers consist of nucleic acids and usually have a length of 12 to 80 nucleic acids. Except for DNA and RNA-type aptamers, peptide aptamers are also developed and selected to bind their target molecules. Peptide aptamers are small combinatorial proteins and consist of a short variable peptide domain [3]. The target molecules of aptamers include metal ions, pathogenic microorganisms, short peptide, micromolecules and macromolecules. While aptamers are analogous to antibodies in their range of target recognition and variety of applications, they possess several advantages over traditional antibodies, such as small molecules, high stability, easy production, possibility for reuse, and unchanged affinity after simple chemical modifications [4]. Because of these advantages, aptamers can be used in biosensor, diagnosis and therapeutic applications.

Considering about the development of PSA-binding aptamer, two types of aptamers were reported by Jeong's [5] and Savory's [6] groups in 2010. Jeong et al. [5] identified specific RNA aptamers against the active PSA with a length of 41 mer. Savory et al. [6] initially obtained five DNA sequences by the systematic evolution of ligands by exponential enrichment (SELEX). After that, the genetic algorithm (GA) was used to generate next-generation sequences for evaluating their PSA-binding ability in the plate assay. Following this, the selection process was repeatedly applied until obtaining a fourth-generation of DNA aptamer which has the highest PSA-binding ability. Generally speaking, a shorter length in sequence could be synthesized much easier for an aptamer. Therefore, Anderson et al. [7] successfully reduced the length of the 35-mer aptamer against cofactor flavin mononucleotide (FMN) to 14 mer by using experimental and computational methodologies, with these truncated aptamer possessing higher binding ability to FMN.

Herein, the main purpose of this study was that we wanted to carry out the selection of truncated DNA aptamers against PSA by using computational and experimental approaches. The original lengths of DNA aptamers reported by Savory et al. [6] were 32, and a total number of 8 consecutive thymine bases existed in the sequences at the 5' and 3' end. We adopted four aptamers from Savory's report as the original source of parent sequence and deleted the consecutive thymine bases at two ends for the subsequent selection processes. By using the selection strategy, we found a new DNA aptamer with a length of 24 mer and the truncated aptamer exhibited high binding capability with PSA in the experiments performed on the quartz crystal microbalance (QCM) biosensor.

2 Materials and Methods

2.1 The Original Sequences of Aptamers

We adopted the four sequences reported by Savory et al. [6], which could bind to PSA and were selected as the original sequences in this study. The names of four sequences were PSap2#1, PSap2#2, PSap2#16 and PSap2#18. Savory et al. [6] found these sequences by using the initial SELEX procedure and subsequent post-SELEX screening with a genetic algorithm (GA). In their study, a fourth-generation aptamer, ΔPSap4#5 (5′-tttttAATTAAAGCTCGCCATCAAATAGCttt-3′), showed the highest PSA-binding ability. Besides, the sequences of four aptamers used in the plate assay contained 5 and 3 consecutive thymine bases at the 5′ and 3′ ends, respectively. For the purpose of truncating the length of aptamer, the consecutive thymine bases at two ends are removed. The sequences of four PSA-binding aptamers used in this study are listed as follows.

- PSap2#1: 5′-GCCTACTGATTTCCTTTTTGAGCC-3′
- PSap2#2: 5′-AATTAAGGATTTCCCGGTTGTATC-3′
- PSap2#16: 5′-ACTGTGAACTCGCCATCAAATATC-3′
- PSap2#18: 5′-AATGTCAACGTTGTTTACTGTCCC-3′

2.2 Research Procedures

The research procedures are shown in Fig. 1, and procedures start from the generation of 200 mutated sequences. The mut function in genetic algorithm toolbox for MATLAB® [8] was utilized to generate random mutated sequences for selection. Firstly, bases A, T, G, and C for the sequences of four PSA-binding aptamers were numbered 00, 01, 10, and 11, respectively. By running the mut function, it could generate new sequences in numerical format and then we used a C# program written by ourselves to convert these sequences back to the original base sequence format. The parameter of mutation probability in the function was set as the default value (0.7/the length of the chromosome structure). With the use of the function, we generated a total number of 200 mutated sequences. However, the mut function produced repeated sequences in different runs. We checked and deleted the repeated sequences, and the final number of new mutated sequences were 152.

After getting these mutated sequences, we utilized RNAfold web server to analyze the secondary structures of sequences. The web server gives the prediction information of the secondary structures of sequences by using Dot-Bracket Notation (DBN). The dot symbol means the position of nucleotide is unpaired. Therefore, the DBN exclusively comprises of periods that indicates such a sequence does not have a clear secondary structure. We deleted these sequences and the rest of the sequences were used to build 3D structural models with the use of RNAComposer web server. For using RNAComposer web server, we changed the letter code T in the DNA sequence to U. After getting the models, we adopted Accelrys Discovery Studio (DS) 3.5 to edit the pyrimidine bases in the 3D structural models of RNA to make Uracil become Thymine. The protein model of PSA numbered "3QUM" was downloaded from Protein Data Bank (PDB). The 3D

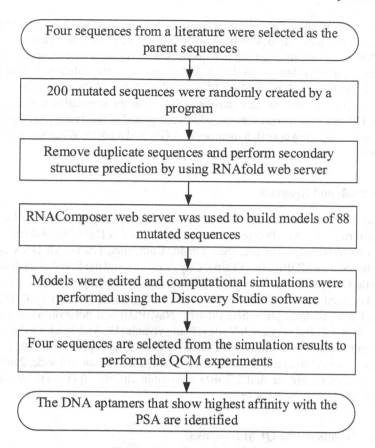

Fig. 1. Flow chart of research

structural model has three biomolecules, including the PSA and two Fab fragments of monoclonal antibodies (antibodies are numbered as "5D5A5" and "5D3D11"), in this file. Except for the structural data of PSA, another two biomolecules were removed by using the DS software. After the cleaning process, the structural file of PSA was ready for using in simulations.

The ZDOCK simulation function in the DS software was adopted to evaluate the interactions between the aptamers and PSA. The ZDOCK function is a Fast Fourier Transform based protein docking program, which uses the shape complementary (SC) to evaluate the interactions between two biomolecules. The fast Fourier correlation techniques that calculate shape complementarity are also applied in another study for investigating protein/DNA complexes [9]. The ZDOCK scoring function also take electrostatics and pairwise atomic potential into consideration. This docking method usually used in studying protein–protein interactions. We chose the ZRANK scoring function to re-rank the ZDOCK results in order to get more precisely prediction results. Because ZRANK program uses an optimized energy function that can remarkably advance the success rate of prediction from docking results of ZDOCK [10]. In the RANK scoring function, van der Waals attractive and repulsive energies, short and long range repulsive and attractive energies, and desolvation are all combined in the calculation. According the ZRANK score, we selected top 4 sequences with higher ZRANK scores and used them to carry out experiments on the QCM biosensor for the validation of simulation results. The aptamer with the highest PSA-binding ability (ΔPSap4#5) discovered by Savory et al. [6] was used as the control group both in the simulation and experiment. We used a HP server to run simulations, which comprised of one Intel® Xeon processors (containing 8 computing cores), a memory of 4 GB, and a 64-bit Windows Server 2008 R2 Operating System.

2.3 Materials and Reagents

The human prostate specific antigen was purchased from MP Biomedicals, LLC. (Santa Ana, California, USA). Poly(ethylene glycol) thiol (Thiol-PEG4-Alcohol) was purchased from Broadpharm Inc. (San Diego, California, USA). All DNA aptamers were synthesized by MDBio, Inc. (Taipei City, Taiwan), and the 5′end of each sequence was modified with 1-hexanethiol (C_6SH). Sodium dihydrogen phosphate monohydrate (KH_2PO_4) and sodium chloride were obtained from Sigma-Aldrich Co. LLC. (St. Louis, Missouri, USA). Sodium phosphate dibasic (Na_2HPO_4) and potassium chloride were purchased from J.T. Baker (Center Valley, Pennsylvania, USA). We adopted phosphate-buffered saline (PBS) solution with 1X concentration in the experiments. The PBS solution contains 10 mM sodium phosphate dibasic, 137 mM sodium chloride, 2 mM potassium dihydrogen phosphate and 2.7 mM potassium chloride (pH 7.4). All chemicals used in this study were reagent grade.

2.4 Experiments on the QCM Biosensor

QCM experiments were conducted after computational analyses for evaluating the binding reactions between the DNA aptamers and PSA. The QCM instrument (Model 410C) is capable of ultrasensitive mass measurements and manufactured by CH Instruments, Inc. (USA). The gold disk on the chip surface was firstly cleaned with pure ethanol once and then blown dry with nitrogen gas. The immobilization of DNA aptamer on the surface of gold disk was achieved by formation of gold-sulfur bond. We used the 1 M KH_2PO_4 solution to dissolve the Thiol-PEG$_4$-Alcohol and synthesized DNAs to the concentrations of 19 mM and 0.2 μM, respectively. Afterwards, adequate amounts of two solutions were took and added together to make the DNA with a final concentration of 100 nM and the PEG thiol with a concentration of 5 μM in the mixed solution. From which, the molar ratio of DNA to PEG was 1 to 50. For the surface functionalization, the chip was placed in a petri dish and then 0.3 ml of the DNA/PEG mixed solution was dropped on the chip surface to cover the area of gold disk. Next, the petri dish was sealed with parafilm and placed in a freezer at 4 °C for 24 h. During the reaction, the sulfur moieties of DNA aptamer and PEG could bind to the gold surface via the formation of gold-sulfur bond and generate a DNA/PEG self-assembled monolayer (SAM) on the surface of gold disk. The PEG molecules could prevent and minimized the nonspecific adsorption of proteins. Additionally, space provided by these PEG molecules could further reduce the steric hindrance for upcoming molecular interactions. The modified chip was rinsed thoroughly with ethanol and blown dry with nitrogen gas. After that, the chip was installed into a QCM cell and ready to perform a measurement. We prepared 10 mM phosphate buffered saline (PBS) solution (pH 7.4) for using in the QCM measurements and dissolving the PSA protein. In order to carry out the experiments, we initially added 3.3 ml of PBS solution to the QCM chamber and started to oscillate the quartz crystal. Until oscillation frequency of the chip reached a stable state, the PSA solution (0.5 ml) with the concentration of 2 ng/ml was injected into the cell to monitor changes in the oscillation frequency caused by binding of the PSA and the immobilized aptamer on the chip surface. The experiments for each aptamer were performed in triplicate.

3   Results and Discussions

3.1   Computational Selection of PSA-Binding Aptamer

For evaluating the interaction of PSA and the mutated DNA sequence, it will take almost 9 to 12 h for the computer in each sequence calculation. Table 1 shows the docking results of PSA with four mutated DNA sequences and ΔPSap4#5. The docking results indicated that PSA-binding ability of these five aptamer in the light of ZRANK score, listed in descending order, were TA13, TA12, TA14, TA87 and ΔPSap4#5. The sequences of TA13, TA12, TA14 and TA87 are 5′- ACCGTGAACTCGCCATCAAA-AATC-3′, 5′-ACCGTGAACCCGCCATCAAAA-ATC-3′, 5′-ACCGTGAACTCGC-CATCAAATATC-3′ and 5′-TCAGTGAACTCGC-CATCAAATATC-3′, respectively. As shown, these four aptamers might have better PSA-binding ability, but further verification on the prediction results using experimental approaches is still required. Because the ZDOCK uses a simple shape complementarity method to study the interaction between two molecules and all six rotational and translational degrees of freedom are fully explored with scoring functions. Sometimes, the best docking pose of aptamer predicted by the ZDOCK algorithm are not in the right location to interact with some amino acids of protein that are important for the binding of aptamer. Therefore, we also examined amino acids involved in aptamer-protein binding interface of best docking pose of each aptamer.

Table 1. The four mutated sequences with the best docking results and one reported PSA binding aptamer (ΔPSap4#5)

Sequence name	ZDOCK score	ZRANK score
TA12	17.7	−98.877
TA13	21.24	−103.239
TA14	20.96	−98.437
TA87	21.06	−93.638
ΔPSap4#5	15.16	−74.673

Figure 2 shows the best docking poses of ΔPSap4#5 against PSA and TA87 against PSA. The binding ability of ΔPSap4#5 had been demonstrated in the literature [6], so the amino acids that were predicted in the binding interface of ΔPSap4#5 and PSA were used to evaluate the accuracy of binding locations of other aptamers. Amino acids involved in binding with aptamers are mentioned below with their abbreviations and position numbers. The amino acids listed with bold and italic styles indicates they also appear in the binding interface between PSA and ΔPSap4#5.

ΔPSap4#5: *S35*, *R36*, *R38*, *A39*, *V41*, *C42*, A56, *H57*, *C58*, *I59*, *R60*, *N61*, K62, *H87*, *S88*, *F89*, *P90*, *P92*, *Y94*, *M95A*, L95C, L95D, *K95E*, *N95F*, *F95H*, M104, *I144*, *E145*, *P146*, *E147*, *E148*, *F149*, *D194*, *S195*, *W215*, *E218*

TA12: I16, V17, W20, E21, C22, E23, K24, L32, V33, A34, *S35*, *R36*, *G37*, *R38*, *A39*, *V41*, *C42*, K62, S63, V64, I65, L66, R69, H70, L72, F73, H75, P76, E77, D78, G80, V82, F83, F149, T151, K153, K154

TA13: V17, G18, G19, L32, V33, A34, G37, *R38*, *A39*, *V41*, *C42*, L66, R69, S71, L72, F73, H75, P76, E77, W141, G142, *I144*, *E145*, *P146*, *E147*, *E148*, *F149*, L150, T151, P152, K153, V171, W186, K188, S189, T190, C191, *D194*, *E218*, P219, C220, A221, L222, P223, E223A, R224

TA14: V17, G18, G19, L32, V33, A34, *R36*, G37, *R38*, *A39*, *V41*, *C42*, L66, R69, S71, L72, F73, H75, P76, E77, W141, S143, *E145*, *P146*, *E147*, *E148*, *F149*, L150, T151, P152, K153, V171, T190, C191, *D194*, *E218*, C220, A221, L222, P223, E223A, R224, P225

TA87: A34, *S35*, *R36*, G37, *R38*, *A39*, *V41*, *H57*, *C58*, *I59*, *R60*, *N61*, *H87*, *S88*, *F89*, *P90*, H91, *P92*, L93, *Y94*, *M95A*, *K95E*, *N95F*, R95G, *F95H*, L95I, R95 J, *I144*, *E145*, *P146*, *E147*, *S195*, *W215*, G216, *E218*, W237

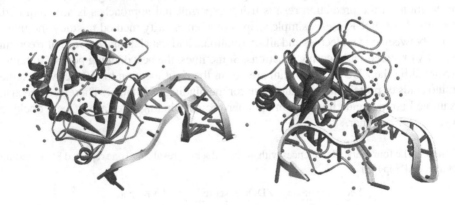

Fig. 2. Docking results between aptamers and PSA (a) ΔPSap4#5 against PSA (b) TA87 against PSA. The DS software can analyze the binding interface between the aptamer and PSA and obtain the information of amino acids and nucleic acids involved in the interaction.

For TA87 aptamer, 75% (27/36) of the amino acids of PSA involved in the aptamer-PSA interaction are the same as ΔPSap4#5. As for the other three aptamers, the percentage values for TA12, TA13 and TA14 are 18.9 (7/37), 26.1 (12/46) and 27.9 (12/43), respectively.

3.2 Sensing of PSA by the QCM Instrument

The four sequences (TA12, TA13, TA14 and TA87) according to the ZRANK scores and ΔPSap4#5 reported in the literature were selected for the aptamer synthesis before QCM experiments were conducted. Representative results of the experiments for these five sequences are shown in Fig. 3. It is worth noting that three of four selected aptamers don't exhibit good binding ability of PSA. By the end of our measurement, only a very slight of or almost no frequency changes were observed for TA12, TA13 and TA14. There is a large difference between the experimental results and the predicted results obtained from the molecular simulation. However, TA87 showed excellent binding ability of PSA as the prediction of simulation. The average value in the decrease of frequency change generated by the binding reaction between the immobilized TA87 aptamer and PSA was 271 (SD = 25.1). For ΔPSap4#5, it could produce an average value of 76.1 (SD = 2.1) in the decrease of frequency variation of QCM chip. TA87 aptamer could generate almost a 3.5-fold greater measured signal than the response produced by ΔPSap4#5.

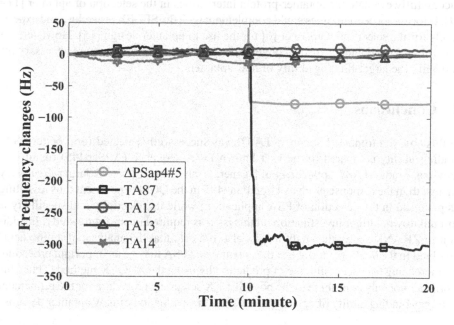

Fig. 3. The QCM experiments for the five aptamer sequences listed in Table 1. The binding reaction between the aptamer and PSA caused mass changes on the surface of a quartz crystal resonator and therefore influenced the oscillation frequency. The aptamer, named as TA87, showed the best binding ability to PSA.

Through the analysis of amino acids involved in binding interface between the aptamer and PSA, the information of amino acids for TA12, TA13 and TA14 revealed that only small portions of amino acids were sharing the same sequences in binding interface of the ΔPSap4#5 and PSA. We speculated that the big difference between the experimental and simulation results was mainly caused by the best docking pose of aptamer predicted by the simulation which didn't interact with the amino acids of PSA that were important for the aptamer-PSA binding. Other factors in this study may also contribute to the inconsistent results between experiments and simulations. The first factor is that the ZDOCK is not the optimal algorithm for the evaluation of aptamer-protein interaction. Secondly, the condition for the immobilization of aptamer on the sensor surface is different from the one used in simulation. In reality, the immobilized aptamer may hinder its further interaction with the target protein while the simulation has no consideration in this problem. The third factor is that the RNAComposer web server is specifically designed to generate RNA models. Although we used DS software to edit the atoms of model to make RNA models which make them become acceptable and reasonable DNA models, the DNA model may be still different from the actual situation. The last factor is that the experimental conditions are more complicated than the conditions of simulation.

The results of several studies revealed that the computational approaches could successfully evaluate the aptamer-protein interactions in the selection of aptamer [11–14]. Information technologies could complement with the SELEX procedure and experiments for the selection of aptamer [6] for the use of aptamer design [15]. However, the experiments conducted by using conventional bioassays or biosensors are necessary for verifying the target-binding ability of new aptamers.

4   Conclusions

In this study, a truncated aptamer, TA87, was successfully selected for a better PSA-binding ability compared to the best known DNA sequence (ΔPSap4#5) found in a previous study. TA87 aptamer could generate an almost 3.5-fold larger frequency change than the response produced by ΔPSap4#5 in the QCM experiment thus revealing its potential in the detection of PSA application; while its selectivity and specificity is still unknown, further investigation on this issue is required in our future study. In addition to ZRANK score, the study results also indicate that the analysis of amino acids involved in the binding interface of the aptamer and PSA may be an important procedure for increasing the successful rate of prediction by using the ZDOCK method. Thus, this work can not only contribute to the post-SELEX screening procedure for the refinement of target-binding ability of aptamer but also to the evaluation of new aptamer designs.

Acknowledgements. The authors gratefully thank the financial support provided by Ministry of Science and Technology, Taiwan under the following contract numbers: MOST 103-2221-E-468-019 and MOST 105-2221-E-468-017.

References

1. Lojanapiwat, B., Anutrakulchai, W., Chongruksut, W., Udomphot, C.: Correlation and diagnostic performance of the prostate-specific antigen level with the diagnosis, aggressiveness, and bone metastasis of prostate cancer in clinical practice. Prostate Int. **2**, 133–139 (2014)
2. Catalona, W.J., Smith, D.S., Ornstein, D.K.: Prostate cancer detection in men with serum PSA concentrations of 2.6 to 4.0 ng/ml and benign prostate examination: Enhancement of specificity with free PSA measurements. JAMA **277**, 1452–1455 (1997)
3. Reverdatto, S., Burz, D.S., Shekhtman, A.: Peptide aptamers: development and applications. Curr. Top. Med. Chem. **15**, 1082–1101 (2015)
4. Thiviyanathan, V., Gorenstein, D.G.: Aptamers and the next generation of diagnostic reagents. Proteomics. Clin. Appl. **6**, 563–573 (2012)
5. Jeong, S., Han, S.R., Lee, Y.J., Lee, S.-W.: Selection of RNA aptamers specific to active prostate-specific antigen. Biotechnol. Lett. **32**, 379–385 (2010)
6. Savory, N., Abe, K., Sode, K., Ikebukuro, K.: Selection of DNA aptamer against prostate specific antigen using a genetic algorithm and application to sensing. Biosens. Bioelectron. **26**, 1386–1391 (2010)
7. Anderson, P.C., Mecozzi, S.: Identification of a 14mer RNA that recognizes and binds flavin mononucleotide with high affinity. Nucleic Acids Res. **33**, 6992–6999 (2005)
8. Chipperfield, A.J., Fleming, P.J.: The MATLAB genetic algorithm toolbox. In: IEE Colloquium on Applied Control Techniques Using MATLAB, pp. 10/1–10/4 (1995)
9. Aloy, P., Moont, G., Gabb, H.A., Querol, E., Aviles, F.X., Sternberg, M.J.: Modelling repressor proteins docking to DNA. Proteins **33**, 535–549 (1998)
10. Pierce, B., Weng, Z.: ZRANK: reranking protein docking predictions with an optimized energy function. Proteins **67**, 1078–1086 (2007)
11. Bini, A., Mascini, M., Mascini, M., Turner, A.P.F.: Selection of thrombin-binding aptamers by using computational approach for aptasensor application. Biosens. Bioelectron. **26**, 4411–4416 (2011)
12. Hu, W.P., Kumar, J.V., Huang, C.J., Chen, W.Y.: Computational selection of RNA Aptamer against Angiopoietin-2 and experimental evaluation. Biomed. Res. Int. **2015**, 1–8 (2015)
13. Kumar, J.V., Chen, W.Y., Tsai, J.J.P., Hu, W.P.: Molecular simulation methods for selecting thrombin-binding aptamers, vol. 253. Lecture Notes in Electrical Engineering, pp. 977–983 (2013)
14. Kumar, J.V., Tsai, J.J.P., Hu, W.-P., Chen, W.-Y.: Comparative molecular simulation method for Ang2/Aptamers with in vitro studies. Int. J. Pharma Med. Biol. Sci. **4**, 2–5 (2015)
15. Shcherbinin, D.S., Gnedenko, O.V., Khmeleva, S.A., Usanov, S.A., Gilep, A.A., Yantsevich, A.V., Shkel, T.V., Yushkevich, I.V., Radko, S.P., Ivanov, A.S., Veselovsky, A.V., Archakov, A.I.: Computer-aided design of aptamers for cytochrome p450. J. Struct. Biol. **191**, 112–119 (2015)

BRIR Refinement with Bone Conduction Headphones to Improve Spatial Sound Reproduction in Conventional Headphones

Shingchern D. You[✉] and Zhi-You Xie

Department of Computer Science and Information Engineering,
National Taipei University of Technology, Taipei, Taiwan
you@csie.ntut.edu.tw

Abstract. Conventional BRIR (Binaural Room Impulse Response) measurement uses only miniature microphones plugged on the entrance of the ear canal to pick up acoustic sound entering the subject's ears. Though simple, this method does not consider the resonance of the ear canal, among other factors. In this paper, the measured BRIRs are refined by using bond conduction headphones to include these factors. Experimental results show that the refined BRIRs produce better distance localization than without the refinement.

Keywords: BRIR · Bone conduction headphones · Sound field localization

1 Introduction

Nowadays multimedia contents are easy to access and consume. In a high-quality video media, such as a blu ray disc, 5.1 or 7.1 audio channels are encoded. To reproduce these channels, multiple loudspeakers are usually used. For many instances, however, it is not possible to use multiple loudspeakers to reproduce audio. Thus, a reasonable alternative is to use headphones. Traditional headphones are unable to reproduce audio with more than two channels. In addition, the reproduced sound does not provide directional cue to the listener, i.e. no surround sound effect. Fortunately, with the use of head-related transfer function (HRTF) [1], it is possible to reproduce spatial sound by headphones to have a similar listening experience as if multiple loudspeakers were in use [2]. In this way, the listener can also enjoy the sound effect of 5.1 channels.

The HRTF datasets are typically measured in an anechoic room with either a dummy [3] or real people [4]. When rendering spatial sound using HRTF, only the localization cue preserves, but not the environment. Taking a video scene with gun shot sound as an example: If the sound of such a scene is reproduced by multiple loudspeakers in a reverberant room, the audience not only hears the direct sound (from the loudspeakers to the ears of the listener), but also the reverberation sound, which gives the listener the aural cue about the size of the room. However, as the HRTF does not contain the room impulse

© Springer Nature Singapore Pte Ltd. 2018
J. C. Hung et al. (Eds.): FC 2017, LNEE 464, pp. 56–63, 2018.
https://doi.org/10.1007/978-981-10-7398-4_6

response, mixing HRTF to reproduce the gun shot sound only provides the direction of the gun shot, but no reverberation sound.

To cope with the mentioned problem, a possible approach is to measure the Binaural Room Impulse Response (BRIR) [5, 6] instead of HRTF, and use BRIR to synthesize the spatial sound. In general, a personalized BRIR is measured in a reverberant room using loudspeakers and miniature microphones placed on both entrances of ear canals of the listener (subject). Once BRIR of an individual is obtained, the BRIR can be used to reproduce better spatial sound for the subject.

When we use BRIR to reproduce spatial sound, the directional cue is well preserved. However, the reproduced sound field does not have sufficient distance cue. That is, the user is unable to distinguish whether the sound source is one meter away from him/her or half a meter away. To address this issue, we refine the measured BRIR by comparing the relative loudness between the headphones and the loudspeakers for multiple frequencies. As direct comparison has some difficulties, we use the bone conduction headphones as a media to achieve this goal. Our experimental results show that the proposed approach has higher distance cue than without refinement.

2   BRIR Measurement with Microphones

This section briefly describes the method we use to measure BRIR using microphones and loudspeakers and how the measured responses can be used in reproducing spatial sound.

2.1   BRIR Measurement

There are several methods to measure impulse response [7]. One widely used method is through the maximum length sequence (MLS). Another one is by using sine sweep signal. In our case, we use the sine sweep method. In general, any signal convolved with its inverse results in an impulse. When a sine sweep is in use, its inverse is obtained by reversing the order of the samples [8]. Therefore, we can easily compute BRIB with the sine sweep.

As mentioned previously, the measurement takes place in a reverberant room. In our case, it is a laboratory room equipped with tables, chairs, and steel cabins. The setup of the measurement environment is shown in Fig. 1. A notebook is used to generate the sine sweep signal, which is converted to analog by an external A/D converter (Babyface). The analog signal drives a loudspeaker to reproduce the sine sweep signal. The signal is then received by a pair of miniature microphones. The received signal is converted back to digital form and recorded by commercial software. The miniature microphones are inserted on the entrance of ear canals, as shown in Fig. 2, to pick up any reflections from the body and auricles. The convolution operation is accomplished with the aid of FFT (fast Fourier transformation) to reduce the required computation. In the measurement, the sample rate is 44,100 s/s and the sine sweep has duration of 15 s.

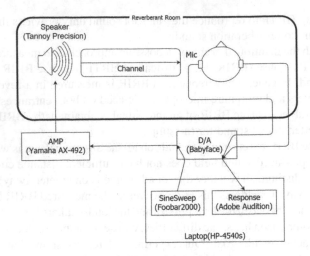

Fig. 1. Experimental setup of BRIR measurement.

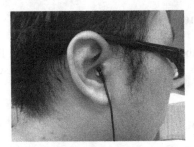

Fig. 2. The miniature microphones used in the experiments.

2.2 Rendering Spatial Sound Using Headphones

If we plan to reproduce spatial sound using headphones for five-channel audio, we need to collect BRIR associated with different loudspeakers. Let x_L, x_R, x_C, x_{LS} and x_{RS} denote the signals in the L, R, C, LS, and RS channels. If the BRIR corresponding to these loudspeakers to the right ear are denoted as $h_{\alpha,R}$, $h_{\beta,R}$, $h_{0,R}$, $h_{\gamma,R}$, and $h_{\delta,R}$, the signal to drive the right transducer (speaker) of the headphones is

$$y_R = (x_L * h_{\alpha,R} + x_R * h_{\beta,R} + x_C * h_{0,R} + x_{LS} * h_{\gamma,R} + x_{RS} * h_{\delta,R})/5 \tag{1}$$

where * denotes convolution. The signal to drive the left transducer can be obtained by a similar method.

In the above equation, the frequency response of the headphones is not considered. To have better performance, we also measure the frequency response of the headphones using the method similar to what we measure BRIR. Specifically, the subject wears the headphones and inserts the miniature microphones inside the ear cups of the headphones. Then, the sine sweep signal is reproduced from the headphones and captured by the microphones. In this way, we are able to find the impulse responses of the headphones. Once the impulse responses are obtained, we use linear predictive coding (LPC) to find the inverse. The order of the LPC filter is 200. The coefficients are computed by using autocorrelation method, which is available in MATLAB, so we simply apply the MATLAB function to obtain the results. The frequency response of the headphones before and after inverse filter is depicted in Fig. 3. It can be seen that the response after compensation is much smoother (and flatter) than before compensation.

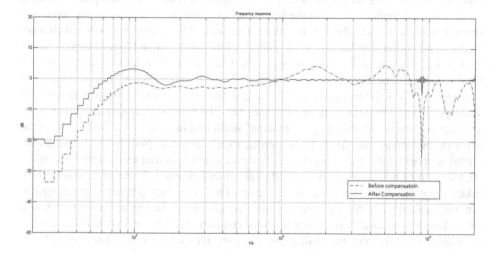

Fig. 3. The frequency responses of the headphones before and after compensation.

3 BRIR Refinement

With the measured BRIR, the listener can sense the sound direction, but not the distance cue, by using headphones. For example, if a loudspeaker is placed on 1 m ahead of the listener, he/she may feel as if the sound source were placed on 50 cm away (instead of 1 m). The causes to this problem include: the influence of the ear canals [9], variations of microphone mounting condition, irregularities in measured frequency characteristics [6], and so on. As the error sources cannot be easily analyzed, we are unable to design a proper inverse filter to offset the errors. To cope with this difficulty, we measure the subjective loudness differences between the sound reproduced by a loudspeaker and by headphones, and then use a software equalizer to synthesize the inverse filter based on the measured loudness differences. In a sense, we add a type of "correction" filter in the signal processing chain, similar to the one used in [6] to improve the processed sound quality.

Conceptually, if we can compare the perceptual loudness differences for sound rendered by a loudspeaker (sound source) and by headphones, we can correct the level differences by a software equalizer. Doing this kind of comparison, however, is not simple. Because we use circumaural headphones, and the sound from the loudspeaker is blocked by the ear cups of the headphones, we will not have correct loudness levels when wearing headphones. Furthermore, human beings can only "memorize" the loudness level of a sound for a very short period of time (a couple seconds). Thus, it is not possible to take off the headphones to hear one tone, and then put on the headphones to hear the same tone to compare the level difference. To walk around this difficulty, we use the bone conduction headphones to aid the comparison.

The proposed BRIR refinement method is shown in Fig. 4. The comparison consists of two phases. In the first phase, the generated pink noise is reproduced by a loudspeaker and one channel (one ear) of the bone conduction headphones. For example, if we want to measure the loudness level of the left front loudspeaker to the left ear, the left ear is open to hear the sound, whereas the right ear is blocked by an ear plug. The left channel of the bone conduction headphone is mute, and the right channel reproduces the same pink noise. The sound alternates between the loudspeaker (sensed by left ear) and the bone conduction headphones (sensed directly by right inner ear) in duration of around one second. Then, the listener is asked to adjust the volume of the loudspeaker until the same perceived loudness achieved for both ears. To have a precise reading, the loudspeaker volume is controlled by the external audio interface capable of adjusting signal level in the step of one dB. In the experiments, (1/3)-octave pink noise is used as the signal source to drive both equipments. To cover as much as frequency span as possible, we use 14 pink noise files each with different center frequency. The entire measurement takes about 80 min to complete one direction. Due to the heavy measurement burden, we only measure one direction for six listeners (subjects). In Fig. 4, the normalization procedure is used to set a reference level for comparison. In the experiments, the reference level is 1 kHz-centered pink noise. The perceived loudness is defined as 0 dB.

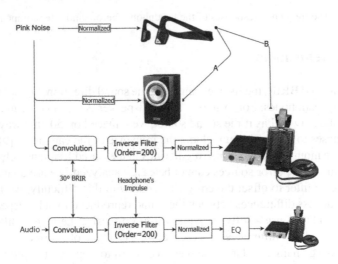

Fig. 4. The proposed BRIR refinement method.

Once the measurement of the first phase is accomplished, the second phase is to measure the level difference between the headphones and the bone conduction headphones. As also shown in Fig. 4, the pink noise is convolved with the personalized BRIR to reproduce the same virtual sound field as the loudspeaker does. In addition, we also use inverse filters to compensate for the responses of the headphones. The level adjustment procedure is the same as in the first phase, so it is not repeated here. One small but useful tip is how to properly wear both bone conduction headphones and circumaural headphones at the same time while ensuring the air-tightness of the headphones. One possible wearing method is depicted in Fig. 5. Correct wearing both headphones is an important factor relating to the measurement accuracy.

Fig. 5. Wearing both bond conduction headphones and circumaural earphones.

Once we obtain both loudness differences, the "actual" level differences are the differences between the measurement results in the first phase and those in the second phase. As also shown in Fig. 4, the level differences will be corrected by an equalizer during the listening evaluation.

4 Results of Subjective Tests

We conduct the experiments in a reverberant laboratory room. The test subjects are six graduate students aged between 20 and 30. In the measurements, the loudspeaker is located on the left front (30° azimuth angle) of the subject in a distance of 1 m. The personalized BRIR for each subject is measured first by using the method mentioned in Sect. 2. Then, loudness compensation of these subjects is measured using the method given in Sect. 3. The average (subjective) loudness difference between left ear and right ear for all subjects is depicted in Fig. 6. Recall that the pink noise is reproduced in the headphones with the BRIR containing the direction and distance cues. If the measured BRIR is fully consistent with the perceived loudness, both curves should coincide together. However, as we can see from Fig. 6, the loudness level of the right ear should be larger than BRIR reproduced, especially at high frequency end (10 kHz).

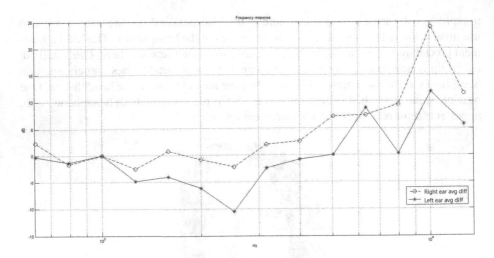

Fig. 6. Average loudness difference between left ear and right ear for six subjects.

We use a commercial equalizer software to make all necessary change for the proposed approach, and then ask the subjects to compare the relative direction and distance sensations with and without BRIR refinement. The subjective distances are reported in the unit of 10 cm. The results are given in Table 1. It is shown that the average distance increases from (around) 6 units to 8 units. In terms of directional cue, four subjects does not have any preference, one subject feels that the direction cue is improved after BRIR refinement. Thus, the experimental results confirms the usefulness of the proposed approach.

Table 1. Results for subjective test.

	Distance	Better dir. cue	Same dir. cue
No refinement	6 unit	1 subject	4 subject
With refinement	8 unit	1 subject	

5 Conclusions

This paper describes a method to refine personalized BRIR by loudness correction. The loudness differences are obtained through the use of bone conduction headphones. The collected data indicate that the average loudness at high-frequency end should be reduced for the ipsilateral ear (the ear close to the sound source), or equivalently, the loudness level of the contralateral ear should be increased. The listening comparison shows that the sensed distance increases with the BRIR refinement, while the directional cue remains. In the future, we plan to carry out more measurements to better test the proposed idea, and how to reduce the loudness measurement procedure for more practical usage.

References

1. Gardner, W.G.: 3-D audio using loudspeakers. Ph.D. thesis, MIT Media Laboratories, Cambridge (1997)
2. You, S.D., Chen, W.-K.: Optimally truncating head-related impulse response by dynamic programming with its applications. Multimedia Tools App. **70**, 2167–2188 (2014)
3. MIT Media Lab: HRTF measurements of a KERMAR dummy-head microphone. http://sound.media.mit.edu/resources/KEMAR.html
4. IRCAM HRTF Database. http://recherche.ircam.fr/equipes/salles/listen/index.html
5. Mickiewicz, W., Sawicki, J.: Headphone processor based on individualized HRTFs measured in listening room. In: Audio Engineering Society Convention, vol. 116 (2004)
6. Mickiewicz, W., Sawicki, J.: Spatial audio reproduction by headphones using binaural room impulse responses measured individually by the listener. Pomiary Automatyka Kontrola **53**, 30–33 (2007)
7. Guy-Bart, S., et al.: Comparison of different impulse response measurement techniques. J. Audio Eng. Soc. **50**, 249–262 (2002)
8. Farina, A.: Simultaneous measurement of impulse response and distortion with a swept-sine technique. Audio Eng. Soc. Convention **108**, 1–24 (2000)
9. Sankowsky-Rothe, T., et al.: Comparison of transfer functions in the ear canal for open-fitting hearing aids. In: AIA-DAGA 2013 Conference on Acoustics, pp. 873–874 (2013)

A Web Crawler Supporting Interactive and Incremental User Directives

Woei-Kae Chen[✉], Chien-Hung Liu, and Ke-Ming Chen

Department of Computer Science and Information Engineering,
National Taipei University of Technology, Taipei, Taiwan
{wkchen,cliu,103598011}@ntut.edu.tw

Abstract. To increase the coverage of states that a Web crawler can explore, most crawlers require their users to define a few crawling directives before the crawler starts crawling. A directive can, for example, assign a special input value to a particular input field so that the application performs a specific action and visits some special states. Note that, a crawler is supposedly capable of exploring an unknown application. But, given an unknown application, how could the user possibly prepare all the required directives in advance? This paper proposes a novel, interactive crawling strategy called GUIDE to address this issue. Instead of passively receiving all directives from the user at once, GUIDE actively asks the user for directives when Web pages containing input fields are found. GUIDE frees the user from knowing the target application in advance. In addition, GUIDE offers a hierarchical directive structure, allowing the user to define multiple values for the same input field. Our experiments indicate that GUIDE is easy to use and can also improve the overall Web page coverage.

Keywords: Web crawler · Coverage · Interactive crawler · Directives

1 Introduction

A Web crawler (called simply crawler hereafter) is an Internet robot that can systematically browse a Web application (or Web pages) and create a state-flow graph for the application. Many types of web analysis and testing techniques can be automated by using crawlers. For example, by analyzing the state-flow graph, tasks such as detecting broken links/images/tooltips, nonfunctional testing (Accessibility, validation, security, …) [3], invariant-based testing [4], cross-browser compatibility testing [6], regression testing [7], and system testing [10] can be automated.

Web crawling technologies have been studied extensively since the advent of the Web [1, 2]. More recently, crawlers are enhanced with the ability to explore AJAX-based Web applications, which use JavaScript and dynamic DOM manipulation on the client side to offer better interactivity and responsiveness. For example, Crawljax [5] is an open source Java tool for crawling modern Web applications and many researches used Crawljax to perform different tasks [3, 4, 6, 7, 9, 10].

Most crawlers (including Crawljax) allow their users to define some crawling *directives (or parameters)* in advance. For example, when a crawler encounters the login page

© Springer Nature Singapore Pte Ltd. 2018
J. C. Hung et al. (Eds.): FC 2017, LNEE 464, pp. 64–73, 2018.
https://doi.org/10.1007/978-981-10-7398-4_7

of a Web application, a valid pair of account and password is needed to gain access to the application. Since it is highly unlikely that the crawler can magically guess the right account and password by simply using randomly generated values, a valid pair of account and password must be provided. This is typically accomplished by defining one or more input directives, which assign specific values to specific input fields. When the crawler is in action, the predefined directives are applied once the target input fields are found.

However, using predefined directives has the following problems that make crawlers less convenient to use and less effective in exploring the target application.

(a) *All directives must be predefined*: one of the most appealing advantage of using a crawler is that exploring a new, unknown Web application can be automated. But, if the target Web application is indeed unknown, how could the user possibly prepare all the required directives in advance? Thus, requiring all directives to be predefined is impractical and inconvenient.

(b) *Reading the source code of the target Web application is necessary*: a directive typically needs an ID (or XPath) that identifies the target input field. Such information cannot be obtained without studying the source code of the Web application. This is not only tedious and inconvenient, but also require the user to possess professional knowledge on how an application is coded.

(c) *Defining two or more input values for the same input field is not possible*: an input directive defines a single input value for a particular input field. However, having a single input value is often not enough to fully explore the target application. For example, given a Web application with two different kinds of users, the administrator user and regular user, what an administrator user can do is often quite different from a regular user. Thus, both of the Web pages designed for the two kinds of users should be explored. But, unfortunately, since a directive defines a single input value, the crawler cannot explore both of the Web pages simultaneously. This limitation makes crawlers less effective in exploring different states of the Web application.

To overcome the above problems, this paper proposes a new crawling strategy along with a crawler, called GUIDE (GUiding crawler with Interactive and incremental DirectivEs). Our approach reveres how a user uses a crawler. A traditional crawler accepts directives from the user and never requests for directives on its own (Fig. 1(a)). GUIDE, on the other hand, actively asks the user for directives (Fig. 1(b)) for Web pages that may need some directives. As discussed in Sect. 3, the reversal frees the user from knowing the target application in advance and the user no longer needs to read the source code of the target application. In addition, GUIDE offers a hierarchical directive structure, allowing the user to define multiple values for the same input field. Our experiments (Sect. 4) indicate that GUIDE is easy to use and can also improve the overall Web page coverage.

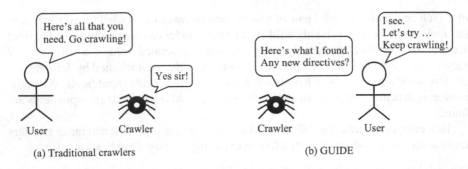

(a) Traditional crawlers   (b) GUIDE

Fig. 1. Traditional crawlers versus GUIDE

2 Related Work

Crawljax [5] is a popular open-source Web crawler that can process AJAX-based applications through automatic dynamic analysis of user-interface-state changes in Web browsers. Crawljax offers a flexible plugin architecture that makes Crawljax very extendable. Many researches took advantage of the plugin architecture and extended Crawljax for different applications [3, 4, 6, 7, 9, 10]. When using Crawljax, the user can define a number of directives before Crawljax starts crawling. This implies that the user needs to be familiar with the target Web application so that the directives can be prepared. In case that the user does not know anything about the target application, the states that can be explored become quite limited. A workaround is to execute Crawljax for several rounds. The user tries to figure out the necessary directives by looking at the resulting state diagram of the previous round, and then set directives for the next round. This is in fact similar to the strategy proposed in this paper (see Sect. 3). However, the workaround requires a lot of manual work and reading source of the target application is unavoidable. In comparison, the proposed strategy is easier to use and also a lot more efficient (see Sect. 4).

Tanida et al. [10] reported an automated approach for the system testing of modern dynamic web applications. In order to facilitate a more comprehensive and scalable crawling behavior, Tanida et al. proposed an extension to Crawljax, called guided crawling. By using *guidance directives*, Web pages that requires user authentication, form data filling, and excluding behavior from the model can be properly explored. This is similar to the problems addressed in this paper. In fact, the objectives of guidance directives are also similar to those of the directives offered in GUIDE. However, like a traditional crawler, guided crawling also requires all guidance directives to be predefined and in addition, each guidance directive must be prepared in the form of text-based instructions. In comparison, GUIDE proposed in this paper offers an interactive crawling environment and allows the user to explore the target application incrementally without the need of writing complicated guidance directives.

3 Guide

GUIDE is an interactive crawler that accepts incremental directives from the user. Initially, GUIDE starts a first-round crawling without using any directives. Upon exploring all the states (pages) that can be found, GUIDE provides state diagrams and screenshots for the user to review. For a Web page that has input fields, GUIDE asks the user for directives. The user decides whether giving directives are necessary and if yes, provides appropriate ones. GUIDE, then continues with a second-round crawling using the new directives, and explores more states. Since some of the newly explored states may also need directives, GUIDE asks the user again for more directives and continues with a third-round crawling. The interaction between GUIDE and the user is repeated for as many rounds as necessary until the user considers that the target application is fully explored.

Two kinds of directives, namely *input directive* and *stop directive*, are available. An input directive assigns a special value to a special input field; a stop directive, on the other hand, stops a specified page from being crawled in the future. The former allows GUIDE to explore more Web pages and the latter informs GUIDE not to waste time continually crawling a particular page that is of no interest to the user. When GUIDE finds a Web page containing some input fields, red rectangles are superimposed on the screenshot to highlight the input fields (Fig. 2). Thus, by reviewing the resulting screenshots, the user can easily determine whether any input directives are necessary. If yes, the user simply keys in the desired input values using GUIDE user interface (Fig. 2). This user interface makes GUIDE a lot easier to use than a traditional crawler that requires users to provide IDs (e.g., XPath) for input directives. GUIDE keeps track of IDs automatically and the user no longer needs to read the source code of the target application.

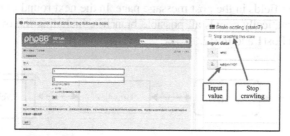

Fig. 2. GUIDE highlights input fields and allows the user to assign input values.

If the user considers that a Web page does not need any additional crawling, the user can also issue a stop directive in the same user interface (Fig. 2). When "Stop crawling this page" is checked, GUIDE ignores any events in this page in the future rounds of crawling.

We now explain the concept of *branch*, the mechanism that allows the user to assign multiple input values for the same input field. This is useful when the application offers different features (pages) for different input values. For example, given a blog (weblog) application, suppose there are two kinds of users, namely blogger and administrator. In this case, we would like to direct the crawler to explore the pages for both kinds of users. That is, the input fields, account and password, in the login page should be assigned with two different sets of account and password combinations. GUIDE accomplishes this requirement by allowing the user to create branches. Initially, GUIDE starts the first-round crawling with a *root branch* that does not contain any directives (Fig. 3). Upon completion, the user will see the login page reported by GUIDE. From this information, the user may create two branches under the root branch, one for the blogger user (User1 branch) and one for the administrator user (Admin branch). Each branch can be assigned with an independent set of directives (either input or stop directives). In the second-round crawling, GUIDE automatically crawls every possible branch using the directives assigned to the branch. The resulting state graph is the union of the state graphs of Root, Admin, and User1 branches. GUIDE automatically merges the state graphs generated in each branch and each round into a single overall state graph.

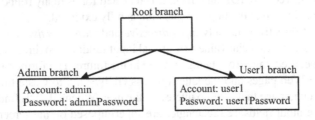

Fig. 3. A root branch with two sub-branches

Note that when the second-round crawling finishes, the user may find that there is a post message page under User1 branch. Suppose, the user decides to post two different messages for User1. In this case, the user can create another two branches under User1 branch, namely Post1 branch and Post2 branch. Each branch simply assigns different input values for the input fields in the post message page. In the next round of crawling, again, GUIDE will automatically crawls all possible branches, including the two different post message branches Post1 and Post2.

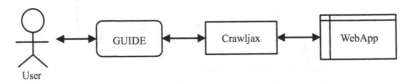

Fig. 4. GUIDE interacts with the user and uses Crawljax to perform crawling

GUIDE uses Crawljax [5] as a sub-component for crawling (Fig. 4). The use of GUIDE is outlined in Fig. 5. The root branch (with no directives) is automatically created at the beginning. GUIDE uses Crawljax to perform the first-round crawling for the root branch. The results is a single state graph along with the screenshots of all the explored states. No merge of state graphs is necessary at this point. GUIDE then searches for HTML input and TextBox from all states by analyzing their DOMs. If an input field is found, an extra red rectangle is superimposed on the corresponding screenshot to highlight the input field. The state graph and screenshots are presented to the user using the user interface shown in Fig. 6. After reviewing the state graph and screenshots, suppose the user decides to provide directives for a state *S*, the user can create one or more branches (on the left of Fig. 6). In each branch, the user can assign a set of different directives (on the right of Fig. 6) for *S*. GUIDE will translate these directives into a format that Crawljax accepts and request Crawljax to perform crawling for each of the new branches. Upon completion, GUIDE merges the resulting state graphs of each branch into a single one, search for new input fields, and presents the results to the user again for the next round of crawling.

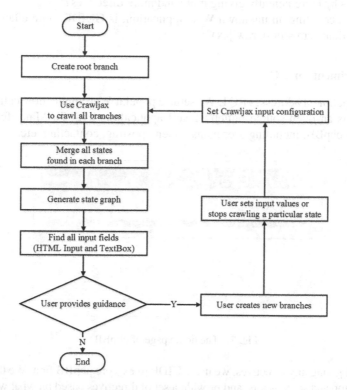

Fig. 5. Flowchart of using GUIDE

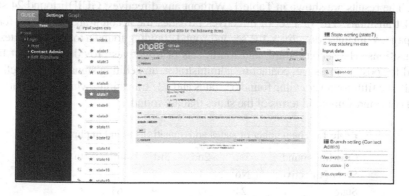

Fig. 6. The user interface of GUIDE

4   Case Study

We conduct two experiments to study whether using GUIDE is practical. The following two research questions are addressed:

RQ1: When using GUIDE, can the user direct the crawler to explore more and more states by incrementally giving more and more directives?

RQ2: When crawling an unknown Web application, is GUIDE more efficient than a traditional crawler (Crawljax)?

4.1   Experiment for RQ1

The first experiment addresses RQ1. We setup a popular Web application, called phpBB [8], which is an open-source bulletin board application (Fig. 7). Many features are available in phpBB, including user management, posting, contacting, etc.

Fig. 7.   The home page of phpBB

Without giving any directives, we use GUIDE to explore phpBB first. We then review the state graph and screenshots, and provide a set of directives based on what we saw. For each round of crawling, we record the number of states and transitions in the resulting state graph. The results are shown in Table 1. Without any directives, GUIDE found 24 states and 44 transitions. The first set of directives provides the account and password for the login page. With these directives, the second-round crawling found a total of 49 states and 87 transitions. The second, third, and fourth sets of directives provided the input fields needed for posting message, contacting administrator, and editing signature, respectively. Overall, the fifth-round crawling found 63 states and 127 transitions, which are about 262% (63/24) of improvement in terms of the states that are found.

Table 1.   The number of states and transitions found in phpBB

Crawling round	1st	2nd	3rd	4th	5th
Sets of directives	No	1	2	3	4
States	24	49	59	63	63
Transitions	44	87	113	125	127

The branch structure used in defining the above directives is shown in Fig. 8. The Login branch was placed under the Root branch, because the login page was discovered after the Root branch was crawled. Similarly, the Post Message, Contact Admin, and Edit signature branches were placed under the Login branch, because they were discovered after the Login branch was crawled.

Root
Login
Post Message
Contact Admin
Edit Signature

Fig. 8. The branch structure used in exploring phpBB

Overall speaking, the results of the first experiment indicated that using GUIDE to provide incremental directives was very effective in increasing the overall coverage of the states and transitions for exploring phpBB. In comparison with no directives, the number of discovered states improved for 262%. Thus the answer to RQ1 is "yes, GUIDE helped the user to provide directives incrementally and also explore more states and transitions." Note that phpBB is an application with rich interactions and login is a prerequisite for some other features such as posting messages, sending contact messages, etc. In addition, it was important to enter appropriate input values for some of the features to be properly activated. Thus, the hierarchical branch structure of GUIDE was very useful for defining and organizing directives for phpBB.

4.2 Experiment for RQ2

The second experiment addresses RQ2. We recruit 8 different participants to act as users. Each participant is requested to use both GUIDE and Crawljax [5] to explore an unknown Web application. We use Crawljax to represent a traditional crawler that accepts all directives in advance, because Crawljax is a modern and well known state-of-the-art crawler. The target application is designed to be unknown to the participants, because we would like to simulate the situation that the participants are exploring a new application, which is a typical application of crawlers.

To make the scale and time of the experiment more manageable, we create a simple, in-house application, called TS, which supports two kinds of users, namely teacher and student. After login, the students can post messages and the teacher can view the posts. Since we are simulating the crawling of an unknown application, the participants are not allowed to use TS directly. They can only review the state graph and screenshots provided by the crawler to determine whether any directives are necessary. The expected branch structure is shown in Fig. 9. However, the participants are not required to come up with the same structure. Instead, the participants are only requested to explore all the features that are available in TS.

Fig. 9. The expected branch structure

For GUIDE, as described in Sect. 3, the participants use GUIDE user interface to perform several rounds of crawling. For Crawljax, this is a bit more complicated, because Crawljax only accepts directives in advance and does not support interactive crawling. The participants end up using a repetitive cycle similar to that of using GUIDE, i.e., requesting Crawljax to get the state diagram, reviewing the state diagram, figuring out the directives to set, setting directives for Crawljax, and requesting Crawljax to get the new state diagram. But, with Crawljax, the cycle can only be done mostly manually.

The participants are divided into two equal-sized groups, namely group A and group B. Group A uses GUIDE first and then Crawljax. Group B, on the other hand, uses Crawljax first and then GUIDE. We record the time needed to complete the exploration of TS for all participants. The results are shown in Table 2. The participants are labeled P1-P8. Each participant spent a different amount of time. On average, using Crawljax required 424 s, while using GUIDE required only 217 s, indicating that GUIDE can reduce the time required for crawling. In this case, the time is reduced to about 51% (217/424). Therefore, the answer to RQ2 is "yes, using GUIDE was more efficient than using Crawljax."

Table 2. The time required to fully explore TS (unit: second)

	Group A				Group B				Average
	P1	P2	P3	P4	P5	P6	P7	P8	
Crawljax	426	476	420	331	453	440	438	406	424
GUIDE	211	293	220	223	209	187	201	188	217

5　Conclusions

This paper proposes an interactive Web crawling strategy as well as a tool called GUIDE, which allows the user to provide directives incrementally. In comparison with traditional crawlers, the advantages of GUIDE include: (a) the user does not need to provide any directives in advance – this is very useful for exploring unknown applications; (b) GUIDE is easier to use – the user does not need to read the source code of the target application; and (c) assigning multiple input values for the same input field is possible – more states of the target application can be explored. Our experiments confirmed that GUIDE can indeed help the user provide directives incrementally and also explore more states and transitions, and using GUIDE is more efficient than using traditional crawlers. Currently, GUIDE supports only two kinds of directives (input and stop directives). In the future, we expect to enhance GUIDE with more kinds of directives.

Acknowledgement. This research was partially supported by the Ministry of Science and Technology, Taiwan, under contract number MOST 105-2221-E-027-085, which is gratefully acknowledged.

References

1. Brin, S., Page, L.: The anatomy of a large-scale hypertextual Web search engine. Comput. Netw. ISDN Syst. **30**(1–7), 107–117 (1998)
2. Burner, M.: Crawling towards eternity: building an archive of the World Wide Web. Web Techn. Mag. **2**(5), 37–40 (1997)
3. Ferrucci, F., Sarro, F., Ronca, D., Abrahao, S.: A crawljax based approach to exploit traditional accessibility evaluation tools for AJAX applications. In: Information Technology and Innovation Trends in Organizations, pp. 255–262 (2011)
4. Groeneveld, F., Mesbah, A., Deursen, A.V.: Automatic invariant detection in dynamic web applications. Delft University of Technology, Software Engineering Research Group, December 2010
5. Mesbah, A., van Deursen, A., Lenselink, S.: Crawling AJAX-based web applications through dynamic analysis of user interface state changes. ACM Trans. Web (TWEB) **6**(1), 3:1–3:30 (2012)
6. Mesbah, A., Prasad, M.R.: Automated cross-browser compatibility testing. In: Proceedings of the 33rd International Conference on Software Engineering (ICSE 2011), pp. 561–570. ACM, New York (2011)
7. Mirshokraie, S., Mesbah, A.: JSART: JavaScript assertion-based regression testing. In: International Conference on Web Engineering, pp. 238–252, July 2012
8. phpBB, an open-source bulletin board application. https://www.phpbb.com. Accessed 1 Apr 2017
9. Silva, C.E., Campos, J.C.: Combining static and dynamic analysis for the reverse engineering of web applications. In: Proceedings of the 5th ACM SIGCHI Symposium on Engineering Interactive Computing Systems, pp. 107–112, June 2013
10. Tanida, H., Prasad, M.R., Rajan, S.P., Fujita, M.: Automated system testing of dynamic web applications. In: International Conference on Software and Data Technologies, pp. 181–196, July 2011

The Design and Implementation
of Cloud-Based Binary Image Customization
System for User-Shooting Photos

Chuan-Feng Chiu[✉], Yu-Chih Hsu, and Ping-Ming Sung

Department of Information Management, Minghsin University of Science
and Technology, Xinfeng, Taiwan
cfchiu@must.edu.tw

Abstract. In recent years, users have more exception for having the cus-
tomizing goods that users can have it by themselves or be a good gift for their
friends. They also want to have an easy system for this purpose and have more
materials for produce goods or gifts. The famous technique is using laser
machine to print photos on cups or steel plate. However the laser machine accept
binary image only, and the photos shoot by users are colored. So converting
color image to binary image is the first issue. But traditional binary image
conversion mechanism will not be against the photos shoot by users because of
the photos will have overexposure beauty shot effect by users. Therefore, in this
paper we propose the binary image mechanism based on lightness histogram
classification that can be against the effects and propose a practical application
for realizing the scenario.

Keywords: Binary image conversion · Lightness histogram · HSL color model

1 Introduction

In recent years users want to have customized goods, so many customization tools
become popular, Beauty shot camera or APP are the such kind of tools. The simple and
popular style is to have a customized picture printing on some products that are just like
cups and so on. To have customized picture on products, stickers is simplest. However
the quality using stickers is not good and easy to damage. Using laser machine to
produce customized picture on goods is more popular and the quality is better than
using stickers. The color will be black and white only when using laser machine to
generate customized pictures. So a binary image that is clear enough to recognize
something important is necessary. On the other hand most laser machine only accept
image sources with vector format. In such scenario, we observe that the color image
can be transformed to binary image is easy. But when transforming binary image to
vector image, the image will lose something important or distort.

Binary image conversion is not new issue. Traditionally, the binary image conversion focus on well-formed image that is not be influenced by lightness or brightness.
So the transformed quality is acceptable. Now because of electronic device is cheap
and users have mobile phone with camera, users can have pictures or photos by

© Springer Nature Singapore Pte Ltd. 2018
J. C. Hung et al. (Eds.): FC 2017, LNEE 464, pp. 74–83, 2018.
https://doi.org/10.1007/978-981-10-7398-4_8

themselves is easily. However, the photos taking by users have many special effects for enhancing user subjective quality including brightness, saturation or the direction of light source etc. So the pictures users shoot will be influenced by lightness or brightness strongly. Therefore, traditional binary image conversion is not applied directly. Because of the detail information that is the notable feature of the object in the picture will be lose, just like hair texture, or wrinkle etc.

Based on the above motivation, we propose an application scenario that users can shoot photos by themselves, upload photos to our designed system, generating binary image that can be put on goods by laser machine in this paper. And in order to produce binary image with less distortion by using laser machine, we also propose the binary image conversion mechanism based on lightness information and customized patching template.

In the following, we will describe the literatures that are related with out work in Sect. 2. In Sect. 3 we reveal the design of binary image conversion mechanism based on lightness information and experimental result. Then we propose the design of the system regarding our application scenario in Sect. 4. Finally we conclude the brief conclusion in Sect. 5.

2 Related Work

In this section we describe the related literatures regarding binary image conversion technique. In the past there exist many literatures for converting gray images to binary images. The basic approach is threshold-based approach. Prewitt et al. [2] use the distribution of the gray color histogram, we can find the peak value from the histogram. The value is the selected threshold. When the gray value of pixel is greater than the threshold, the value pixel is set to black, otherwise the value of the pixel is set to white. So the binary image is produced. But when the peak value of the gray histogram is not clear, this approach is not working well. So the average value of gray value of all pixels of the image is proposed to select the threshold. Otsu [3] proposed another threshold selection mechanism and the mechanism can select the threshold adaptively. The basic idea is to calculate the threshold based on statistic approach that is the smaller variance called within-group variance in same group and the larger variance called between-group variance between different groups. By calculating the minimum of aggregated within-
group variance from weighted within-group variance and between-group variance with respect to each gray value $(0 \sim 255)$, the gray value of minimum of aggregated within-group variance is selected as threshold. Entropic method [11, 12] is another approach for converting images to binary image. This kind of approach use information theory as the basis. By estimation the entropy or probability of pixel value, the binary image can be produced based on the estimation values.

Halftone image [4–6] is another related research for binary image conversion. The halftone technique use dots to describe the continue tone images. When the dots in halftone image is closed enough, the image for human eyes is similar with original images. However, halftone image is not suitable for laser machine to produce the customized image on goods because of using dots tone not for continue tone and will decrease the life-time of laser machine. Figure 1 shows the halftone image examples. Image segmentation is another application by using binary image. Car license recognition [8, 10] is the popular application by using binary image conversion as the preprocessing of the original image.

Fig. 1. Examples of halftone image

3 Proposed Binary Image Conversion Mechanism

In this section, we reveal the proposed mechanism for converting user-shooting photos to binary image without losing user subject feeling. Figure 2 is the illustration of our proposed mechanism. The process includes two major phase including training phase and testing phase. The training phase is to train the model that includes different clusters with different feature of images. Each cluster have different patching template used to produce binary image. The testing phase is to generate the binary image via the mode from training phase and the generated binary image will satisfy user subject felling.

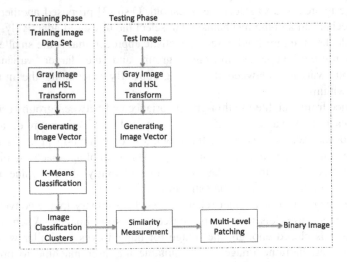

Fig. 2. Flowchart of proposed binary image conversion mechanism

3.1   Training Phase

In the training phase, the color image data set input to generate gray level image and produce the HSL (Hue, Saturation, Lightness) model information first. Based on the lightness value of HSL model of each image, the image Feature Vector (FV) is generating according the lightness histogram. Equation (1) shows the format of Feature Vector (FV) with ten fields.

$$FV = \{v_0, v_1, \ldots, v_9\} \tag{1}$$

The lightness histogram is divided into ten parts equally and each part is the number of pixel that the lightness value belongs to the interval of the part of the lightness histogram showed in Fig. 3. Equation (2) shows the calculation of the lightness-based feature vector.

$$v_i = \begin{cases} the\ number\ of\ pixel\ with\ i \times 10 \leq lightness\ value \leq i \times 10 + 9, 0 \leq i \leq 8 \\ the\ number\ of\ pixel\ with\ i \times 10 \leq lightness\ value \leq (i+1) \times 10, i = 9 \end{cases} \tag{2}$$

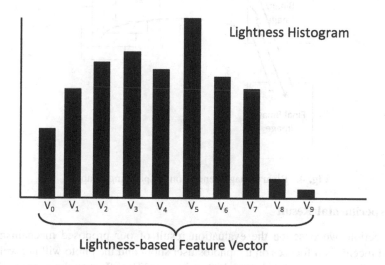

Fig. 3. Illustration of lightness-based histogram Feature Vector (FV)

After having all Feature Vectors (FVs) from the input image data set, we apply K-Means classification algorithm to produce 20 different groups and calculate the centroid of each group. The centroid is the Cluster Feature Vector (CFV). Equation (3) is the designed classification method based on K-Means.

$$arg\ min_S \sum_{i=1}^{20} \sum_{x \in S_i} \sqrt{\sum_{j=1}^{10} (x_j - CFV_j)^2} \tag{3}$$

3.2   Testing Phase

In the testing phase, we input the test image and generate the gray level image and HSL (Hue, Saturation, Lightness) model information. We also calculate the feature vector with respect to the lightness histogram. Then we calculate the similarity between the test image feature vector and Cluster Feature Vector to determine which group the test image belongs to. Finally we apply the Patch Template to generate the binary image of test image. The Patch Template includes the binary image threshold and patching level information. The test image will generate the original binary image according the patch threshold first. Then we select the pixel from the binary image according the pixel position of each patch level of patching template to produce the final binary image. Figure 4 shows the process.

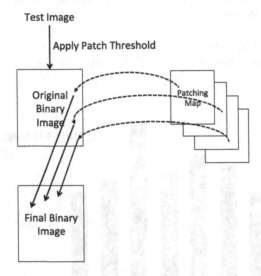

Fig. 4. Binary image generation by patching template

3.3   Experimental Result

In this section, we describe the evaluation result of our proposed mechanism. Our proposed mechanism focuses on the photos user shoot and the photo will not well-form that means the photo may have special effect by user. The effect on the photo will cause un-satisfied result. In the following we use 213 photos as the training image data set and shows the result by applying our method with two kinds of such user shooting photos and compare with Otsu algorithm.

Figure 10 show the result by using beautyshot effect photo. In the figure, our proposed method can shows the user subject feeling regarding the photo. Otsu method will lose some information so that the result cannot satisfy user's feeling. Figure 10 shows the result of overexposure photos. The overexposure photo will have more brightness so that the user feeling is good on color. However, in order to put on good we can see the Otsu algorithm still can not show the user subject feeling. On the contrary our proposed method have better user subject feeling on binary format.

In general our proposed mechanism can keep the user subject feeling after converting binary image. And the converting result is suitable for laser machine to produce customized goods. Otsu had benefit on threshold selection, but the method might be influenced by light information so that the result will lose some important information regarding user subject feeling. Figures 5, 6, 7, 8, 9, 10, 11, 12, 13, 14, 15, 16, 17, 18, 19, 20, 21, 22, 23, 24, 25, 26, 27 and 28 shows the result by original image, Otsu approach and our proposed method.

Fig. 5. Original image (1)   **Fig. 6.** Binary image by Otsu (1)   **Fig. 7.** Our proposed method (1)

Fig. 8. Original image (2)   **Fig. 9.** Binary image by Otsu (2)   **Fig. 10.** Our proposed method (2)

Fig. 11. Original image (3)   **Fig. 12.** Binary image by Otsu (3)   **Fig. 13.** Our proposed method (3)

Fig. 14. Original image (4)   **Fig. 15.** Binary image by Otsu (4)   **Fig. 16.** Our proposed method (4)

Fig. 17. Original image (5)   **Fig. 18.** Binary image by Otsu (5)   **Fig. 19.** Our proposed method (5)

Fig. 20. Original image (6)   **Fig. 21.** Binary image by Otsu (6)   **Fig. 22.** Our proposed method (6)

Fig. 23. Original image (7)   **Fig. 24.** Binary image by Otsu (7)   **Fig. 25.** Our proposed method (7)

Fig. 26. Original image (8)   **Fig. 27.** Binary   image   **Fig. 28.** Our proposed method (8)
by Otsu (8)

4 Application and System Design

Based on the proposed binary image conversion for laser machine with maintaining user subject feeling on goods, we also design a practical application for realizing the proposed mechanism. Figure 29 shows the architecture and scenario. Users can use their owned camera to shoot pictures and upload the picture they want to print on the cup or steel plate etc. to our designed system. When the system receives the picture request, the system will convert the uploaded picture to binary images as the input for laser machine. The resulting binary image will be replied to users and user can make a decision regarding the binary image. If users agree the resulting binary image, the users can select the goods that users want to print on. However it users do not agree and want to enhance some part of the result binary image, users can mark the part in rectangle or circle area and feedback to our system. Our system will make binary image conversion process regarding the marking part only. Finally the system will start the process for produce the customized goods after user conform the result. The application is useful for users because the user can customize their goods on cup or steel plate not only print on t-shirt now. And our proposed application can receive the feedback and reproduce the result according to the feedback.

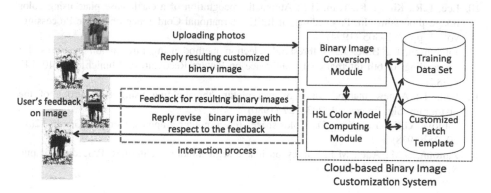

Fig. 29. The design of cloud-based binary image customization system

5 Conclusion

In this paper we proposed an application for users to make customized photos on special material like cups or steel plates. The users can interact with our system and make feedback to proposed application with boring process. On the other hand, we also propose a binary image conversion mechanism based on lightness histogram. By using lightness histogram, our proposed mechanism can against the photos that have beautyshot effect or overexposure in converting nature photos to binary image.

However there still have some improvement that we can enhance. The proposed binary image conversion mechanism relies on the patching template to convert images. The automatic process for patching template generating is still to be considered. In the future we will study the learning mechanism to produce the patching template that is more close to users exception.

References

1. Doyle, W.: Operation useful for similarity-invariant pattern recognition. J. Assoc. Comput. **9** (2), 259–267 (1962)
2. Prewitt, J.M.S., Mendelsohn, M.L.: The analysis of cell images. Ann. N. Y. Acad. Sci. **128**, 1035–1053 (1966)
3. Otsu, N.: A threshold selection method from gray-level histogram. IEEE Trans. Syst. Man Cybernet., 62–66 (1978)
4. Lieberman, D., Allebach, J.: Digital halftoning using direct binary search. In: Proceedings of 1996 1st IEEE International Conference on High Technology, pp. 114–124 (1966)
5. Allebach, J.P.: Selected Papers on Digital Halftoning. SPIE Milestone Series (1999)
6. Bares, J.: Algorithms and procedures for digital halftone generation. In: Proceedings of Color Hard Copy and Graphic Arts, vol. 1670, pp. 26–40 (1992)
7. Sezgin, M., Sankur, B.: Survey over image thresholding techniques and quantitative performance evaluation. J. Electron. Imaging **13**(1), 146–165 (2004)
8. Poon, J.C.H., Ghadiali, M., Mao, G.M.T., Sheung, L.M.: A robust vision system for vehicle license plate recognition using grey-scale morphology. In: Proceedings of the IEEE International Symposium on Industrial Electronics (ISIE 1995), vol. 1, pp. 394–399 (1995)
9. Gonzalez, R.C., Woods, R.E.: Digital Image Processing. Prentice Hall (2002)
10. Lee, E.R., Kim, P.K., Kim, H.J.: Automatic recognition of a car license plate using color image processing. In: Proceedings of IEEE International Conference on Image Processing, vol. 2, pp. 301–305 (1994)
11. Johannsen, G., Bille, J.: A threshold selection method using information measures. In: Proceedings of 6th International Conference on Pattern Recognition, Munich, pp. 140–143 (1982)
12. Pun, T.: A new method for gray-level picture thresholding using the entropy of the histogram. Sig. Process. **2**, 223–237 (1980)
13. Joblove, G.H., Greenberg, D.: Color spaces for computer graphics. Comput. Graph. **12**(3), 20–25 (1978)
14. Zaitoun, N.M., Aqel, M.J.: Survey on image segmentation techniques. Procedia Comput. Sci. **65**, 797–806 (2015)

15. Atmaja, R.D., Murti, M.A., Halomoan, J., Suratman, F.Y.: An Image processing method to convert RGB image into binary. Indonesian J. Electr. Eng. Comput. Sci. **3**(2), 377–382 (2016)
16. Samopa, F., Asano, A.: Hybrid image thresholding method using edge detection. Int. J. Comput. Sci. Netw. Secur. **9**(4), 292–299 (2009)
17. Liu, D., Yu, J.: Otsu method and K-means. In: Proceedings of the 2009 Ninth International Conference on Hybrid Intelligent Systems (HIS 2009), vol. 1, pp. 344–349 (2009)
18. Chaudhari, R., Patil, D.: Document image binarization using threshold segmentation. Int. J. Innovative Res. Comput. Commun. Eng. **3**(3), 1873–1876 (2015)
19. Bhargava, N., Kumawat, A., Bhargava, R.: Threshold and binarization for document image analysis using otsu's Algorithm. Int. J. Comput. Trends Technol. (IJCTT) **17**(5), 272–275 (2014)

Digital Watermarking Scheme Enhancing the Robustness Against Cropping Attack

Ching-Sheng Hsu[1] and Shu-Fen Tu[2(✉)]

[1] Department of Information Management, Ming Chuan University,
Taoyuan City, Taiwan
cshsu@mail.mcu.edu.tw
[2] Department of Information Management, Chinese Culture University,
Taipei City, Taiwan
dsf3@ulive.pccu.edu.tw

Abstract. Most digital watermarking schemes using QR factorization suffer from being unable to fully utilize the elements of the R matrix. Thus, these schemes are neither secure nor robust to resist the cropping attack. Besides, these schemes do not deal with the allowable modification ranges of the R elements, thereby causing the damage to the hidden watermark. In this paper, we designed an algorithm to redundantly embed the four copies of the watermark bits to enhance the ability to against the cropping attack. During the embedding process, the property of sign wave is employed to ease the modification of real number coefficients. After the four copies of a watermark bit are extracted, they may be different due to possible attacks. Therefore, we designed a weighted strategy to resolve the watermark bit. The experimental results show that our scheme satisfy the requirements of imperceptibility and robustness. Particularly, our scheme has prominent robustness against cropping attacks.

Keywords: Digital watermarking · QR decomposition · Cropping attack

1 Introduction

Digital watermarking is a common way to protect digital images. The digital image is watermarked with a watermark, which may be a binary image or just a binary stream. Depending on the purpose of the watermarking scheme, the watermark is either robust or fragile. The purpose of a robust watermarking scheme is to protect the copyright of the digital image. The copyright of the digital image is proved by the extracted watermark, so the extracted watermark should be robust enough even if the watermarked image undergoes slight modification. On the other hand, the purpose of a fragile watermarking scheme is to protect the integrity of the digital image. The extracted watermark should be fragile when the watermarked image has been tampered with. The working domain of the digital images may be spatial or frequent [5]. Recently, some researches tried to work on another domain which is derived from matrix factorization, such as SVD [1, 2, 8], QR decomposition, or LU decomposition. The main idea is to utilize matrix factorization to decompose the image and embed the watermark into one of the decomposed matrices. Among these matrix factorization methods, QR

© Springer Nature Singapore Pte Ltd. 2018
J. C. Hung et al. (Eds.): FC 2017, LNEE 464, pp. 84–93, 2018.
https://doi.org/10.1007/978-981-10-7398-4_9

decomposition is a good choice for digital watermarking and image steganography. The reasons are as follows. Firstly, in comparison with SVD, QR decomposition has lower computational complexity and can avoid false positive problems [12, 14]. Secondly, in comparison with LU factorization, QR decomposition is more accurate for least square problems. Thirdly, LU factorization can only applied to square matrices while QR decomposition can be applied to rectangular and square matrices. Fourthly, the first row elements of the R matrix obtained by QR decomposition are able to resist to several image processing operations, such as lossy compression, noise addition, and filtering. Finally, the first row elements of the R matrix are likely to be greater than those elements in other rows, thereby allowing a greater modification range [12, 14]. Because of the above mentioned properties of QR decomposition, some researches use the first row elements for watermark embedding and image steganography [4, 7, 11–14].

Most QR-based digital watermarking schemes suffer from being unable to fully utilize the coefficients of the R matrix. Therefore, such methods may have low data hiding capacity and watermark security. For example, Su et al. [12] adopted QR decomposition to hide a robust color watermark into a color image. They use the last element of the first row in the R matrix to hide a watermark bit, but the allowable modification range of this coefficient was not discussed. Thus, the embedded watermark may be damaged by the reverse QR decomposition without any image processing operation or any malicious attack. This problem also occurs in Subhedar and Mankar's image steganography scheme [14]. The main concern of a QR-based digital watermarking scheme is that the modification to the decomposed matrix may cause the pixel values out of range. In order to fully utilize all the coefficients in the R matrix, we have to know the allowable modification range of each coefficients.

To cope with this issue, we have built formulas to compute the allowable modification ranges, thereby increasing the capability of information hiding and watermark security [6]. Our watermark embedding scheme employ the sine wave function in trigonometry and achieve the possible minimal modification within the allowable bounds. The sign wave simplifies the modification to real number coefficients by its nature. Like Su et al.'s scheme [12], the host image is decomposed by QR decomposition, and the watermark is embedded in the R matrix. But unlike their scheme, we use all elements of the first row of matrix R and adopt Householder reflections [15] for QR decomposition because it is numerical stable relative to Gram-Schmidt process [3, 10], which is adopted by Su et al.'s scheme [12]. In addition, most QR-based watermarking scheme embeds one watermark bit in an image block; therefore, their watermarks are prone to be fragile to cropping attacks. Take the example of Su et al.'s scheme. When the watermarked image is cut 25% off, the correct rate of the extracted watermark is about 87.75%. However, when the watermarked image is cut 50%, the correct rate of the extracted watermark decreases to 62.64% [12]. Therefore, we design an embedding strategy to enhance the robustness against cropping attacks. The rest of this paper is organized as follows. In Sect. 2, we will analyze the allowable modified ranges of the elements and explain the watermark embedding and extraction scheme in detail. In Sect. 3, we will demonstrate the experimental results and give some discussions. Finally, we will give conclusions in Sect. 4.

2 The Proposed Scheme

2.1 The Allowable Modification Range

Suppose the host image is a gray-level image. Thus, the integer pixel values of the host image range from 0 to 255. In this research, 4×4 image blocks are used to embed watermark. Thus, the matrix A is denoted as a 4×4 matrix. Each matrix A is factorized to matrix Q and R via QR decomposition, which are illustrated as follows:

$$A = \begin{bmatrix} a_{00} & a_{01} & a_{02} & a_{03} \\ a_{10} & a_{11} & a_{12} & a_{13} \\ a_{20} & a_{21} & a_{22} & a_{23} \\ a_{30} & a_{31} & a_{32} & a_{33} \end{bmatrix}, \ Q = \begin{bmatrix} q_{00} & q_{01} & q_{02} & q_{03} \\ q_{10} & q_{11} & q_{12} & q_{13} \\ q_{20} & q_{21} & q_{22} & q_{23} \\ q_{30} & q_{31} & q_{32} & q_{33} \end{bmatrix}, \text{ and } R = \begin{bmatrix} r_{00} & r_{01} & r_{02} & r_{03} \\ 0 & r_{11} & r_{12} & r_{13} \\ 0 & 0 & r_{22} & r_{23} \\ 0 & 0 & 0 & r_{33} \end{bmatrix}$$

If the modification applies to the first row elements of the R matrix, then the reconstructed values in matrix A should also range from 0 to 255. Thus, the allowable modification range of the R matrix should be identified to accommodate the valid pixel values. For $x = R[0, j]$, where $j = 0..3$, we define the allowable upper bound ub and lower bound lb of the modified x by Eqs. (1) and (2) [6].

$$lb = \max_{i \in \{0..3\}} (LB[i]) \tag{1}$$

And

$$ub = \min_{i \in \{0..3\}} (UB[i]) \tag{2}$$

where

$$LB[i] = \begin{cases} -\infty & \text{if } Q[i, 0] = 0, \\ \min(a, b) & \text{otherwise.} \end{cases} \tag{3}$$

$$UB[i] = \begin{cases} \infty & \text{if } Q[i, 0] = 0, \\ \max(a, b) & \text{otherwise.} \end{cases} \tag{4}$$

$$a = R[0, j] - A[i, j] / Q[i, 0] \tag{5}$$

and

$$b = R[0, j] + (255 - A[i, j] / Q[i, 0]). \tag{6}$$

2.2 Watermark Embedding

Suppose that the host image is an $M \times N$ gray-level image and the watermark is an $M/4 \times N/4$ binary image, where both M and N are multiples of four. At first, the host image is divided into n non-overlapping 4×4 blocks, where $n = M/4 \times N/4$. Then, each block A, which can be seen as a 4×4 matrix, is factorized to Q and R matrices by means of Householder reflections, and all R matrices are used to embed the watermark. Let $W = (w_0, w_1, \ldots, w_{n-1})$ denote the binary watermark, where $w_i \in \{0, 1\}$ for $i = 0, 1, \ldots n - 1$. Let $(R_0, R_1, \ldots, R_{n-1})$ denote the array of n R matrices constructed from the n image blocks. For the sake of increasing the survival of the watermark, we adopt a two-pronged strategy. Firstly, the original array $(R_0, R_1, \ldots, R_{n-1})$ is shuffled according to a pseudo-random number generator seeded by the secret key SK. Secondly, four copies of each watermark bit are redundantly inserted into the first row elements of the R matrix. Let $(R'_0, R'_1, \ldots, R'_{n-1})$ denote the shuffled array, and $R_i[0, 0]$, $R_i[0, 1]$, $R_i[0, 2]$, and $R_i[0, 3]$ respectively denote the first row elements of R_i, for $i = 0..(n - 1)$. To embed a watermark bit w_i, the four copies of w_i are redundantly inserted into the four R elements: $R'_{i \bmod n}[0, 0], R'_{i+1 \bmod n}[0, 1], R'_{i+2 \bmod n}[0, 2]$, and $R'_{i+3 \bmod n}[0, 3]$. Finally, the reverse operation of the QR decomposition is used to generate the watermarked image.

The rule for embedding a watermark bit $w \in \{0, 1\}$ to an element x of a matrix R depends on a sine wave function:

$$f(x) = sin(k \cdot x), \tag{7}$$

where $k > 0$ and is a real number. Accordingly, the wavelength λ of the function f is $(360/k)$. If $w = 1$, then x is modified to x' such that $f(x') \geq 0$; otherwise, if $w = 0$, then x is modified to x' such that $f(x') < 0$. Note that the modification of x to x' is restricted to the following three constraints:

1. The range of x' needs to be within $[0, 255]$.
2. The modification of x should be as less as possible so that x' is near to x to ensure the imperceptibility of our scheme.
3. The value of $|f(x')|$ should be as large as possible, hence our scheme can tolerate an acceptable alteration on the watermarked image.

Therefore, the proposed scheme follows the three steps below to modify x.

Step 1: For $x = R[0, j]$, where $j = 0..3$, define the allowable upper bound ub and lower bound lb of x' by Eqs. (1) and (2).

Step 2: Complying with the following rules, modify x to x' such that $|f(x')| = 1$ and the difference between x and x' is as small as possible.

Rules for $w = 0$:

> **If** $(x \geq 0 \wedge r \leq 0.25\lambda)$ **then** $x' = (x - r) - 0.25\lambda$.
> **If** $(x \geq 0 \wedge r > 0.25\lambda)$ **then** $x' = (x - r) - 0.75\lambda$.
> **If** $(x < 0 \wedge r \geq -0.5\lambda)$ **then** $x' = (x - r) - 0.25\lambda$.
> **If** $(x < 0 \wedge r < -0.5\lambda)$ **then** $x' = (x - r) - 1.25\lambda$.

Rules for $w = 1$:

> **If** $(x \geq 0 \wedge r \leq 0.75\lambda)$ **then** $x' = (x - r) + 0.25\lambda$.
> **If** $(x \geq 0 \wedge r > 0.75\lambda)$ **then** $x' = (x - r) + 1.25\lambda$.
> **If** $(x < 0 \wedge r \geq -0.25\lambda)$ **then** $x' = (x - r) + 0.25\lambda$.
> **If** $(x < 0 \wedge r < -0.25\lambda)$ **then** $x' = (x - r) - 0.75\lambda$.

The remainder r of x divided by λ is defined by

$$r = s\left(|x| - \lambda\left\lfloor\frac{|x|}{\lambda}\right\rfloor\right), \tag{8}$$

where $s \in \{1, -1\}$ represents the sign of x.

Step 3: If x' is out of allowable range, perform the following procedure to adjust x' so that x' is able to be within $[lb, ub]$, the difference between x and x' is as close as possible, and $|f(x')|$ is as large as possible.

```
Procedure ValueAdjust(x', lb, ub, λ):
low ← x' - 0.25λ
high ← x' + 0.25λ
If x' < lb then
  If lb ≥ high and ub ≤ high + 0.5λ then
    Indicate that "No feasible solution for x'."
  If ub ≥ x' + λ then x' ← x' + λ
  Else if lb < high then x' ← lb
  Else if ub < x' + λ then x' ← ub
Else if x' > ub then
  If ub ≤ low and lb ≥ low + 0.5λ then
    Indicate that "No feasible solution for x'."
  If lb ≤ x' - λ then x' ← x' - λ
  Else if ub > low then x' ← ub
  Else if lb > x' - λ then x' ← lb
Else
  Indicate that x' is valid and do nothing
```

2.3 Watermark Extraction

To extract the robust watermark W, the secret key SK is used to generate the shuffled array $(R'_0, R'_1, \ldots, R'_{n-1})$. In principle, the extracted bit w' of an element of the R matrix is determined according to Eq. (9).

$$w' = \begin{cases} 1 & \text{if } f(x) \geq 0, \\ 0 & \text{otherwise.} \end{cases} \tag{9}$$

where $x = R[0, j]$ for $j = 0..3$.

Remember that each watermark bit is redundantly embedded in four elements of the R matrix. Therefore, the four copies of the watermark bit w_i are extracted from coefficients $R'_{i \bmod n}[0, 0], R'_{i+1 \bmod n}[0, 1], R'_{i+2 \bmod n}[0, 2]$, and $R'_{i+3 \bmod n}[0, 3]$, and each copy is determined by Eq. (9). Note that the extracted four copies may be different due to the possible alteration to the watermarked image, it is necessary to have a rule to reach a consensus about the extracted watermark bit w_i'. Intuitively, the majority voting principle is practical. That is, if two or more copies are '1', then the extracted watermark bit w_i' is determined as '1'; otherwise, the extracted watermark bit w_i' is determined as '0'. Because different elements of the R matrix may have different ability to resist image processing operations or malicious attacks [12], more sophisticated way can be used, which puts different weights on the four extracted copies. Assume that the ability to resist image processing operations of the R element r_{0i} is weighted as α_j, for $j = 0, 1, 2, 3$ and

$$\sum_{j=0}^{3} \alpha_j = 1$$

Also assume that $c_i(0)$, $c_i(1)$, $c_i(2)$, and $c_i(3)$ are the extracted four copies of the watermark bit w_i. Then, the extracted watermark bit w_i' is determined according to the following rule.

$$w_i' = \begin{cases} 1 & \text{if } \sum_{j=0}^{3} \alpha_j \cdot c_i(j) \geq 0.5, \\ 0 & \text{otherwise.} \end{cases} \tag{10}$$

In fact, the majority voting principle is a special case of the weighted scheme. Generally, the weight α_j can be determined according experimental analysis.

3 Experiment Results and Discussions

In general, a robust watermarking scheme is evaluated by its imperceptibility and robustness. The imperceptibility means that the difference between the host image and the watermarked image should not be perceived by human eyes, and the robustness means that the extracted watermark should be similar to the original watermark. In this paper, we use $PSNR$ to evaluate the imperceptibility of our scheme [9].

$$PSNR = 10 \times \log \frac{255^2}{MSE}, \tag{11}$$

where

$$MSE = \frac{1}{M \times N} \sum_{i=1}^{M} \sum_{j=1}^{N} (p_{i,j} - p_{i,j}')^2. \tag{12}$$

The notation M and N in Eq. (12) denote the width and height of an image, respectively, and $p_{i,j}$ and $p_{i,j}'$ denote the pixel of the original and watermarked image, respectively. Generally, the difference between the watermarked image and the host image is visually imperceptible if *PSNR* is greater than 30. The robustness is evaluated by the indicator *NC* as follows.

$$NC = \frac{1}{W \times H} \sum_{i=1}^{W} \sum_{j=1}^{H} \overline{(w_{i,j} \oplus w'_{i,j})} \tag{13}$$

The notation W and H in Eq. (13) denote the width and height of the watermark image, respectively, $w_{i,j}$ and $w'_{i,j}$ denote the bit of the original and extracted watermark, respectively, and '$\oplus$' represents the logical XOR operation. The larger the *NC* is, the more robust the extracted watermark is.

Figure 1(a) is our watermark, which is embedded into the host image (Fig. 1(b)). The embedding rule (Eq. (7)) needs a parameter k, but it is not necessary to use the same value of k for each element of the R matrix to embed a bit. In this experiment, we use four values (10, 90, 90, 90, 90) respectively corresponding to the four elements (R[0,0], R[0,1], R[0,2], R[0,3]). Figure 1(c) is the watermarked image, and the PSNR is 38.0081. Obviously, our scheme is qualified for the imperceptibility. With regard to the robustness, we simulate two kinds of attacks, namely cropping and JPEG lossy compression, on the watermarked image and inspect the survival of the extracted watermark. As mentioned in the above section, we can use a set of weights (α_0, α_1, α_2, α_3) for the extracted four copies to determine the extracted watermark bit. In this experiment, we test two different sets (1.0, 0.0, 0.0, 0.0) and (0.25, 0.25, 0.25, 0.25) on the robustness of the watermark.

We performed experiments on the watermarked image under cropping attack with different cropping ratios (CR) from 0.05 to 1.00. Figure 2(a)–(t) are the extracted watermarks and their respective NC values, and the corresponding cropping ratios are listed as well. Figure 2 shows that the similarity between the extracted watermark and the original watermark is greater than 80% even though the cropping ratio reaches as high as 0.6. Compared to the NC value against the attack with 25% cropping ratio of Su et al.'s scheme [12], our scheme gets 97.81%, which is higher than Su et al.'s 87.7%. When the watermarked image is cut 50% off, the NC value of our scheme is 88.59% while that of Su et al.'s is 62.64%. Thus it can be seen that our embedding strategy largely enhances the robustness against cropping attacks. The other experiments were performed on the watermarked image under JPEG lossy compression with different compression qualities, *i.e.* compressed image quality, from 0.05 to 1.00. The compression is implemented with Java Image I/O API provided in Java SE 8. Figure 3(a)–(t) are the extracted watermarks and their respective NC values, and the corresponding JPEG qualities are listed as well. Figure 3 shows that the similarity between he extracted watermark and the original watermark is greater than 80% even though the compression quality decreases as low as 0.45.

We have mentioned earlier that two sets of weights are used to test the robustness of the watermark. Actually, different weights have different effects on the robustness of the watermark. Figure 4(a) demonstrates the correlations between the NC values and cropping ratios, and Fig. 4(b) demonstrates the correlations between the NC values and JPEG compression qualities. It is observed that the two weights perform vary and each has its own strengths. The set with equal weights outperforms the set (1.0, 0.0, 0.0, 0.0) on extracting watermarks from cropped watermarked images; on the contrary, the set (1.0, 0.0, 0.0, 0.0) outperforms the extracting scheme with equal weights on extracting watermarks from compressed watermarked image. Accordingly, the set with equal weights is suitable for being against the alterations that concentrates in a region. The set (1.0, 0.0, 0.0, 0.0) means that the decision of the extracted watermark bit depends on the first element only and is suitable for being against the alterations that spread over the watermarked image. This distinct fact gives our scheme flexibility. We can juxtapose the watermarks extracted by our scheme with different weights and display the best one. Correspondingly, Fig. 2 shows the extracted watermarks with the weights (0.25, 0.25, 0.25, 0.25), and Fig. 3 shows the extracted watermarks with weights (1.0, 0.0, 0.0, 0.0).

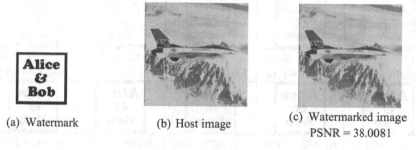

(a) Watermark   (b) Host image   (c) Watermarked image
PSNR = 38.0081

Fig. 1. The watermark, the host and watermarked images

Alice & Bob	Alice & Bob	Alice & Bob	Alice & Bob	Alice & Bob
(a) NC = 99.98 CR = 0.05	(b) NC = 99.87 CR = 0.10	(c) NC = 99.43 CR = 0.15	(d) NC = 98.80 CR = 0.20	(e) NC = 97.81 CR = 0.25
Alice & Bob	Alice & Bob	Alice & Bob	Alice & Bob	Alice & Bob
(f) NC = 96.83 CR = 0.30	(g) NC = 95.3 CR = 0.35	(h) NC = 93.4 CR = 0.40	(i) NC = 91.11 CR = 0.45	(j) NC = 88.59 CR = 0.50
Alice & Bob	Alice & Bob	Alice & Bob	Alice & Bob	Alice & Bob
(k) NC = 85.89 CR = 0.55	(l) NC = 82.81 CR = 0.60	(m) NC = 79.67 CR = 0.65	(n) NC = 76.31 CR = 0.70	(o) NC = 73.66 CR = 0.75
Alice & Bob	Alice & Bob			
(p) NC = 70.76 CR = 0.80	(q) NC = 68.12 CR = 0.85	(r) NC = 66.00 CR = 0.90	(s) NC = 64.71 CR = 0.95	(t) NC = 64.15 CR = 1.00

Fig. 2. The NC values of extracted watermarks for cropped images with different cropping ratio values.

(a) NC = 53.78 Quality = 0.05	(b) NC = 54.57 Quality = 0.10	(c) NC = 54.30 Quality = 0.15	(d) NC = 61.18 Quality = 0.20	(e) NC = 65.60 Quality = 0.25
(f) NC = 70.49 Quality = 0.30	(g) NC = 75.61 Quality = 0.35	(h) NC = 79.40 Quality = 0.40	(i) NC = 81.73 Quality = 0.45	(j) NC = 84.19 Quality = 0.50
(k) NC = 86.32 Quality = 0.55	(l) NC = 88.63 Quality = 0.60	(m) NC = 90.61 Quality = 0.65	(n) NC = 92.60 Quality = 0.70	(o) NC = 95.14 Quality = 0.75
(p) NC = 97.71 Quality = 0.80	(q) NC = 99.59 Quality = 0.85	(r) NC = 100 Quality = 0.90	(s) NC = 100 Quality = 0.95	(t) NC = 100 Quality = 1.00

Fig. 3. The NC values of extracted watermarks for JPEG compressed images with different quality values

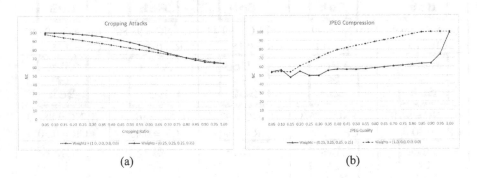

(a)   (b)

Fig. 4. The NC values relative to different parameters

4 Conclusions

Most QR-based digital watermarking schemes suffer from being unable to fully utilize the coefficients of the R matrix. Therefore, such methods may have low data hiding capacity and watermark security. To cope with this issue, we have built formulas to compute the allowable modification ranges and employ the sine wave function in trigonometry and achieve the possible minimal modification within the allowable bounds. In the embedding scheme, the host image is decomposed by QR decomposition, and each watermark bit is redundantly embedded in the R matrix. We adopt Householder reflections for QR decomposition because it is numerical stable relative to Gram-Schmidt process. Redundantly embedding a watermark bit can increase the robustness and security of our scheme. Observing from the experiment results, our scheme indeed succeeds in enhancing the robustness against cropping attacks However, after the four copies of a watermark bit are extracted, they may be different due to possible attacks. Therefore, we design a weighted strategy to resolve the watermark bit. The experimental results show that our scheme satisfy the requirements of imperceptibility and robustness. In the future, we will test more sets of weights on extracting the watermark and give analysis of the relationship between the set of weights and the type of attacks.

References

1. Ansari, I.A., Pant, M., Ahn, C.W.: SVD based fragile watermarking scheme for tamper localization and self-recovery. Int. J. Mach. Learn. Cybern. 7(6), 1225–1239 (2015). https://doi.org/10.1007/s13042-015-0455-1
2. Byun, S.C., Lee, S.K., Tewfik, A.H., Ahn, B.H.: A SVD-based fragile watermarking scheme for image authentication. LNCS, vol. 2613, pp. 170–178. Springer, Heidelberg (2003)
3. Cheney, W., Kincaid, D.: Linear Algebra: Theory and Applications. Jones and Bartlett, Sudbury (2009)
4. Gao, J., Fan, L., Xu, L.: Solving the face recognition problem using QR factorization. WSEAS Transl. Math. 11(8), 712–721 (2012)
5. Hsu, C.S., Tu, S.F.: Probability-based tampering detection scheme for digital images. Opt. Commun. 283(9), 1737–1743 (2010)
6. Hsu, C.S., Tu, S.F.: Image authentication based on QR decomposition and sinusoid. In: 11th International Conference on Computer Science and Education. IEEE Press, pp. 479–482 (2016). https://doi.org/10.1109/iccse.2016.7581627
7. Huang, H.F.: A fragile watermarking algorithm based on Chaos and QR decomposition. Intl. J. Adv. Comput. Tech. 5(4), 117–124 (2013)
8. Kang, Q., Li, K., Chen, H.: An SVD-based fragile watermarking scheme with grouped blocks. In: 2nd International Conference on Information Technology and Electronic Commerce. IEEE Press (2014). https://doi.org/10.1109/icitec.2014.7105595
9. Katzenbeisser, S., Petitcolas, F.A.P.: Information Hiding: Techniques for Steganography and Digital Watermarking. Artech House Inc., MA, USA (2000)
10. Stoer, J., Bulirsch, R.: Introduction to Numerical Analysis, 3rd edn. Springer, New York (2002)
11. Su, O., Niu, Y., Zou, H., Zhao, Y., Yao, T.: A blind double color image watermarking algorithm based on QR decomposition. Multimed. Tools Appl. 72(1), 987–1009 (2014)
12. Su, Q., Niu, Y., Wang, G., Jia, S., Yue, J.: Color image blind watermarking scheme based on QR decomposition. Signal Proc. 94, 219–235 (2014)
13. Su, O., Wang, G., Zhang, X., Lv, G., Chen, B.: An improved color image watermarking algorithm based on QR decomposition. Multimed. Tools Appl. 76(1), 707–729 (2017). https://doi.org/10.1007/s11042-015-3071-x
14. Subhedar, M.S., Manbkar, V.H.: Image steganography using redundant discrete wavelet transform and QR factorization. Comp. Elec. Eng. 54, 406–422 (2016)
15. QR decomposition. https://en.wikipedia.org/wiki/QR_decomposition

The Practical Research of the Computer-Based Courses in University

The Teaching Model of Flipped Class Based on SPOC

Dongmei Zhao and Xiaofan Liang[(⊠)]

College of Information Technology, Hebei Normal University,
Shijiazhuang 050024, China
1406501316@qq.com

Abstract. Along with the development of multimedia system and its relevant technology, the researchers began to explore and attempt to use multimedia technology to assist teaching which is in order to improve the quality of education. The computer-based course is a public compulsory course which has significant implications for cultivating the students' information literacy. However, there is a big difference between learners' Computer level and other factors. Since all these unfavorable factors, the Teaching Practice of computer-based courses is not ideal and there is a variety of conflict. This research is trying to use the multimedia system and related technology to serve the daily teaching activities of the computer-based course after sorting out the related research results. This study built the flipped class mode based on SPOC. The study combined two rounds of teaching practice to improve the teaching mode and the study use the questionnaire survey and interview technique to prove the effectiveness of the model. Finally, the study suggested that the teaching model could be improved from three aspects—strengthening the construction of network teaching platform, enhancing the teachers teaching skills and focusing on the creative application of multimedia technology in teaching.

Keywords: Multimedia system · The computer-based courses in the university
SPOC · Flipped class · Teaching model

1 Introduction

The computer-based course is a public compulsory for all the freshmen in the university. As a public foundation course, whether the course could carry out effectively is ever more important for nowadays society which is developing rapidly in the field of information technology. The Basic knowledge about computer, operational skills and

---

D. Zhao—Received her M.S. and Ph.D degree in College of Computers from Xidian University, China in 1998 and 2006 respectively. Currently she is a professor at Hebei Normal University, Shijiazhuang, China. Her research interests include network and information security.
X. Liang—Studying for a master's degree at Hebei Normal University, Shijiazhuang, China. Her research interests include educational informationization and information security.

© Springer Nature Singapore Pte Ltd. 2018
J. C. Hung et al. (Eds.): FC 2017, LNEE 464, pp. 94–103, 2018.
https://doi.org/10.1007/978-981-10-7398-4_10

the practical ability have already been the most basic information literacy for the modern college students [1]. Meanwhile, "the ten years' development planning of Educational Informationization (2011–2020)" mentioned many major concepts, such as combining information technology and education deeply and paying attention to the development of students' cultivation of the information literacy [2]. The information literacy has already been the most essential ability for the contemporary college students. While, one of the main ways of improving the university students' information literacy is setting up the Computer-based Courses [3]. Researching and practicing the teaching activities in the computer-based courses is not only important to the cultivation of students' information literacy, but providing reference for transformational work in other subjects to combine the information technology and education deeply.

The study which on the basis of having read a large amount of relevant documents and having summarized the teaching experience for many years in the computer-based course found that there is large difference between freshmen's information literacy when they just begin their college life; Teachers need to provide corresponding content because students whose academic background are different have differential learning needs; Since the number of students in the computer-based course is huge, more effective ways are to be used creatively in the teaching activities [4]; There are a great deal of conflict between the great class capacity and the very limited teaching hours; The evaluation approach of this course could not reflect students' information literacy and related ability effectively [1]; These problems mentioned above exist truly in the teaching activities of the computer-based course, so the studies by researchers on related field are urgently needed.

One of the most typical cases is the practice research made by professor Fox and his colleagues at the University of California, Berkeley. The research team has used their existing MOOC course named "Software Engineering" to orient SPOC course to their students inside their campus. The teaching practice which was conducted by the teaching team leading by professor Fox, to a great extent, demonstrated the effectiveness of SPOC teaching model. This research obtained good teaching effect in many respects such as the students' learning experience and the students' evaluation. Similarly, the teaching model based on SPOC was designed in more detail At MIT in Boston. The successful of their research conducted by Massachusetts institute of technology was not only reflected in the students' consistent high praise but reflected in the very ideal average result of the students [5]. Most studies on SPOC in China was focused on the following aspects, such as the comparative study of MOOC with SPOC, the analysis of the advantages on SPOC teaching model, the design of teaching model and the teaching practice on SPOC. Many well-known domestic universities, such as Tsinghua university, Zhejiang university, Beijing Institute of Technology and so on, are devoted to the practical teaching application of the SPOC teaching model. While there is a lack of the applied researches on using SPOC teaching model in the college computer basic courses.

At the same time, researchers began to explore and attempt to use multimedia technology to assist education and teaching in order to improve the quality of teaching with the rapid development of multimedia system and its related technology. Using the multimedia systems and service to assist teaching has become an indispensable part of teaching activities in college classrooms [6]. This study based on the summary of research findings from related researchers and the firsthand teaching experience for many years is an exploration of the computer-based courses' reform. Using multimedia technology to service the education and teaching better is regarded as the research ideas. This study based on SPOC has conducted a design for the teaching model of Flipped class in detail and used this model in the practical teaching activities. Finally, researching its effectiveness with empirical study. The reflection is based on the teaching effect and feedback information collected from the first round of teaching practice. Then, optimizing the design of the original model on the second round of the teaching practice. Now Stating the research ideas, the process of experiment, feedback data and research conclusion to provide a certain reference for other researchers who also conduct the same researches on the transformational attempt of the computer-based course. This study is also expected could provide ideas and thoughts for the teaching practice which use multimedia system to service other disciplines.

2   The Theoretical Basis of Flipped Class Teaching Model on SPOC

The emergence of SPOC (Small Private Online Course) is due to the reflection of MOOC (Massive open online course) which exists many defects such as the high rate of dropout and the lack of deep learning behaviors. SPOC was proposed by professor Armando Fox [7] at the earliest. He believed that the Emergence of SPOC is in order to better use MOOC and better play the potential of MOOC, so that better quality teaching resource could be used in small groups such as school. SPOC which mainly by setting the access condition to participate in forming a small-scale private online course could be considered as the product of the "MOOC development" Times [8].

"The significance of SPOC, a more flexible and effective teaching method, is to make the teaching process jump out of the repetitive teaching stage, so liberating the limited labor of teachers into more valuable and creative teaching activities", said professor Robert [9] of Harvard University. Yeqin Kang as a Chinese scholar has summarized some advantages about SPOC after having researched the teaching model explore about SPOC conducted by Harvard University, University of California Berkeley and Massachusetts Institute of Technology. One of the most important thing is that SPOC closely bound up with teaching content could reinvest the learners with complete experience of deep learning [10]. The conclusion based on summarizing the research achievements of many scholars is that SPOC is effective improvement and reasonable supplement to realize that MOOC could merge with the traditional classroom deeply [11]. This study compared SPOC with MOOC from twelve dimensions such as the openness of the course, the number of the students, the completion rate of the course, interaction between teachers and students, the timely feedback, the freedom degree of learning, the way students learning, the form of evaluation, the efficiency of teaching, the idea of teaching, the diversity of teaching model and the learning cost. As shown in Fig. 1, setting up a "12×5" analysis model in this study. "12" is the twelve dimensions such as the openness of the course and so on and "5" is the evaluation order of 5 levels with the outermost ring getting the highest score and excellent performance while the innermost ring getting the lowest score and poor performance. As the proposer professor

Fox of SPOC pointed out, using SPOC in the school teaching not only could enrich the teachers' teaching methods, but improve the degree of students' participation, the throughput of the class and the students' mastery degree of knowledge [12].

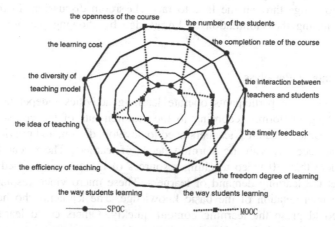

Fig. 1. The comparison between SPOC and MOOC from twelve dimensions

3 The Concrete Construction of the Flipped Class Teaching Model Based on SPOC

It is a teaching model which mixes the online learning on the web-based teaching platform with the face-to-face learning in the classroom together naturally [13]. The analysis of pretest was conducted on the information literacy of the students who are to learn with the flipped class teaching model based on SPOC. And the analysis of pretest mainly included the statistical analysis on the e-Learning ability in the experimental class. The results show that: Nearly 91.3% of the students who can use the Internet media resource and social software has certain information literacy and the experience of using information technology to learn. Only a small number of students is short of digital learning experience. Then, this study interviewed these learners and learned that they received basic education in the remote areas. The advantage is that these learners show the strong learning motivation generally. The teachers should focus on these students who are lack of the information literacy and give more guidance at the beginning of the flipped class teaching model based on SPOC to prevent these students missing the interest in learning content.

The new model, combined both the feature of the computer-based courses in the university and the analysis of pretest on the subjects, is based on the flipped class teaching model raised by Gerstein with four stages which include the stage of experiential learning, the stage of Concept exploration, the stage of meaning construction and the stage of display and application. In this research, the flipped class teaching model based on SPOC is divided into three stages which include pre-class(on-line), in-class(Offline), post-class (the combination of on-line and Offline). The stage of pre-class is mainly experiential learning that participants are asked to check the learning tasks assigned by the teachers and learn the micro video about basic knowledge on the network teaching platform, and then complete the test in order to comprehend the knowledge; During the Stage of in-class, the learners should explore the concepts and complete significant construction of knowledge to achieve the internalization of knowledge through the face to face classroom discussion. In the stage of post-class, finishing the sublimation of knowledge by showing the production and application.

3.1 The Stage of Pre-class

During this stage, the participants operate learning activities independently on the network teaching platform. According to the requirements of the course's teaching goals and the learners' features, teachers need to develop, design, and provide the study resource which accords with the learning needs of learners. The research team has developed more than 80 video teaching resources of the computer-based courses in order to meet the learning demand of learners. These micro video resources mainly aimed at the interpretation of the basic knowledge. The learners who have a good foundation could grasp the learning content quickly; Others could learn the video resources repeatedly. This has a good flexibility. The learners understand the instructional objectives by reading the task modules of course, and then the students learn independently by watching the video resources on the network teaching platform.

3.2 The Stage of In-class

This stage is mainly conducted in face to face offline classroom. In this stage, the role of teachers is not only teaching knowledge, but organizing teaching activities, constructing the learning situation and guiding the students' learning behaviors. At first, teachers learn the students' overall learning situation on the basis of their behaviors in stage of pre-class and explain the common problems in the stage of in-class. Then teachers assigned the well-designed and meaningful tasks(problems) to guide students to learn collaboratively. In the process of inquiry learning, students could solve the problem through personalized inquiry, group investigation, collaborative inquiry and the deep communication with teachers. Teachers could explain the universal problems in class.

In the process of inquiry learning, learners comprehend and grasp the basic knowledge learned in their stage of pre-class and realize the internalization and migration of knowledge in a variety of changing situations. The teaching goal is improving the students' information literacy and cultivating the students' ability to solve the complicated problems of ill-structured domain. In the end of the course, teachers guide students to extract the knowledge map of this course and cultivate the students' structural thinking to restructure and migrate the knowledge.

3.3 The Stage of Post-class

Considering that the feature of the computer-based courses are strong applicability and strong practicality, learners need to complete a comprehensive teamwork in this stage.

Each team can choose the theme according to their own interests within the fixed time and then conduct demand analysis and finish the works' design and implementation. Knowledge and skills can be comprehensively used through completing the teamwork cooperatively. Most of the students can get a sense of achievement from their finished products to gain a greater learning motivation. Many studies have shown that there is a mass of students who have the psychological need to learn others' learning outcomes. Therefore, setting up a module to display products. In this module, evaluating is by the way of peer review each other. Learners are tutting praise and emerging to try when they see perfect products in the practice teaching process.

4 The Investigation of Practical Effect and the Analysis of Data

The practice research is based on the computer-based courses in the university. During two rounds of teaching practice, the researchers collect experimental data continuously through class observation, interview and questionnaire survey to the students. After the first round of teaching activities, modifying and improving the teaching model.

4.1 The Preliminary Exploration: Teaching Activities in the First Round

The content of the first round of teaching activities is about the module of word typesetting. During the process of teaching activities in the first round, the researchers have arranged two observers trained before to observe the situation of the implementation of teaching activities and record the class scene as objective as possible. At the end of the teaching, the questionnaire survey was conducted on all subjects and the interview investigation was conducted on 15 students selected randomly. The researchers consider that the records of the observers have a certain objectivity, but may not be enough comprehensive. While the data collected from the questionnaire survey and the interview survey of learners has a certain subjectivity. So the researchers study comprehensively on the data collected from both the two channels in order to reflect the research process objectively.

The study based on comprehensive analysis of the data collected by the above ways found that there were the following problems in the first round of the teaching activities. In the teaching process of the in-class stage, teachers organize the teaching activities difficultly because the capacity of the computer-based courses is huge. If using the fixed voice tube, the teachers only could speak in fixed position instead of interacting well with students among them. If using the wireless microphone, there also are many issues such as the poor battery performance, noisy and poor effect. In the process of collaborative inquiry learning in-class, some teams encountered problems which hindered the process of group collaboration and they all hoped could be given guidance by teachers.

Researchers have proposed the following constructive improvement measures based on the above problems. Building the intellective multimedia platform based on the smart mobile phone in the stage of in-class (offline) through incorporating the smart phones into the multimedia system in classroom. Using the screen transmitter to receive the audio and video information from smartphones. Thus the information could been displayed in the previous multimedia devices of the classroom in near real-time. The public address equipment is of a great significance to reduce the work intensity of teachers, especially in the huge-capacity courses. But the research results based on the first round of teaching activities show that if using the fixed voice tube, the teachers only could speak in fixed position instead of interacting well with students among them. If using the wireless microphone, there also are many issues such as the poor battery performance, noisy and poor effect. In the intellective mobile multimedia system based on smartphones, the wireless microphone system with excellent performance is built by using smartphones, headset and the pick-up APP cooperatively. Building the intellective mobile multimedia platform based on smartphones also has a good effect on solving the second problem. Camera which has the good performance on video is the most important part of smart phones. Using the camera of smartphone can realize intelligent, multi-dimension and omni-directional video [14]. Based on this, the collaborative inquiry progress of the groups could be recorded conveniently. The constructive suggestions would be provided for carrying forward the team cooperation by reproducing the scene, analyzing the details and finding the problems. Meanwhile, the possible reason of the second problem is the lack of cooperative learning skills. The improve strategy is that providing the related teaching resource which could improve the learners' collaborative learning skills in the stage of pre-class online. These resources should guide the learners effectively to understand that how to communicate in the process of cooperative learning, how to discuss and how to treat the internal conflicts of the groups correctly.

4.2   The Improvement: Teaching Activities in the Second Round

The improvement of the flipped class teaching model based on SPOC is through rethinking the problems existed in the teaching activities of the first round. On the basis of this, expanding the research cycle of teaching activities in the second round. The evaluation of the teaching effect in the second round mainly includes two aspects: On the one hand, the questionnaire survey was implemented to learn the satisfaction of the students about this teaching model and the interview survey was conducted to learn the improvement suggestions from the students; On the other hand, analyzing the final result of the experimental class.

The survey on students' degree of satisfaction
To learn the students' recognition of the flipped class teaching model based on SPOC, the questionnaire survey was conducted on the experimental class's students. The research recycled the 97 effective questionnaires. The survey was mainly from the three following dimensions. Are you satisfied with this teaching model which bring the new learning experience? Do you think that the flipped class teaching based on SPOC has a promoting effect to learning? Do you desire to use the flipped class teaching based on SPOC to study in the future learning life? The three above dimensions all adopt the scoring system with five rating levels.

About the 81% students gave positive evaluations which showed the advancement of this teaching model to some extent. The students reflected generally that the flipped class teaching model based on SPOC could bring their new learning experience. Compared with the traditional teaching model, their learning autonomy, anytime and anywhere, was given by watching the micro-video which could better meet the needs of their personalized learning needs. The immediate feedback mechanism could make it possible for students' self-adaptive learning behaviors. The module of displaying the final productions also makes them gain a great sense of achievement. Then, stimulating the learning motivations eventually.

The statistics indicate that almost all the learners in experimental class think that this teaching model play a stimulative role. After interviewing with the students, the learners think that their deep learning experience can be given by this model. At the same time, they also show that the deep interaction in-class is based on completing the study task seriously in the stage of pre-class.

The experiment result shows that there are 8% learners expressed clearly that they did not desire to use the flipped class teaching based on SPOC to study in the future learning life while they recognize the effectiveness of this model. The subsequent research for this problem reveal that there are a few learners who are lack of the digital learning ability and most of the students fear that this teaching mode would take up their too much energy. Considering this point, the teaching mode need to be optimized in the future practice by simplifying the too heavy learning tasks.

The analysis of the final result in the experimental class
The flipped class teaching model based on SPOC utilized other researchers' work and used the developmental teaching evaluation model based on the method of comprehensive index [15]. In order to evaluate the students' learning behaviors objectively and scientifically, this study use the multidimensional evaluation ways which include prepositive evaluation, formative evaluation and summative evaluation. The prepositive evaluation is set for the learners' learning behaviors between the stage of in-class; The formative evaluation exists in the whole learning process of learners, in order to feedback in time and adjust the learning behaviors; The summative evaluation is the reflection on the study results at the end of the learning activities and its form is usually the final test.

Because the other part of the results is hard to compare, the study analyzes contrastively the average score in the experimental class with the average score of the traditional teaching class. They have the same final test questions, the same professional background and almost the same average scores before. The average score in the experimental class which used the flipped class teaching model based on SPOC is 9.87 points higher than the traditional teaching class in the same period. This illustrates the flipped class teaching model based on SPOC has the validity and advantage to some extent.

4.3   The Conclusion and Reflection

The study has designed the flipped class teaching model based on SPOC concretely by setting the stage of pre-class, in-class and post-class. Confirming the validity and advantage of this model to some extent based on the empirical research. At the same time the study found that there are some aspects to improve: First, the function of the network teaching platform need to be improved. The intelligent learning platform is the important support for building the well interactive learning situation. The network teaching platform, open, smart and social, should be based on the user experience; Second, improving the teachers' knowledge structure and teaching skills.

The requirement of the flipped class teaching model based on SPOC will be higher for teachers because teachers need to layout the learning tasks, choose teaching resources, organize teaching activities carefully and evaluate the students' learning behaviors reasonably [16]; Third, building the advanced and intelligent multimedia teaching system requires using the advanced multimedia technology to assist teaching.

5   The Conclusion

The flipped class teaching model based on SPOC making the multimedia technology and education mixed together deeply. The survey found that this model was praised by the students in the experimental class and the teaching effect was improved distinctly. The ways of teaching are clear, while the practical teaching situation was not [17]. The concrete implementation method also need to be explored by teachers. What need to be pointed is that the research on this field is still at the primary stage. The development and improvement of the flipped class teaching model based on SPOC need to expand the research sample and extend the research cycle by researchers.

References

1. Zhong, Q., Wu, Z.-Y.: Research on 'micro-lecture' teaching mode of computer basic courses in college. Mod. Educ. Technol. 02, 26–33 (2014)
2. The ministry of education. The ten years' development planning of educational informationization (2011–2020) [EB/OL]. http://www.moe.edu.cn/publicfiles/business/htmlfiles/moe/s5892/201203/133322.html. Accessed 13 Feb 2012
3. Yun, X., Li, Z.: Exploration and reflection of SPOC: based teaching model in flipped classroom. China Educ. Technol. 05, 132–137 (2016)
4. Zhang, J.-L.: The path analysis of the teaching effectiveness on the liberal art's computer-based courses based on SPOC in the University. Dist. Educ. China 05, 71–75 (2016)
5. White, B.: An edX SPOC as the online backbone of a flipped college course [OL]. https://www.edx.org/blog/edx-spoc-online-backbone-flipped-college
6. Nie, Z., Yang, Y., Liu, Z.: The perspective of twenty years' changes of the national education technology plan of America. China Educ. Technol. 02, 132–139 (2017)
7. Fox, A.: From MOOCs to SPOCs. Commun. ACM 12, 38–40 (2013)
8. Luo, J.-T., Sun, M., Gu, X.-Q.: The innovation research of MOOC from the perspective of blended learning: case study of SPOC. Mod. Educ. Technol. 7, 18–25 (2014)
9. Coughlan, S.: Harvard plans to boldly go with 'Spocs' [EB/OL]. http://www.bbc.com/news/business-24166247. Accessed 24 Sept 2013
10. Kang, Y.-Q.: An analysis on SPOC: post-MOOC era of online education. Tsinghua J. Educ. 1, 85–93 (2014)
11. Li, H.-L., Li, X.-L.: The discussion on the era of "MOOC development" based on the experiential learning of SPOC's distribution to flip. e-Education Res. 11, 44–50 (2015)

12. Fox, A.: From MOOCs to SPOCs[DB/OL]. (2013-12-16). http//cacm.acm.org/magazines/
    2013/12/169931-from-moocs-to-spocs/fulltext. Accessed 20 Feb 2015
13. Naisheng, Y.: Classroom situations under the perspective of deep learning. Res. Educ.
    Devel. **12**, 76–79 (2013)
14. Shenglan, X.: The construction and application base on smart phone's teaching support
    platform. China Educ. Technol. **03**, 127–131 (2017)
15. Zhang, Z.-Y., Ren, J.-P., Wu, Y., Zhou, B.-B.: Research on computer basis teaching
    reformation based on computational thinking. Mod. Comput. **1**, 16–19 (2016)
16. Terry, A., Wang, Z.-J.: Hope or adventure: Massive Open Online Courses (MOOCs) and
    open distance education. China Educ. Technol. **1**, 46–51 (2014)
17. Huang, G.-F., Wu, H.-Y., Jin, Y.-F.: The practice and research on SPOC effective teaching
    in the U-Learning environment. e-Education Res. **05**, 50–57 (2016)

A Research on the IOT Perception Environment Security and Privacy Protection Technology

Xinli Zhou[✉], Liangbin Yang, and Yanmei Kang

School of Information Science and Technology, University of International Relations,
Beijing 100091, China
zhouxinli001@126.com

Abstract. The Internet of Things, which is called the IOT for short, is the third wave of the information industry. Some simple superposition of the existing technology of privacy protection cannot meet the new demand for privacy protection of the Internet of Things in the third wave of the information industry. A timely overview of the development of the Internet of Things security and privacy technology, from numerous existing research results, will serve as a cornerstone of future research. Despite some literatures have given us a detailed summary of the relevant issues of the Internet of Things according to its perception or network, previous studies fail to discuss the terms of the overall framework of the Internet of Things or simply apply the security and privacy protection of the internet to the Internet of Things, which will inevitably affect the directions of future research more or less. This paper first analyzes the system structure and features of the Internet of Things, and then focuses on the security and privacy protection technology of the perceived environment by presenting a systematic overview of the current privacy protection technology concerning the Internet of Things, with a view to laying a foundation for future research.

Keywords: The Internet of Things · Perception environment · Privacy protection Location privacy

1 Introduction

The concepts of sensor network, ubiquitous network, M2M, CPS, and IOT [1, 2], etc. have been put forward. They all realize the high integration of physical and information world. In China, IOT (the Internet of Things) is a network that can connect sensor devices and RFID with Internet by near field network or mobile network. The IOT can intellectually and automatically perceive and manage everything, so that all the people and objects will be included in the ubiquitous network, and information of the users will be perceived, transmitted, and saved while the users are unaware. For example, intelligent medical system has to real-timely collect the data of physiological signals, such as heart rate, temperature, and blood pressure, of the users. Similarly, private data of the users is exposed to the open, connective, and transparent physical and information world as well, which challenges the privacy security of IOT [3]. How to protect the privacy security of IOT becomes a hot issue.

© Springer Nature Singapore Pte Ltd. 2018
J. C. Hung et al. (Eds.): FC 2017, LNEE 464, pp. 104–115, 2018.
https://doi.org/10.1007/978-981-10-7398-4_11

The privacy has three essential attributes which is personal related, personal governed, and confidentiality. This concept has similar aspects with information security. Information security protects the validity and security of data by firewall, IDS, and encryption; while privacy security pays more attention to ensure that other people cannot get the private information. In some cases, we can protect private data by approaches of information security field. In others, we permit that data can be anonymously opened. Protecting privacy has its own algorithm and scheme, which have interdisciplinary combinations with information security.

2 Correlation Studies

2.1 The Problem of Protecting Privacy on Internet

IOT still needs the help of Internet to transmit data over long distances, so that security issues on Internet will appear on IOT as well. Because the cornerstone of the Internet TCP/IP was based on trust, users of Internet will provide a great deal of private information to the Internet service providers. So, if Internet has information security issues or the management is ineffective, there will be a lot of events of privacy disclosure.

The Table 1 shows the user privacy content classification and disclosure model.

Table 1. User privacy content classification and disclosure model

Disclosure model	Content	Explanation
Directly unauthorized acquiring: such as personal information or records of Cookies is illegally acquired, private data is collected by Trojan or embedded software [4]	Identity information	Which can uniquely identify data of entity, such as user's name, ID card, mobile phone, etc.
	Private data	Which relates to person or department information, such as hobbies, belief, log, legal person, business scope, etc.
	Users' address	Which can track IP address and MAC address
Indirectly acquiring by DM: Data on Internet is related to each other. Some information is used to determine the unique network node, so although personal private information is protected, users' privacy can still be mined by cross relation and cause privacy disclosure	Social relations	Which includes private information between a user and another user by DM, when user uploads information, interacts with others, and uses "friend recommendation" function
	Trace information	Which includes sensitive information of user's visiting and browsing Internet, personal information which is inferred by user's behavior, such as living habits, personal hobbies, and health condition, etc., and the living trace presented by user's locating
	Location	Which includes user's past or current position information and personal information which is inferred by position information

2.2   Traditional Privacy Protecting Technology

According to different privacy protecting object, there are two kinds of traditional privacy protecting technology.

1. Analysis-oriented Data Dissemination

Analyzing data based on statistics can protect privacy from DM. This technology can acquire general statistics rules without individual information by data perturbation and query restriction. Researches of the technology pay most attention to DM and data dissemination of privacy protecting, which anonymize data before issue the data to third-party to analyze.

There are two common technologies in privacy protecting of DM. One technology disconnects user's unique identity from other information, so that the attacker cannot determine target users. The scheme of this technology mainly includes anonymous mechanism, which is a common approach, and pseudonym mechanism. Anonymous mechanism means that information of the object is unrelated to any identity markers. There are many classical privacy protecting models, such as k-anonymity model [5], 1-diversity model [6], and t-closeness model [7], came up by scholars. K-anonymity needs generalizing, modifying, or disturbing data to hide an object in k users. Even though information disclosure happens, the possibility of finding the privacy reduces to 1/k. The disadvantage of k-anonymity is that if k objects have the same sensitive attributes, the privacy will leak out. Besides, k-anonymity cannot protect privacy from homogeneity attack and background knowledge attack. 1-diversity has improved k-anonymity. It ensures that every same identifier has one different sensitive data at least. However, it is difficult to realize 1-diversity, and 1-diversity will be under skewness attack and similarity attack. T-closeness requires that the distribution of sensitive attributes in equivalence class should be like that in integrated data. Users can protect their information by anonymity, because the private information cannot correspond to specific user.

The other technology is called data interleaving which generalizes information or adds noise into information, so that attackers cannot acquire target users' accurate information. For example, if a user's location information is private information, the accurate location will be replaced with another region. This technology costs less and simply practices, but it cannot avoid information loss and sensitive information disclosure entirely.

2. Transport-based Data Encryption

We use cryptological technique, such as traditional public-key cryptography, to encrypt data and use decryption algorithm to access and use the data on Internet. Using cryptological technique in data transmission can prevent data from privacy disclosure and protect the original data, but it needs large calculation amount, and it is difficult to deploy it.

Traditional privacy protecting technology cannot apply to IOT directly, because the opening tendency of IOT makes attacks have more chance to acquire more diversified background knowledge which causes privacy disclosure. Moreover, current privacy protecting technologies, such as data encryption and publication limits, are static, and they does not consider background knowledge of attackers. The data in IOT has altitudinal dynamic which includes changes of data attribute, adding new data, and deleting old data. So it is essential to research new security and protection mechanism to be fit for current business scenarios.

3 The Architecture of IOT and New Feature

3.1 The Architecture of IOT

AS the application fields of IOT is increasing continuously, the security vulnerability and privacy security problem of IOT become more serious. Before study the protection of safety and privacy problem of IOT, we need to make clear the system architecture, corresponding functions and key technology of IOT.

IOT concerns many key technologies. We can divide IOT into 3 layers:

Perception layer. Perception layer is the core technology of IOT, which include Radio Frequency Identification (RFID), Sensing device and other automatic information collected equipment, the WSN which consist of sensing device by self-organization and multi-hop paths.

Network layer. The role of this layer put the data of sensing layer into internet, which provide to Application layer that can use. The features of this layer can achieve based on basic network of existing internet.

Application layer. This layer provides massive amount of data through intermediate software to efficiently and reliably aggregate and storage. Through the treatment of big data, Cloud computing, Intelligent Data. Finally, the layer provide service for Intelligent Transportation, Smart home, Intelligent Building, Intelligent Electric Grid, Intelligent school, intelligent city, intelligent agriculture and other field.

3.2 The Feature of IOT

IOT system has features like open, Spontaneous Interoperability, Smart actuators and so on, which challenge to the Privacy and safety of IOT system. In the perception layer of IOT, electronic tag, Wireless Communication and unattended ground sensor are widely used so that attackers may be easier to obtain private information, the users potential Safety Hazard is increasingly outstanding.

While the safety problems of IOT system itself, are mainly reflected in perception layer. In IOT, the perception layer which connects the information with physical world is the essential part of IOT so that it is most vulnerable to be attacked. Because network layer and application layer are based on internet and we have explored a number of safety and privacy problem related to this two layers, so this paper mainly discusses the perception of environmental security and privacy protection, which in Perception layer of IOT. The characteristic of IOT will bring new privacy issues which focus on the following aspects (Table 2).

Table 2. The issue of privacy security in the Internet of Things

Characteristics	Phenomenon	New privacy security problem
Opening access point and opening visiting	Communication, data collaboration and other operations will be more frequent while different safety performance of perceived devices can track user behavior	Third-party application will know more about our behavior and bobbies than us through data mining, which will threaten the sensitive data and the privacy of information
Constrained devices and complex network security management	Labels or intelligent sensor device of IOT have several characteristic like low-cost, weak in computing ability, limited size in energy	Traditional information security system such as firewall, IDS systems that cannot apply to equipment of IOT are deployed independently
Intelligent execution	Equipment is unattended	Cannot monitor
Spontaneously interoperability relationship between objects	The relationship between objects is more complex than Internet. In addition to the network connection, there is also are relationship between common vendor, location, owner and so on	Under the complex environment of IOT, the trust issues between the multiple subjects need to be considered. In addition, information tracking and data mining based on the relationship between objects let privacy more easier to be Threatened

4   The Key Issues: The Security and Privacy Protection of Perceived Environment

The connection between objects are achieved by the sensor and RFID tags of perception layer as well as this construct the sensor networks and we will discuss from this several point of view. When users use the mobile terminal devices by wireless communication, such as sensors and tags, to communicate when they obtain service, their location and other information can easily be exposed. RFID technology is widely used in various fields. Because RFID sent information to reader by the way of no-contact through the wireless channel, so it is easier to get information of RFID flag by eavesdropping reader or using the forged reader to communicate with RFID flag. Due to the nature of opening in wireless communication of sensor network, opponents can easily capture the transmission signal from the sensor node and decipher which can achieve the privacy information like the secret key storage in sensor node, so this raises higher requirements for the key management of the wireless sensor networks.

4.1 Location Privacy Protection

1. The definition of the Location-Based Service LBS and divulge the privacy location

The development of the perception positioning technology allows people informed of their location quickly and accurately. People can use LBS to search for nearby restaurants, attractions, entertainment and so on. The continuous development of IOT will let application based on location services become more frequent to appear.

The applications based on LBS will obtain location information automatically, therefore it can intercept the transmission of location through eavesdropping on the communication line or software existing vulnerabilities will lead to disclosure the information of user.

Position information recorded have when, where, who, what or other factors, therefore disclosure the location information, on one hand will let the user's tracks are leaked and monitored. On the other hand, attackers can synthetically infer user behavior, vocational, social and personal information by observing the user location information. For example, parent often appear in the vicinity of the school at a fixed time explain that their child attend in this school. In addition, due to the location information leaked, you may receive spam messages based on location or even be attacked and harassed.

2. The existing privacy protected methods of location

States have begun to focus on the location privacy problems, and actively carry out the research related to legal systems and solutions. In 2004, PCP (Privacy-Conscious Personalization), developed by Bell Labs software architecture, will start the "Houdini" rules to determine whether to provide location information to others in a series of background knowledge.

K-anonymity technology which is a classical algorithm was first used to protect the privacy of the relational database. Gruteser put forward position k-anonymity technologies [9], with use the k-1 redundancy to construct the fuzzy region so it can hide the true location of anonymous users. Since the value of k is one of the key factors to affect the service results, Gedik [10] proposed the value of k which can be defined by the users themselves, use a large amount of computation to in exchange of narrowing down the fuzzy region. Fake name is a special anonymous method. However, this method has its limitations. Sometimes the user's location may reveal his real identity, such as a private house or a café.

SIGSPATIAL, founded in 2008 by ACM, begin to study the protection of privacy information based on LBS that ensure the safety of privacy services is an important indicator to evaluate whether the entire IOT network is mature

3. The protection of location in WSN

Currently, There are four main technical routes to protect privacy location of WSN [12] (Table 3).

Existing technical solutions cannot prevent all of attacks, there are also many disadvantages. In the future, the development of IOT should be combined with physical layer to consider how to design a project which can protect location privacy. At the same time, meet the requirements of protection granularity, communication quality and low power.

Table 3. Location privacy protecting technology of WSN

Route	Explanation	Main protocol	Disadvantage
Path camouflage	Sensor nodes choose camouflage path as a communication link	PRS, IRL, WRSE	Large computing and bigger time delay
Traps Induced	Set up a fake data source or base station to let attacker fall into a loop trap	TARP, DBT	High consumption of energy
Anonymous policy	Hide sensitive sensor node	EAC, MQA	Complex to realize
The control of communication	Modify the communication protocol. There are silent mechanism, cross-layer routing, network coding, data transfer	ALBS, CLS, SUNC, MSS	Bigger time delay, unstable communication performance

4.2 RFID

1. Privacy protection of RFID

   Radio frequency identification (RFID) is one of the key technologies of the IOT, which consist of the label moiety containing memory chips, better storage and security capabilities of the reader, antenna and other parts. Identify electronic tags automatically by using Radio Frequency signals of the reader. The following are the related security and privacy issues:

(1) The information transfer of the reader and the tag is automatically identified via wireless communication. The label owner doesn't grasp the situation;
(2) Wireless communication signals can pass through ordinary obstructions and then leak information.
(3) Wireless communication may be affected by signal interference, which could affect the normal operation of the RFID system.
(4) The adversary will implant the malicious code or malicious scripting language through RFID tags. This kind of attacks may cause a fatal loss to back-end database.
(5) A close distance is required when the RFID tag chip and the reader are working, which will reveal the location of readers and the privacy of user's location.

2. Privacy issues and Solutions of RFID (Table 4)

Table 4. Security threat of RFID

	Name	Explanation	Solution
Passive mode: Steal entity information privacy; speculate entity information, tracking entity identity	Cloning label	Counterfeiting or copying the same RFID tags as the original target	Perform authentication protocol; Data encryption; Faraday cage method; Blocked tagging
	Illegal authorization	Reading RFID tag data illegally in unauthorized situation	
	Counterfeit labels	The attacker steals the data of RFID tags posing as a legal reader	
	Physical attacks	Nodes are damaged physically, resulting in the leakage of privacy information	Blocking labels; active interference method and so on
Active attacks	Channel blocking denial of service	Attackers keep sending malicious requests through long-time occupied channel, with the result that legal traffic cannot be transmitted	Encrypt radio signal
	Information tampering	The attacker eavesdrops the information and then transmits to receiver after modified	
	Malicious tracking	Using social engineering methods to hide RFID tags around the user, realizing Malicious tracking when the label moves along with the user	Social engineering methods
	RFID virus	Tags don't have firewall or anti-virus software. For example, the virus code is written to the tag, attacker could attack the system through SQL injection or buffer overflow [13]	Smart Label Standard Guide of the US NIST, which recommend isolating the RFID databases from other IT systems by using firewall

We can solve the problems above in four ways which have both advantages and disadvantages:

(1) The anonymity of RFID tag

Early label discard method is simple but not very good programs. Now many methods destroy labels after the sale of goods by use of KILL, which ensure that the user is not illegally tracked. These methods are simple and effective. However, the irreversibility of KILL command makes the label not to be used, which actually increases the cost of tags. We can also make tags dormant, and then activate when needed.

(2) The implement of RFID privacy protection by the use of blocking communication method

This method is to prevent the information leakage by Faraday outside nets, legal tags and establishing metal mesh outside readers, which ensures that the attacker cannot be fake or intercept to protect user privacy. The disadvantage of the method is metal mesh damages the openness and portability of the IOT, and large-scale implementation is not guaranteed. We can interfere the signal actively to prevent label information from being read by nearby attacker, but interfering signals may affect legal signal.

The key point of the Internet of Thing is openness, low-cost, creating information-physical link unattended intelligently. Using the method of blocking communication above would violate the features of the development of the IOT.

(3) The method of privacy protection based on the distance measurement

This method is proposed by Intel's [14], which mainly consider that the channel distance of attackers generally is longer than that of normal users. So we can measure the distance by use of triangle analysis based on time or signal strength, transmitting different amounts of information to visitors at different distances.

This method has high effectiveness, large computation, complex selection and determination of suitable distance parameters.

(4) The method of authentication and data encryption

We can protect privacy safe from authentication and data encryption point of view. Whether information transmission between the reader and the tag or between the reader and the back-end server, we should strengthen the security and privacy protection from authentication and data encryption point of view.

Hash-Lock decides to use metalD or real ID in response to a query. The protocol achieves the simple access control by use of one-way cryptographic hash function, which can guarantee a lower cost tags. Randomized Hash Lock designs a pseudo-random number generator [15] in tags and readers

Some experts propose smart tags that using a variety of encryption technologies for access control, but this method will request for the storage size, craftsmanship and price of label, resulting in the higher cost. In addition, massive pirate data of the IOT are bound to make encryption more onerous, which can also cause the delay of communication interaction.

4.3   Sensor Network

Abundant applications of the IOT need to collect a lot of fine-grained information. Application projects often deploy a large number of sensor nodes, even spread by aircrafts. These nodes exposed to the public can only remote control. Sensitive information will leak because of signal interception or node capture.

Sensing nodes will appear multi-source heterogeneous due to different principles and functions, which don't have the ability of complicate security protection due to storage size and less energy. IOT data and network standards are not unified lead to the security system cannot be unified (Table 5).

Table 5. Security and privacy issues of sensor nodes and sensing networks

Type	Name	Content
Physical security operation: sensor nodes have simple hardware, limited secrecy and fragility of brute force	Node capture	The key nodes such as gateways are controlled by the attacker, which will result in the leakage of key or information. In the process of using encrypt and authentication technologies to solve node capture problem, we should pay attention to forward privacy. Utilizing periodic key update
	Cloning node	Fake nodes for additional information
Data security	Sensor information eavesdropping	Ubiquitous wireless communication in sensor networks is easy to be tapped and intercepted, so as to analyze the sensitive data, or to infer the role of sensor nodes through the analysis of network traffic
Network security control	DOS attacks	Gateway nodes have limited energy, which will lose operational capability because of energy depletion after being subjected to DOS attacks
	Integrity attacks	Attackers intercept and modify the transmitting information of sensing network was intercepted modification, forgery integrity attacks, resulting in poor decisions based on the information network
	False routing information	The attacker can increase the network delay and exhaust the sensor node energy by false routing information. For example sinkhole attack: the influence of distance vector routing algorithm based on forged node to other nodes to transmit zero distance announcement, causing the neighbor nodes of all communication data flow of forged nodes to form a routing black hole (sinkhole)
Trust security	Massive node authentication problem	Wireless sensor network has a large number of unstable mass nodes. Identity management and authentication issues become more prominent. We need to design flexible, low overhead authentication protocol

5   The Next Stage of Development

The characteristics of the IOT technology bring new problems to the security and privacy protection technologies. This will affect the ability of the IOT to develop quickly. We should pay attention to the following question in the latter study:

(1) The IOT has the characteristics of openness, heterogeneity, low storage capacity, low energy efficiency and so on. We must take full account of the relevant protocols in the design.
(2) Due to the diversity of user roles in the IOT, privacy protection needs are different. It may overprotect some of the user privacy information in the open IOT, and leak another part of the user privacy information. So we should pay attention to granulation in the privacy protection mechanism.
(3) The application scenarios of the IOT are complex, which have huge communication flow or transinformation, limited equipment resources and computing, storage capacity. Many encryption algorithms cannot be directly applied to the perception layer. Therefore, it is needed to design the encryption algorithm according to the characteristics of the sensing layer. The encryption algorithm [16] has fast adaptive, low communication and communication.

In this paper, we discuss on the structure and characteristics of the IOT, and then focus on security and privacy protection technology in the perceptual environment. We make a systematic review of the existing privacy protection technologies of the IOT, which will lay a foundation for the future reasonable security and privacy protection protocols and models of the IOT.

References

1. Sun, Q.B., Liu, J., Li, S., Fan, C.X., Sun, J.J.: Internet of Things: summarize on concepts, architecture and key technology problem. J. Beijing Univ. Posts Telecommun. **33**(3), 1–9 (2010). (in Chinese)
2. International Telecommunication Union (ITU): ITU Internet Reports 2005: The Internet of Things. World Summit on the Information Society (WSIS), Tunis (2005)
3. Weber, R.H.: Internet of Things new security and privacy challenges. Comput. Law Secur. Rev. **26**(1), 23–30 (2010)
4. Zhou, C.L.: Protection strategy of privacy data for the Internet of Things. Ms.D. thesis, Harbin Engineering University, Harbin (2012). (in Chinese)
5. Sweeney, L.: k-anonymity: a model for protecting privacy. Int. J. Uncertainty Fuzziness Knowl. Based Syst. **10**(5), 557–570 (2002)
6. Machanavajjhala, A., Gehrke, J., Kifer, D., et al.: L-diversity: privacy beyond k-anonymity. In: Proceedings of the 22nd IEEE International Conference on Data Engineering (ICDE 2006), pp. 24–36 (2006)
7. Li, N., Li, T., Venkatasubramanian, S.: t-closeness: privacy beyond k-anonymity and l-diversity. In: Proceedings of the 23rd IEEE International Conference on Data Engineering (ICDE 2007), Istanbul, Turkey, pp. 106–115 (2007)
8. Yu, R.X.: Research and application of location service privacy protection technology. Ms.D. thesis, Nanjing University of Science and Technology, Nanjing (2013). (in Chinese)

9. Wang, L., Meng, X.F.: Location privacy preservation in big data era: a survey. Ruan Jian Xue Bao/J. Softw. **25**(4), 693–712 (2014). (in Chinese). http://www.jos.org.cn/1000-9825/4551.htm
10. Kalnis, P., Ghinita, G., Mouratidis, K., et al.: Preserving anonymity in location based services. Bt Technol. J. **21**(1), 34–43 (2006)
11. Medaglia, C.M., Serbanati, A.: An overview of privacy and security issues in the Internet of Things. In: Proceedings of the 20th Tyrrhenian Workshop on Digital Communications, pp. 389–395. Springer, Sardinia (2010)
12. Peng, H., Chen, H., Zhang, X.Y., Fan, Y.J., Li, C.P., Li, D.Y.: Location privacy preservation in wireless sensor networks. Ruan Jian Xue Bao/J. Softw. **26**(3), 617–639 (2015). (in Chinese). http://www.jos.org.cn/1000-9825/4715.htm
13. Rieback, M.R., Crispo, B., Tanenbaum, A.S.: Is your cat infected with a computer virus? In: Proceedings of IEEE International Conference on Pervasive Computing and Communications (PerCom), Pisa, Italy, 10–18 (2006)
14. Liu, X.L., Qi, H., Li, K.Q., Wu, J., Xue, W.L., Min, G.Y., Xiao, B.: Efficient detection of cloned attacks for large-scale RFID Systems. In: Lecture Notes in Computer Science, vol. 8630, pp. 85–99 (2014)
15. Yan, T.: Research the key problems of privacy protection and key management in the Internet of Things. Ph.D. thesis. Beijing University of Posts and Telecommunications, Beijing (2012). (in Chinese)
16. Yang, G., Geng, G.N., Du, J.: Security threats and measures for the Internet of Things. J. Tsinghua Univ. (Sci. Tech.) **51**(10), 1335–1340 (2011). (in Chinese)

Traffic Sign Classification Base on Latent Dirichlet Allocation

Lei Song[1], Zheyuan Liu[1(✉)], Xiaoteng Zhang[1], Huixian Duan[1,2,3(✉)],
Na Liu[1], and Jie Dai[1]

[1] The Third Research Institute of the Ministry of Public Security,
Shanghai, China
f6489701l@hotmail.com, hxduan005@163.com
[2] The Key Laboratory of Embedded System and Service Computing,
Ministry of Education, Tongji University, Shanghai, China
[3] Shanghai International Technology and Trade United Co., Ltd,
Shanghai, China

Abstract. Traffic sign classification is a significant issue in the intelligent vehicle domain, which helps vehicles to follow the traffic rules and ensure the safety. Feature selection and description are very important and difficult for classification. In this paper, a novel traffic sign classification method is proposed which is based on the Latent Dirichlet Allocation (LDA) model. Feature topics are modeled based on various traffic signs by the LDA automatically. And traffic signs captured onboard are classified according to the modeled features. The experiment results show the efficiency of our work.

Keywords: Traffic sign · LDA model · Feature topic · Visual word

1 Introduction

Traffic sign detection and recognition are required for intelligent vehicles for understanding the traffic rules and navigating. Therefore, related researches have been done since 1984 [1, 2]. After that, more countries started the research, i.e. the Saint-Blancard system of France focus on the red traffic signs detection [3]. The ADIS from U.S. detected the stop signs [4]. And the system developed by Germany detected traffic signs by color and shape [5]. Then binocular camera was used by Japan for sign detection. In the year of 2005, Sweden and Australia jointly developed a system to recognize traffic signs. At the same year, Siemens produced a traffic sign tracking and recognition system [6]. Germany exploded a real-time traffic sign detection system based on FPGA [7]. However, the methods used in traffic classification include template matching [8, 9], SimBoot [10, 11], SVM (Support Vector Machine) [12–14], HOG (Histograms of Oriented Gradients) [15, 16], wavelet [17]. neural networks [18, 19], SURP descriptor [20], Laplacian of Gaussian (LOG) filter [21], image matching [22] and so on.

Considering the complicated situation of the traffic scenes (i.e. weather variations, illumination changings, obstruction), all the researchers are still working on the development of robust algorithms of traffic sign detection and classification. The most

J. C. Hung et al. (Eds.): FC 2017, LNEE 464, pp. 116–123, 2018.
https://doi.org/10.1007/978-981-10-7398-4_12

important issue is feature, which leads to some problems, including how to select the features, which features to adopted, and how to extract the features.

In this paper, a novel feature extraction method is proposed and applied for traffic sign classification. The rest of the paper is organized as follows: the LDA model adopted in our work is introduced in Sect. 2. In Sect. 3, the main algorithm of traffic sign classification is introduced. Experiment results are shown and discussed in Sect. 4. And we conclude the paper in Sect. 5.

2   LDA Model

LDA is a generative probabilistic model for collections of discrete data (e.g., text corpora) [22]. Its graphical model is shown in Fig. 1. In the LDA model, the corpus is a collection of documents over a word vocabulary; each document is a sequence of unordered words. Given the documents, LDA modeling can find out groups of co-occurring words, which are called "topics". The number of topics is assumed known and fixed. Then, each topic is represented by a multinomial distribution over the word vocabulary, while each document is represented as a multinomial distribution over topics. In LDA, topics are discovered as frequent co-occurring words shared by all documents.

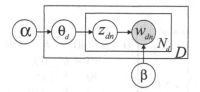

Fig. 1. LDA model for fist-level processing

3   Traffic Sign Classification

3.1   Visual Word and Document

Traffic signs are first detected and segmented by the bounding rectangles as introduced in Ref. [23]. Traffic signs in different images captured by the on-board cameras are with various sizes. Therefore, normalization is applied for the modeling processing. Besides the coordinates of pixels, color is also an important feature, and is quantized in the HSV color space. In our work, a visual word contains spatial location and color. Consequently, a visual word is represented by the vector (x, y, clr). Then, traffic sign areas (rectangular regions) are considered as documents. Referring to the codebook, each pixel is numbered. And the processing flow is shown in Fig. 2.

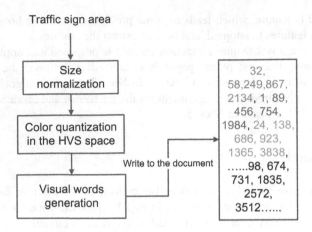

Fig. 2. The generation of a document (Color figure online)

Particularly, the size normalization is achieved by the down sampling to large images and the nearest interpolation to small images. Assume that the traffic sign regions have the normalized size as $X \times Y$. And the color quantization is preceded as Eq. (1) described in the HVS color space. There are totally CLR levels. So the size of the codebook is $X \times Y \times CLR$.

$$
\begin{cases}
v \prec 0.25, & \text{black} \\
v \geq 0.25 \& \begin{cases}
s \prec 0.2, & \text{white} \\
s \geq 0.2 \begin{cases}
0 \leq h \leq 30, or, h \succ 330, & \text{red} \\
30 \prec h \leq 90, & \text{yellow} \\
90 \prec h \leq 150, & \text{green} \\
150 \prec h \leq 210, & \text{cyan} \\
210 \prec h \leq 270, & \text{blue} \\
270 \prec h \leq 330, & \text{purple}
\end{cases}
\end{cases}
\end{cases} \tag{1}
$$

3.2   Feature Extraction

Traffic sign regions are documents, and the topics modeled by the LDA model can be considered as the common features shared by traffic signs, which are adopted to classify the traffic signs.

Taking the standard traffic sign library for example (partially shown in Fig. 3), there are 92 different traffic signs, and 3 categories which are prohibition, warning and indication signs. Particularly, the library contains 31 prohibition signs, 45 warning signs and 16 indication signs. All the images have the same illumination and size, which is 50×50 pixels. And the background is simply white. Comparing to the real traffic situation, the feature exaction processing is much easier in the library. 36 images in the library are shown in Fig. 3. The topic number is set to be 3 in the LDA model, which represents indication, warning and prohibition signs. And the color level is 3, which indicates red, yellow and blue. So the size of the codebook is $50 \times 50 \times 3$.

Fig. 3. Examples of traffic signs in the library (Color figure online)

The 3 feature topics modeled by the LDA are shown in Fig. 4. It is obviously to see that Fig. 4(1) shows the feature of indication signs. However the contents of different indication signs are various, so the feature only describes the comment features which are blue color and circular shape. Similarly, Fig. 4(2) shows the warning signs' features with yellow triangle, and Fig. 4(3) shows the prohibition features. In Table 1, the top 10 visual words in each feature topic are shown.

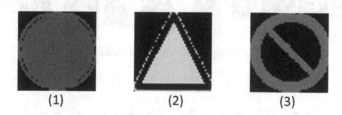

(1)   (2)   (3)

Fig. 4. Feature topics (Color figure online)

Table 1. Feature topics and their visual words

Categories	Top 10 visual words in each topic (probability from high to low)									
Indication	210	213	216	219	222	225	228	231	234	237
Warning	1127	1142	1274	1277	1280	1424	1427	1430	1571	1574
prohibition	76	199	202	205	208	211	214	217	220	223

3.3 Classification

Assuming the traffic sign image contains V visual words $\{w_v\}(v = 1,\ldots, V)$ and K feature topics. In topic k, the word w_v appears w_v^k times. Then in all the images, the times word w_v appears is $W_v = \sum_{k=1}^{K} w_v^k$. Therefore, the probability of word w_v belongs to topic k is $P_v^k = \frac{w_v^k}{W_v}$, and $\sum_{k=1}^{K} P_v^k = 1$.

The classification of a traffic sign is processed as follows:

Step 1: Normalize the traffic sign image and number the pixels according to the codebook;

Step 2: Calculate the probabilities of each visual word belongs to each feature topic, and get $P_k = \sum_{v^f} P_v^k$, where v^f represents all the visual words in the traffic sign image.

Step 3: Choose the topic k as the classification result, corresponding to $\max\{P_k\} (k = 1, \ldots, K)$.

4   Experimental Results

In our experiment, the database is from GTSRB (German Traffic Sign Recognition Benchmark) [24]. Some of the images are shown in Fig. 5. The images in the database contains perspective transform, illumination changes, various weather conditions, partial blocking, rotation, and blurry.

Fig. 5. Examples of segmented traffic signs

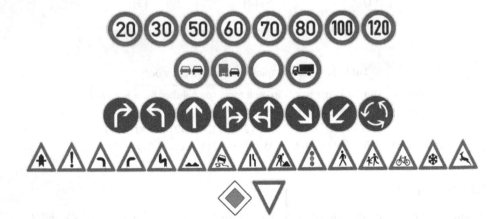

Fig. 6. 37 Traffic signs used in feature extraction

200 images are randomly sampled from the database and used to extract features. These images mainly include 37 kinds of traffic signs as shown in Fig. 6. Traffic sign segmentation is applied to the images to obtain the traffic sign regions. And the regions are normalized to 50×50 pixels. The traffic signs are classified into 5 categories, thus the feature topic number is 5. By the LDA modeling, the 5 topics are extracted and shown in Fig. 7. And the top 10 visual words in each topic are shown in Table 2. The probabilities of each visual word belonging to each topic are recorded.

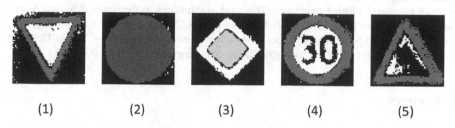

(1)　　　　(2)　　　　(3)　　　　(4)　　　　(5)

Fig. 7. 5 feature topics extracted by the LDA

Table 2. Visual words in 5 feature topics

Number of traffic sign	Top 10 visual words in each topic (probability from high to low)									
1	3424	3900	6824	2433	3664	3904	6584	7064	2669	3908
2	2967	2491	2727	2739	2971	3439	12099	12343	12619	2979
3	7644	4488	7156	2764	5952	6924	7884	6288	7408	7640
4	10456	10672	11164	10932	1577	3793	10220	11168	11941	3472
5	12497	12029	13017	2757	8089	12261	12273	12757	13005	13245

12000 images are classified by our method. And the accuracy is 86.74%. Particularly in Table 3, we analyzed all the incorrect classifications. Numbers in the last column are calculated as:

$$R(i,j) = \text{false negative of traffic sign } i/\text{all the false results} \qquad (2)$$

And the numbers in the first 5 columns are calculated as:

$$r(i,j) = \text{traffic sign } i \text{ incorrectly classified as sign } j/\text{false negative of traffic sign } i \quad (3)$$

Table 3. Analysis of the false classifications

	TS 1	TS 2	TS 3	TS 4	TS 5	R
TS 1	0	0.83	0	0.13	0.04	0.0487
TS 2	0.22	0	0	0.73	0.05	0.0792
TS 3	0.12	0.62	0	0.14	0.12	0.0823
TS 4	0.03	0.90	0.04	0	0.03	0.5462
TS 5	0.34	0.59	0	0.07	0	0.2436

5 Conclusion

In this paper, a traffic sign classification method is introduced. We utilize the LDA model to extract the features of different kinds of traffic signs. Based on these features, traffic signs are classified. The experiment results are discussed in the end. And the efficiency of our method is verified.

Acknowledgement. The authors of this paper are members of Shanghai Engineering Research Center of Intelligent Video Surveillance. This work is sponsored by the National Natural Science Foundation of China (61402116, 61403084, and 61300028); by the Project of the Key Laboratory of Embedded System and Service Computing, Ministry of Education, Tongji University (ESSCKF 2015-03); and by the Shanghai Rising-Star Program (17QB1401000).

References

1. De La Escalera, A., Armingol, J.M., Salichs, M.A.: Traffic sign detection for driver support systems. In: International Conference on Field and Service Robotics (2001)
2. Paclik, P.: Road sign recognition survey. http://euler.fd.cvut.cz/research/rs2/files/skoda-rs-survey.html
3. de Saint Blancard, M.: Road sign recognition: a study of vision-based decision making for road environment recognition. In: Vision-Based Vehicle Guidance. Springer-Verlag New York, Inc., New York (1992)
4. Kehtarnavaz, N., Griswold, N.C., Kang, D.S.: Stop-sign recognition based on color/shape processing. Mach. Vis. Appl. **6**(4), 206–208 (1993)
5. Priese, L., Klieber, J., Lakmann, R., Rehrmann, V., Schian, R.: New results on traffic sign recognition. In: Proceedings of the Intelligent Vehicles 1994 Symposium, pp. 249–254. IEEE (1994)
6. Bahlmann, C., Zhu, Y., Ramesh, V., Pellkofer, M., Koehler, T.: A system for traffic sign detection, tracking, and recognition using color, shape, and motion information. In: 2005 Proceedings of the Intelligent Vehicles Symposium, pp. 255–260. IEEE (2005)
7. Muller, M., Braun, A., Gerlach, J., Rosenstiel, W., Nienhuser, D., Zollner, J.M., Bringmann, O.: Design of an automotive traffic sign recognition system targeting a multi-core SoC implementation. In: 2010 Design, Automation and Test in Europe Conference and Exhibition (DATE), pp. 532–537. IEEE (2010)
8. Khan, J.F., Bhuiyan, S.M.A., Adhami, R.R.: Image segmentation and shape analysis for road-sign detection. IEEE Trans. Intell. Transp. Syst. **12**(1), 83–96 (2011)
9. Chen, L., Li, Q., Li, M., Mao, Q.: Traffic sign detection and recognition for intelligent vehicle. In: 2011 IEEE Intelligent Vehicles Symposium (IV), pp. 908–913. IEEE (2011)
10. Lim, K.H., Seng, K.P., Ang, L.M.: Intra color-shape classification for traffic sign recognition. In: 2010 International Computer Symposium (ICS), pp. 642–647. IEEE (2010)
11. Ruta, A., Li, Y., Liu, X.: Robust class similarity measure for traffic sign recognition. IEEE Trans. Intell. Transp. Syst. **11**(4), 846–855 (2010)
12. Sathiya, S., Balasubramanian, M., Sivaranjini, R.: Image based detection and recognition of road signs. IJREAT Int. J. Res. Eng. Adv. Technol. **1**(1), 1–5 (2013)
13. Liang, M., et al.: Traffic sign detection by ROI extraction and histogram features-based recognition. In: International Symposium on Neural Networks, pp. 1–8 (2013)
14. Mammeri, A., et al.: North-American speed limit sign detection and recognition for smart cars. In: Local Computer Networks, pp. 154–161 (2013)

15. Qingsong, X., Juan, S., Tiantian, L.: A detection and recognition method for prohibition traffic signs. In: 2010 International Conference on Image Analysis and Signal Processing (IASP), pp. 583–586. IEEE (2010)
16. Creusen, I., Wijnhoven, R.G.J., Herbschleb, E.D.: Color exploitation in hog-based traffic sign detection. In: International Conference on Image Processing, pp. 2669–2672 (2010)
17. Fatmehsan, Y.R., Ghahari, A., Zoroofi, R.A.: Gabor wavelet for road sign detection and recognition using a hybrid classifier. In: 2010 International Conference on Multimedia Computing and Information Technology (MCIT), pp. 25–28. IEEE (2010)
18. Zhu, Z., Liang, D., Zhang, S.-H., Huang, X., Li, B., Hu, S.-M.: Traffic-sign detection and classification in the wild. In: IEEE Conference on Computer Vision and Pattern Recognition, pp. 2110–2118. IEEE (2016)
19. Surinwarangkoon, T., Nitsuwat, S., Moore, E.J.: Traffic sign recognition system for roadside images in poor condition. Int. J. Mach. Learn. Comput. 3(1), 121–126 (2013)
20. Solanki, D.S., Dixit, G.: Traffic sign detection using feature based method. Int. J. Adv. Res. Comput. Sci. Softw. Eng. 5(2), 340–346 (2015)
21. Laguna, R., Barrientos, R., Blázquez, L.F., Miguel, L.J.: Traffic sign recognition application based on image processing techniques. In: The 19th World Congress the International Federation of Automatic Control Cape Town, South Africa, August 24–29, 2014, pp. 104–109 (2014)
22. Shah, D.M., Sindha, P.D.: Traffic sign detection and recognition system using translation of images. Int. J. Adv. Res. Comput. Sci. Softw. Eng. 4(10), 433–435 (2014)
23. Griffiths, T.: Gibbs sampling in the generative model of latent Dirichlet allocation. Unpublished note (2002). http://citeseerx.ist.psu.edu/viewdoc/summary?doi=10.1.1.7.8022
24. Song, L., Liu, Z.: Color-based traffic sign detection. In: 2012 International Conference on Quality, Reliability, Risk, Maintenance, and Safety Engineering (ICQR2MSE), pp. 353–357. IEEE (2012)
25. http://benchmark.ini.rub.de

Secure Cyber Physical System R&D Project Issue in Korea

Donghyeok Lee[1], Won-chi Jung[2], and Namje Park[1(✉)]

[1] Department of Computer Education, Teachers College, Jeju National University,
61 Iljudong-ro, Jeju-si 63294, Korea
{bonfard,namjepark}@jejunu.ac.kr
[2] Jeju Free International City Development Center, Jeju Science Park,
Elite Bldg. 3F. Cheomdanro 213-4, Jeju City,
Jeju Special Self-Governing Province, Republic of Korea
jwonchi@jdcenter.com

Abstract. In this secure smart grid R&D project issue in Korea, we aimed to develop the smart grid security technology which is the type of support of security vulnerability analysis through the convergence of IT technology. In addition, through securing such cutting edge technologies as smart meter monitoring, vulnerability analysis modeling, privacy mining, light-type mutual authentication service mechanism, and contents-based approach controls, we are going to pursue our research and development activities to make the core technology suitable for domestic environments, and also plan to have these technologies used by the industries by transferring the technology in a timely manner.

Keywords: Smart grid · R&D project · Smart grid security
Smart grid privacy

1 Introduction

The smart grid is a technology intended to enhance the stability and security of power systems by providing optimal electric power to the right place at the right time, thereby protecting people and property from significant incidents caused by electrical faults, and increasing the dependability and quality of electricity. The smart grid works by integrating information technology including computer, communication, and network technologies into conventional electrical power systems as shown in Fig. 2. Numerous smart grid studies have been recently conducted by industry and academia. One of critical issues in smart grid related to information and communications technologies is cyber physical systems security. Cyber physical systems security includes the protection of networks and servers from unauthorized accesses and malicious attacks. Cyber physical systems security also covers the protection of compromised control and measurement units from doing harm to the cyber physical infrastructures. Although such critical security issues of smart grid still remain unaddressed, there is a strong need to address and solve those important questions (Fig. 1). [1–5]

© Springer Nature Singapore Pte Ltd. 2018
J. C. Hung et al. (Eds.): FC 2017, LNEE 464, pp. 124–130, 2018.
https://doi.org/10.1007/978-981-10-7398-4_13

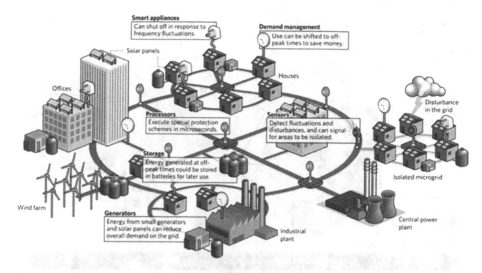

Fig. 1. Potential framework to define scope of smart grid (Source: DataCenterPulse2009)

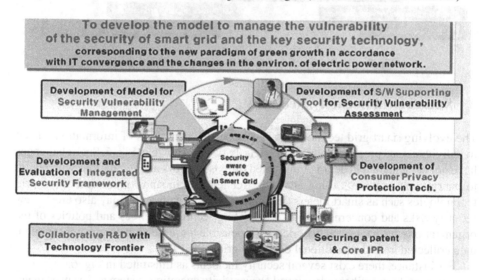

Fig. 2. Goal and scope of R&D project in Korea

In this R&D project in Korea, we will analyze security weaknesses of smart grid and propose corresponding security countermeasures. Without diminishing the benefits of smart grid, it is desirable to inject security-aware services directly into smart grid by accommodating such weaknesses seamlessly. In addition, this R&D project will attempt to realize the envisioned economic, environmental, and societal benefits of smart grid in providing a reliable and robust security environment for its services as depicted in Fig. 3. [6–13]

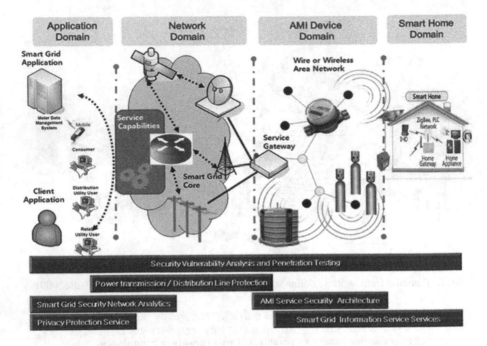

Fig. 3. R&D project's research contents and scope definition

2 Smart Grid Security Necessity

The evolving smart grid technologies and associated new types of information related to individuals and premises may create security risks as shown in Fig. 5. Such challenges have not been studied in depth or mitigated by existing laws and regulations with regard to energy consumption, billing and other related data in smart grid. New smart grid technologies such as smart meters and similar types of endpoints, may also create new security risks and concerns beyond the existing security practices and policies of the organizations that have been historically responsible for protecting energy consumption data collected from the traditional electrical grid. [14–19]

For instance, there exist several security incidents as illustrated in Fig. 6:

Smart grid can deliberately control an electricity operation system and a smart meter with illegal data forgery and falsification during the two-way communication between home smart meter, an electricity supplier, and a management enterprise.

When an electricity control system is interlocked with the Internet, there is the possibility that an external element makes use of the vulnerabilities of the Internet and penetrates into the control system. Smart meter vulnerabilities and worm virus epidemic POC has been introduced at the Blackhat conference 2009, where ZigBee based smart meter hacking technique has been introduced at the control system security conference (Fig. 4). (Black Hat 2009, '09.7.30)

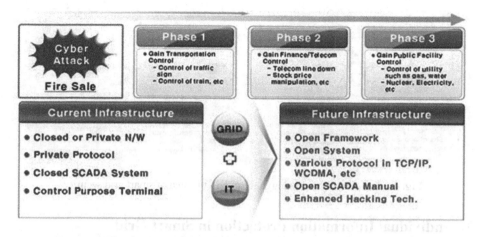

Fig. 4. Growing hacking probability to control system

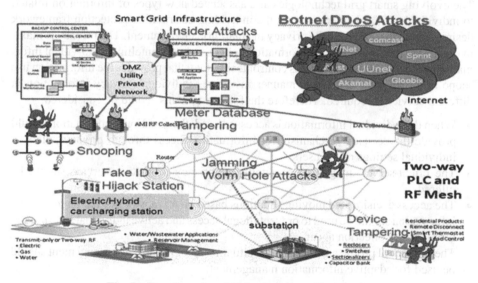

Fig. 5. Security vulnerabilities in smart grid network

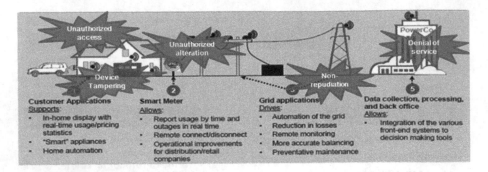

Fig. 6. Key element technology research in the area of smart grid security

3   Individual Information Protection in Smart Grid

The evolving smart grid technologies and associated new types of information related to individuals. This chapter analyzes the individual information protection framework design factors that consider the privacy in smart grid environment. The privacy is maximally guaranteed in individual information collection but an intelligent individual information protection agent and access control technology that provides the information use scope in an adaptive and active manner in various individual privacy information in different service is required. Therefore the following items should be considered.

- When the individual information is accessed/used, the situational information should provide the decision mechanism for information access through privacy policy in individual settings of the users.
- The client-server integration of intelligent privacy protection service should be offered.
- The access decision and moving situations based on privacy policy on the individual information should be used to prevent illegal access and active individual information through the collection agent.
- The individual profile and the separate situation recognition data management should be used for adaptive information management.

4   The Main Contents of R&D Project in Korea

This project includes three major tasks, (1) articulation of the core technologies and relevant issues in smart grid security, (2) design and development of the integrated security reference model and risk management framework, and (3) evaluation and testing of the selected technologies in our framework.

- Task 1. Articulation of the core technologies and relevant issues in smart grid security

    This task first focuses on the rigorous analysis of potential threats and weaknesses of existing security solutions in smart grid. As shown in Fig. 6, the critical security requirements for each communication and system component in smart grid environments will be examined. Also, we will articulate security services to meet the identified security requirements. The analysis results will be eventually accommodated in the integrated security. [21–25]

5 Conclusions

In this secure smart grid R & D Project in Korea, we aimed to develop the smart grid security technology which is the type of support of security vulnerability analysis through the convergence of IT technology. In addition, through securing such cutting edge technologies as smart meter monitoring, vulnerability analysis modeling, privacy mining, light-type mutual authentication service mechanism, and contents-based approach controls, we are going to pursue our research and development activities to make the core technology suitable for domestic environments, and also plan to have these technologies used by the industries by transferring the technology in a timely manner.

Acknowledgments. This work was partly supported by Institute for Information & communications Technology Promotion (IITP) grant funded by the Korea government (MSIP) (No. 2017-0-00207, Development of Cloud-based Intelligent Video Security Incubating Platform) and Basic Science Research Program through the National Research Foundation of Korea (NRF) funded by the Ministry of Education (NRF-2016R1D1A3A03918513).

References

1. Heer, T., Garcia-Morchon, O., Hummen, R., Keoh, S., Kumar, S., Wehrle, K.: Security challenges in the IP-based Internet of Things. Wirel. Pers. Commun. **61**, 527–542 (2011)
2. Park, J., Shin, S., Kang, N.: Mutual authentication and key agreement scheme between lightweight devices in internet of things. J. Korea Inf. Commun. Soc. **38**, 707–714 (2013)
3. Park, N.: Implementation of terminal middleware platform for mobile RFID computing. Int. J. Ad Hoc Ubiquitous Comput. **8**, 205–219 (2011)
4. Park, N.: Secure data access control scheme using type-based re-encryption in cloud environment. In: Katarzyniak, R. et al. (eds.) Semantic Methods for Knowledge Management and Communication, vol. 381, pp. 319–327. Springer, Berlin (2011)
5. Park, N.: Security scheme for managing a large quantity of individual information in RFID environment. In: Zhu, R. et al. (eds.) Information Computing and Applications, vol. 106, pp. 72–79. Springer, Berlin (2010)
6. Park, N.: Mobile RFID/NFC linkage based on UHF/HF dual band's integration in U-sensor network era. In: Park, J.H. et al. (eds.) Information Technology Convergence, Secure and Trust Computing, and Data Management, vol. 180, pp. 265–271. Springer, Berlin (2012)
7. Park, N., Ko, Y.: Computer education's teaching-learning methods using educational programming language based on STEAM education. In: Park, J. et al. (eds.) Network and Parallel Computing, vol. 7513, pp. 320–327. Springer, Berlin (2012)
8. Park, N., Cho, S.; Kim, B., Lee, B., Won, D.: Security enhancement of user authentication scheme using IVEF in vessel traffic service system. In: Yeo, S. et al. (eds.) Computer Science and Its Applications, vol. 203, pp. 699–705. Springer, Berlin (2012)

9. Kim, K., Kim, B., Lee, B., Park, N.: Design and implementation of IVEF protocol using wireless communication on android mobile platform. In: Kim, T., et al. (eds.) Computer Applications for Security, Control and System Engineering, vol. 339, pp. 94–100. Springer, Berlin (2012)

10. Kim, G., Park, N.: Program development of science and culture education tapping into Jeju's special characteristics for adults. In: Kim, T., et al. (eds.) Computer Applications for Security, Control and System Engineering. CCIS, vol. 339. Springer, Berlin, Heidelberg (2012)

11. Park, N., Kim, M.: Implementation of load management application system using smart grid privacy policy in energy management service environment. Clust. Comput. **17**, 653–664 (2014)

12. Sultan, U., Zheng, X.: TCLOUD: a trusted storage architecture for cloud computing. Int. J. Adv. Sci. Technol. **63**, 65–72 (2014)

13. Kim, N., Jing, C., Zhou, B., Kim, Y.: Smart parking information system exploiting visible light communication. Int. J. Smart Home **8**, 251–260 (2014)

14. Park, N.: Implementation of privacy policy-based protection system in BEMS based smart grid service. Int. J. Smart Home **7**, 91–100 (2013)

15. Kim, J., Park, D.-H., Bang, H.-C., Park, N.: Development of open service interface's instructional design model in USN Middleware platform environment. Lect. Notes Electr. Eng. **279**, 411–416 (2014)

16. Park, N., Park, J., Kim, H.J.: Hash-based authentication and session key agreement scheme in Internet of Things environment. Adv. Sci. Technol. Lett. **62**, 9–12 (2014)

17. Park, J., Kang, N.: Entity authentication scheme for secure WEB of Things applications. J. KICS **38**, 394–400 (2013)

18. Park, N., Kwak, J., Kim, S., Won, D., Kim, H.: WIPI mobile platform with secure service for mobile RFID network environment. In: Shen, H.T.et al. (eds.) Proceedings of the APWeb Workshops 2006. LNCS, vol. 3842, pp. 741–748, Harbin, China, 16–18 January 2006. Springer, Heidelberg (2006)

19. Park, N.: Customized healthcare infrastructure using privacy weight level based on smart device. Commun. Comput. Inf. Sci. **206**, 467–474 (2011)

20. Park, N.: The implementation of open embedded S/W platform for secure mobile RFID reader. J. Korea Inf. Commun. Soc. **35**, 785–793 (2010)

21. Park, N., Bang, H.-C.: Mobile middleware platform for secure vessel traffic system in IoT service environment. Secur. Commun. Netw. **9**(6), 451–582 (2014)

22. Zhou, L., Chao, H.-C.: Multimedia traffic security architecture for the internet of things. IEEE Netw. **25**, 35–40 (2011)

23. Susanto, H., Muhaya, F.: Multimedia information security architecture framework. In: Proceedings of the 2010 5th International Conference on Future Information Technology, FutureTech 2010, Busan, Korea, 21–23 May 2010

24. Zhou, L., Wang, X., Tu, W., Muntean, G., Geller, B.: Distributed scheduling scheme for video streaming over multi-channel multi-radio multi-hop wireless networks. IEEE J. Sel. Areas Commun. **28**, 409–419 (2010)

25. Wu, D., Cai, Y., Zhou, L., Wang, J.: A cooperative communication scheme based on dynamic coalition formation game in clustered wireless sensor networks. IEEE Trans. Wirel. Commun. **11**, 1190–1200 (2012)

An Empirical Study on the Clickbait
of Data Science Articles in the WeChat
Official Accounts

Shuyi Wang[⊠] and Qi Wu

School of Management, Tianjin Normal University, Tianjin 300387, China
nkwshuyi@gmail.com, cathyqi055@gmail.com

Abstract. In the Internet age, clickbait is an effective method to attract people's attention, which usually uses some ways to achieve, such as exaggerating, omitting the details or using punctuation exceedingly. In order to attract the readers to click on the link, some news aggregator sites or social media will choose to use the clickbait. If the clickbait applied to scientific articles, not only will affects the article quality, but also will affects the development of relevant subjects. Thus, the purpose of this research is to explore whether there is a clickbait in the data science articles of WeChat official accounts. This paper collects the relevant data by using the shenjianshou platform, and then uses some steps to analyze data, including cleaning data, doing word segmentation, extracting keywords and building a regression model. According to the adjusted r-square value in the regression model, the model can only explains the change of 3.17% page views, which means that the clickbait phenomenon is not prominent in the data science articles of WeChat official accounts. Finally, the regression analysis results are discussed from subject perspective, writer perspective and reader perspective.

Keywords: Clickbait · WeChat official account · Data science
Keyword · Page view

1 Introduction

The clickbait refers to some very arresting titles on the Internet, and its purpose is to attract readers to click, in order to get a lot of benefit [1]. Writers will use some methods to achieve, including exaggerating, omitting key details or using punctuation exceedingly in the title [2]. For example, "Relying on this secret, he earned 10 billion in the stock market for 1 year", "She uncovered the sofa mat to see that ...... heinous!" But these titles tend to hide a lot of advertisements or junk information, wasting people's valuable attention.

This article is one of the research results of the National Social Science Fund Youth Project "Research on the privacy protection of social media users based on information price dynamic disclosure" (Project Approval No.: 15CTQ017).

© Springer Nature Singapore Pte Ltd. 2018
J. C. Hung et al. (Eds.): FC 2017, LNEE 464, pp. 131–140, 2018.
https://doi.org/10.1007/978-981-10-7398-4_14

In recent years, social media have gradually become the main channel for people to access information, and have become a platform for self-media to promote, such as Facebook, WeChat and so on. This trend has accelerated the transformation from traditional media to new media. Facing with the fierce competition in the online news market, many media regard the titles as a key factor in attracting people's attention [3]. At the same time, with the advent of fast-paced lifestyles and the spread of fast food culture, people are forced to develop a fast-food reading habit. Fewer readers have time and patience to read the entire article, most readers decide whether to read according to the article title. Plus, the article title is often limited by space in the Sina Weibo, WeChat official accounts and so on. This situation has accelerated the production of the clickbait.

At present, the degree of user activity in the WeChat is very high. It is reported that the number of WeChat official accounts exceeded 12 million in 2016, 52.3% of users use the WeChat official accounts to get the latest information via WeChat official accounts. However, the articles pushed in the WeChat official accounts only show the title, but do not display the content, so the title has become an important factor to make a difference in clicking the article link [4]. In order to obtain high page views, each WeChat official account will makes effort on the title. However, if the clickbait applied to scientific articles, not only will affect the article quality, but also will affect the development of relevant subjects. Thus, this paper focuses on the field of data science. Under the development rush of big data, data science has gradually become a popular research field [5]. In general, the purpose of this study is to explore whether there is a "clickbait" in the data science articles of WeChat official accounts.

The following is organized as follows: Sect. 2 is related work. Section 3 is data collection and processing. Section 4 is the part of building model. Section 5 is discussion. Section 6 is conclusion, which discusses the limitations and future prospects of this paper.

2 Related Work

In recent years, the emergence of clickbait on social media has been on the rise, this phenomenon is partly caused by people's curiosity. Readers click on the article link just to satisfy their curiosity and fill the knowledge gaps, which is called "information gap" theory of curiosity [6]. George Loewenstein explained the concept of "information gap", he thought that curiosity would be produced when people realized the knowledge gaps. The knowledge gaps will led to the feeling of pain and deprivation. At this time, people will be impulsive because they want to get back their lost knowledge at all hazards [6]. Clickbait just takes advantage of this psychological, create a "curiosity gap" to stimulate the reader's curiosity.

Besides, in the "attention economy" era, if the industry can catch the public's attention, it is easy to stand out from the competition. American economist Michael Goldhaber once said: "The information in the network is not only rich, but even flooded. With the development of information, the value is not information, but your attention." [7] Thus, these articles with ultra-high clicks have successfully attracted public's attention with the help of the article title.

At present, there are many methods to solve the clickbait problem. As the world's leading content recommendation platform, Taboola introduces a tool that allows people to remove articles they do not like, including clickbait articles [8]. Facebook has always attached great importance to the users' experience. As early as August 2014, Facebook proposed to boycott the clickbait, and built a system to detect the clickbait [9]. In August 2016, Facebook once again announced to ban the clickbait and made adjustments to the detection system, clickbait articles would be filtered out of the reader's news feed by using the sorting algorithm. In order to solve the clickbait phenomenon in Twitter, Potthast et al. [1] used the top 20 most prolific publishers on Twitter as the data set, and built a related detection model which can help readers reduce the clickbait articles in their news feed.

There are some scholars who have proposed other methods to detect the clickbait. Chakraborty et al. [9] aimed at detecting the clickbait from all online news media by using different characteristic between the "clickbait" articles and "unclickbait" articles. They created a browser extension that readers can filter out clickbait articles. Biyani et al. [2] extracted the various indicators from the article titles, body and web links of the news aggregators. According to these indicators, a machine learning model is created to detect the clickbait phenomenon automatically.

In addition, some scholars have studied the clickbait phenomenon from other angles. Blom et al. [10] have conducted an in-depth analysis of online news headlines from the pragmatics perspective. The results found that some commercial and tabloid media are more likely to use forward referring headlines to lure readers click. Pengnate [11] has studied from the psychology perspective. By using eye tracking devices to measure the change of pupil and evaluate the clickbait article on the degree of reader's emotional awakening. Chen et al. [12] have explored other potential methods from four aspects to detect clickbait automatically, including language, grammar, image and user's reading behavior.

By summarizing the literature, we found that few scholars have studied the clickbait of WeChat official accounts. In this paper, we choose the WeChat official accounts of data science field as the study object, and use tools to collect data about data science articles, including article title, article content, page view and upvote number. By building a regression model to verify whether there is a "clickbait" in the data science articles of WeChat official accounts.

3 Data Collection and Processing

3.1 Data Collection

We mainly use the shenjianshou platform to collect data in the process of data collection. By using the crawler interface of the WeChat articles collection (multi-keyword crawling) which provided by shenjianshou platform, to obtain data about data science articles. In the crawler keyword setting, we enter 12 keywords about data science, including data analysis, data mining, big data, machine learning, depth learning, artificial intelligence, data visualization, data collection, database, data warehouse, data retrieval and crawler. After starting the crawler, there are 8947 articles related to these keywords.

3.2 Data Cleansing

There may be some missing values in the data set. If we retain these missing values, they will affect the accuracy of research result, so data cleansing is required. First, we need to filter out all the articles that posted on the crawling day, because most WeChat official accounts are updated in a limited number in one day. Readers usually make a decision to read or not read the article within one day after updating. Therefore, the page views of articles will be changed significantly after the first day's uploading. Some articles we collected that may include just released less than one day or even less than an hour. These articles may be very popular ultimately, but the page view at the current crawling time is poor. If these articles calculate together, it will produce data perturbation. Next, we need to filter out the missing data of article title, article content, page view and upvote number. In addition, it is necessary to remove some problematic data, such as the upvote number is more than page view, duplicate articles, etc. After data cleaning, the number of data has been reduced to 6698.

3.3 Preparation Work

It is necessary to analyze whether the article quality has an impact on the page view before the regression analysis. We sort the articles by page view, and pick out the top 20 articles, as shown in Table 1. Since the original data is in Chinese, so the tables and figures in this paper are translated from Chinese into English. Figure 1 is a histogram of twenty articles' page view. As can be seen from Fig. 1, the gap between these 20 articles' page view is big. The page view of ranked No.1 article is more than 100,000, while ranked No. 20 article's page view is only more than 60,000. The page views of these 20 articles are far more than the mean and median of the total page views. The mean is indicated by a solid line, and the median is represented by a dashed line. It is important to notice that the top six articles' page views are the same in Fig. 1. This is because that the WeChat in order to avoid the clickbait, plagiarism and other bad behaviors, decided to limit the page view into 100,000.

Table 1. The top 20 articles' titles

No.	Article title
1	Magical magic formula database
2	Artificial intelligence tearing human, just a matter of time? A little bit of fear…
3	Big data definition "city new owner", you find your place?
4	Artificial intelligence will be the next winning point that Baidu promoted?
5	Must read database maintenance announcement
6	Choosing voluntary by "Big Data" need offer 39,800 yuan, is "sky-high price" consultation reliable to help you fill in the voluntary college entrance examination?
7	Looking at unknown Beijing through big data: about salary, rich and poor, house, the Fifth Ring Road
8	Tonight dry goods: deep learning private placement
9	Sogou CTO Yang Hongtao: what kind of position to participate in artificial intelligence?

(continued)

Table 1. (*continued*)

No.	Article title
10	(The new regulations) The Supreme Court, the Supreme Procuratorate, the Ministry of Public Security "on the stipulation of electronic data collection and judgment"
11	Big data: get your insurance before 40 years old!
12	It is said that annual salary of more than 500,000 people, are hopelessly in love with the "data visualization"
13	Human lost! Artificial intelligence defeats the world championship, its strength is far more than that
14	The world's largest movie database IMDb selected the highest score movie of 25 years!
15	Huang Xiaoming PK artificial intelligence? Angelababy into a dilemma!
16	Database|08 card Chinese database is online, the player recommended recently launch
17	QQ space broadcast fighting of the iteration artificial intelligence, Alpha dog or has become a past
18	Encryption	Netease user database was leaked, Alipay, etc. may also be hidden
19	Li Yanhong: Do not panic for artificial intelligence, at best, it is " a wolf wearing skin sheep"
20	Artificial intelligence is fully rolled, is human thinking really superiority?

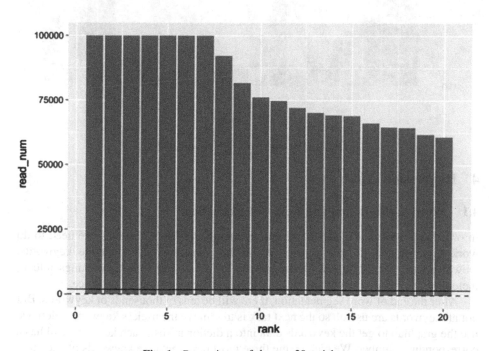

Fig. 1. Page views of the top 20 articles

The quality of article is expressed by the ratio of dividing the upvote number by the page view. Likewise, the mean is represented by a solid line, and the median is represented by a dashed line. There was no one articles whose quality values exceeded the mean, and there were seven articles that did not reach the median, as shown in Fig. 2. Thus, the article quality is difficult to guarantee, even the most popular article. Readers do not know what the quality of this article is, so the article title is the only information to based on.

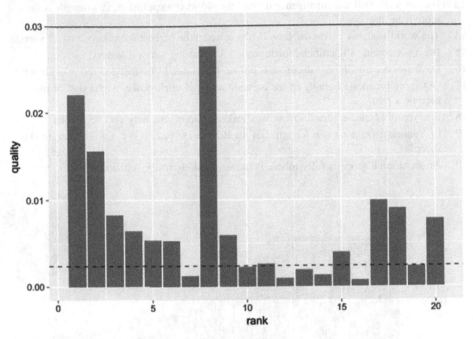

Fig. 2. The quality of the top 20 articles

4   Regression Analysis

4.1   Word Segmentation and Keyword Extraction

In order to remove some stop words and extract the keywords of titles, we need to do work segmentation. In this paper, we use the jieba tool to extract the keywords. Download the list of stop words from jieba tool, and add some new things into it, including common punctuations and the text sentence separated with space.

After the end of word segmentation, there will be tens of thousands of keywords. But not all keywords are useful, so the next task is to remove the useless keywords. Here we use the graphlab to get the keywords data into a dictionary list, each keyword will has a corresponding number. We pick up the three most representative keywords of each title according to the number, and count the frequency of each keyword. Then we sort out the top 200 keywords with higher frequency, and use these 200 keywords as feature variables to construct matrix. Table 2 lists the top 20 keywords with higher frequency.

Table 2. The top 20 keywords with higher frequency

Keyword	Frequency	Keyword	Frequency
Data	90	Learning	29
Database	64	Deep	27
Crawler	59	Application	26
Era	46	MySQL	22
Data analysis	43	Future	21
Data mining	39	Change	21
Visualization	37	Algorithm	21
What	37	Do	21
Artificial intelligence	35	Develop	21
Machine	34	Summary	19

4.2   Regression Model

In the previous section, we got the top 200 keywords with higher frequency. The independent variables are the top 200 keywords, the dependent variable is the page view. We build a regression model to explore the relationship between keywords and page views. To prevent the model from being over-fitting, we need to optimize the model. The more independent variables the model have, the more likely to be over-fitting. In this paper, we use the AIC (Akaike information criterion) rule to simplify the model. By constantly adjusting the AIC parameters to eliminate the lower correlation keywords, and improve the fitting quality of the model. Finally, after several adjustments, the number of independent variables is reduced to 30, including data mining, data analysis, learning, etc.

Since we mainly study Chinese keywords, so we translated the Chinese keywords into English, is to facilitate the understanding of the results of regression analysis, as shown in Table 3. In the regression model, the most critical value is adjusted r-squared. In general, the adjusted r-squared is one of the most important indicators to measure the quality of model. However, this conclusion does not apply to this model. Because this regression model is not used to predict the page view, but used to test whether using keywords is dominant in the contribution of page views. As can be seen from Table 3, adjusted r-square value is 0.0317, indicating that the model can explain the variance of 3.17% page views. Therefore, the keywords of data science articles' titles do not have a major effect on the page views, which means that the "clickbait" phenomenon in the data science articles of WeChat official account is not prominent.

Table 3. Regression analysis results

Keyword		Estimate	Std.Error	T value	Pr(>|t|)	Significance
Chinese	English					
(Intercept)		1955.7	267.3	7.317	2.82e-13	***
数据	Data	-704.1	281.2	-2.504	0.012297	*
数据库	Database	1416.2	287.5	4.927	8.56e-07	***
时代	Era	-753.2	364.9	-2.064	0.039032	*
数据分析	Data Analysis	637.2	254.0	2.509	0.012144	*
数据挖掘	Data Mining	1092.4	399.3	2.736	0.006243	**
可视化	Visualization	1392.6	454.7	3.063	0.002203	**
人工智能	AI	1371.6	334.6	4.099	4.19e-05	***
学习	Learning	576.4	302.3	1.907	0.056619	.
应用	Application	-860.2	402.4	-2.139	0.032477	*
通知	Notice	-1167.5	786.6	-1.484	0.137802	
期权	Option	-2658.3	1532.2	-1.735	0.082791	.
生活	Life	-2544.3	863.8	-2.945	0.003236	**
检索	Retrieve	-1475.7	456.5	-3.232	0.001234	**
领域	Field	-1109.0	712.1	-1.557	0.119426	
机器人	Robot	-1539.2	852.1	-1.806	0.070895	.
一篇	A piece of	6544.7	1436.1	4.557	5.27e-06	***
Python	Python	6102.5	2617.7	2.331	0.019773	*
到	To	979.0	471.2	2.078	0.037775	*
推荐	Recommend	1442.9	750.0	1.924	0.054429	.
关键词	Keyword	-3835.1	2448.1	-1.567	0.117264	
种	A kind of	1160.1	727.5	1.595	0.110832	
带来	Bring	-1975.5	1217.7	-1.622	0.104776	
三个	Three	4509.8	1857.2	2.428	0.015199	*
报告	Report	1187.5	644.1	1.844	0.065293	.
日	Day	-818.8	495.9	-1.651	0.098771	.
自己	Myself	5158.8	1401.8	3.680	0.000235	***
人类	Human	4030.9	816.7	4.936	8.18e-07	***
产业	Industry	-1135.2	614.8	-1.847	0.064859	.
干货	Dry Goods	950.9	665.8	1.428	0.153302	
施行	Implement	7352.8	2084.9	3.527	0.000424	***

Residual standard error: 6403 on 6667 degrees of freedom

Multiple R-squared: 0.03604   Adjusted R-squared: 0.0317

F-statistic: 8.308 on 30 and 6667 DF，   p-value: < 2.2e-16

5   Discussion

In this section, we will discuss the research results from three aspects, including the perspective of subject, the writer's perspective and the reader's perspective.

From the perspective of subject, the core of data science is to extract potential and valuable information from massive data by mixing different elements, theories and techniques in different fields, and even transforming into products. Therefore, according to the characteristics of this subject, data science articles emphasize the authenticity and practicality, professional and technical aspects are also more prominent.

From the writer's perspective, the authors who will write data science articles, basically have a certain depth of professional knowledge, and be able to master professional skills about data science. Therefore, such articles will not blindly pursue the click rate and access traffic, but will pay more attention to the article content. The research process of data science articles is generally inseparable from the data, so author's attitude towards the article will be more rigorous.

From the reader's perspective, there are three kinds of readers who will choose to read data science articles. One is the beginner who is interested in data science field, one is engaged in the relevant professional staff, and the other one is scholar whose research direction is data science. Reading these articles is generally to fill the knowledge gap, strengthen the technological advantages, and understand the development trend of data science. These readers will not easily waste their attention, but pay more attention to the value of article content.

6   Conclusion

This paper mainly discusses whether there is a "clickbait" in the data science articles that published in the WeChat official accounts. By constructing the regression model, it is found that the clickbait phenomenon in the data science articles of WeChat official accounts is not prominent. Although this paper has made some research results, but there are still many shortcomings. Firstly, there are omissions of setting keywords in the data collection process, resulting in an incomplete data set, because the related keywords of data science are far more than twelve. Secondly, this article ignores the influence of the keywords on the clickbaits in the abstract. If the WeChat official accounts only publish one article within one day, the reader will also see a portion of the abstract in addition to the title. In the future study, we will improve the completeness of the data set, and try to use other methods to simplify the data analysis process.

References

1. Potthast, M., Köpsel, S., Stein, B., Hagen, M.: Clickbait detection. In: The 38th European Conference on Information Retrieval, pp. 810–817. Springer International Publishing (2016)
2. Biyani, P., Tsioutsiouliklis, K., Blackmer, J.: 8 amazing secrets for getting more clicks: detecting clickbaits in news streams using article informality. In: Proceedings of the Thirtieth AAAI Conference on Artificial Intelligence, vol. 16, pp. 94–100 (2016)
3. Chen, Y., Conroy, N.J., Rubin, V.L.: News in an online world: the need for an 'automatic crap detector'. In: Proceedings of the Association for Information Science and Technology, vol. 52, no. 1, pp. 1–4 (2015)
4. Fang, J., Lu, W.: A study on influential factors of Wechat public accounts information transmission hotness. J. Intell. **35**(2), 157–162 (2016)
5. Wang, Y.F., Xie, Q.N., Song, X.K.: Review and prospect of overseas research on data science. Libr. Inf. Serv. **60**(14), 5–14 (2016)
6. Loewenstein, G.: The psychology of curiosity: a review and reinterpretation. Psychol. Bull. **116**(1), 75–98 (1994)
7. Jia, F.H.: Information awareness and attention economy. J. Intell. **1**, 89–90 (2002)
8. Bashir, M.A., Arshad, S., Wilson, C.: Recommended for you: a first look at content recommendation networks. In: Proceedings of the 2016 ACM Conference on Internet Measurement, pp. 17–24 (2016)
9. Chakraborty, A., Paranjape, B., Kakarla, S., Ganguly, N.: Stop clickbait: detecting and preventing clickbaits in online news media. In: The 2016 IEEE/ACM International Conference on Advances in Social Networks Analysis and Mining (ASONAM), pp. 9–16 (2016)
10. Blom, J.N., Hansen, K.R.: Click bait: forward-reference as lure in online news headlines. J. Pragmat. **76**, 87–100 (2015)
11. Pengnate, S.F.: Measuring emotional arousal in clickbait: eye-tracking approach. In: Twenty-Second Americas Conference on Information Systems (2016)
12. Chen, Y., Conroy, N.J., Rubin, V.L.: Misleading online content: recognizing clickbait as false news. In: Proceedings of the 2015 ACM on Workshop on Multimodal Deception Detection, pp. 15–19 (2015)

Object Measurement System Based on Surveillance Video-Images

Hui-Xian Duan[1,2,3], Jun Wang[1(✉)], Lei Song[1(✉)], and Na Liu[1]

[1] Cyber Physical System R&D Center, The Third Research Institute of Ministry of Public Security, Shanghai 201204, China
wangjun_darwin@163.com, songlei9312@126.com
[2] The Key Laboratory of Embedded System and Service Computing, Ministry of Education, Tongji University, Shanghai 200092, China
[3] Shanghai International Technology and Trade United Co., Ltd., Shanghai 200031, China

Abstract. Object measurement based on surveillance video-images is a critical task in the field of intelligent video surveillance. In this paper, we proposed an object measurement system based on surveillance video-images. Firstly, considering the scene of the urban video surveillance, a novel calibration tool is designed for such scene, which is portable, collapsible and operational. Next, based on the portable calibration tool, the intrinsic and extrinsic parameters of the surveillance camera can be estimated. Then, an object measurement system based on surveillance video-images is developed, which include the acquisition module of calibration data, the computation module of camera parameters, the selection module of measurement endpoint and the computation module of object. Finally, in the traffic checkpoint, experiments are carried out to verify the effectiveness of the object measurement system. The results show that as long as the surveillance camera is one-time calibrated based on the portable calibration tool, this system can measure the object on the surveillance video-images captured by the calibrated camera. What's more, experimental results have shown that the object measurement system can be used to estimate the vehicle speed and measure the body height within the acceptable error range, and then have demonstrated the effectiveness of the object measurement system.

Keywords: Object measurement system · Portable calibration tool
Surveillance video-images

1 Introduction

In recent years, with the wide application of the surveillance video equipment and the continuous development of image processing technology, image processing has been widely applied to traffic analysis and city security, especially to the calculation and measurement of object (such as height, speed and so on). Object measurement based on surveillance video-images is a critical task in the field of intelligent video surveillance. That is, it can estimate the physical quantities of object in video-images.

As we all known, camera calibration and pose estimation are major issues in object measurement. There are a lot of literatures [1–9] to calibrate camera parameters. In

© Springer Nature Singapore Pte Ltd. 2018
J. C. Hung et al. (Eds.): FC 2017, LNEE 464, pp. 141–151, 2018.
https://doi.org/10.1007/978-981-10-7398-4_15

practice, the method based on the black and white checkerboard is widely used. Considering the actual situation of city video surveillance, the surveillance camera is usually mounted in a high position, thus a larger checkboard with the rigid structure is necessary to obtain the high-precision camera calibration results, which are necessary to estimate the vehicle speed [10, 11] and the body height [12, 13]. However, the larger checkboard is very difficult in the actual operation. In this paper, we design a portable calibration tool to solve this problem, and develop an object measurement system based on the designed calibration tool.

Firstly, in order to adapt to the scene of the urban video surveillance, a novel calibration tool is designed, which is portable, collapsible and operational. Secondly, through the designed calibration tool, the surveillance camera parameters are estimated using the method based on the plane pattern [1]. What's more, we develop an object measurement system based on surveillance video-images. This object measurement system consists of four modules: the acquisition module of calibration data, the computation module of camera parameters, the selection module of measurement endpoint and the computation module of object. At last, experimental results on traffic checkpoint have shown that the object measurement system can be used to estimate the vehicle speed and the body height within the acceptable error range. From the experiment results, it can be seen that the designed calibration tool and the developed object measurement system are effective and acceptable.

This paper is organized as follows: Sect. 2 is some preliminaries. Section 3 describes the portable calibration tool and estimates the surveillance camera parameters. Section 4 develops the object measurement system. Section 5 shows the experimental results of the calibration tool and the object measurement system on traffic checkpoint. Finally, Sect. 6 presents some concluding remarks.

2   Preliminaries

A bold letter denotes a vector or a matrix. Without special explanation, a vector is homogenous coordinates. In the following, we briefly review the image formation for camera introduced in [2].

Let the intrinsic parameter matrix of the pinhole camera be

$$K_c = \begin{pmatrix} r_c f_c & s & u_0 \\ 0 & f_c & v_0 \\ 0 & 0 & 1 \end{pmatrix},$$

where r_c is the aspect ratio, f_c is the focal length, s is the skew factor, and $\begin{pmatrix} u_0 & v_0 & 1 \end{pmatrix}^T$ denotes as p is the principle point.

As shown in Fig. 1, under the pinhole camera model, a space point M is projected to its image point m by:

$$\alpha m = K_c \begin{bmatrix} R & t \end{bmatrix} M, \tag{1}$$

where α is a scalar, R is a rotation matrix, t is a translation vector, K_c is the camera intrinsic matrix and $P = K_c \begin{bmatrix} R & t \end{bmatrix}$ is 3×4 transformation matrix.

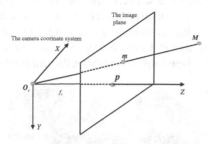

Fig. 1. The image formation of a point under a pinhole camera.

3 Surveillance Camera Calibration

In this section, we design a portable calibration tool for the urban scene, and then estimate the surveillance camera parameters.

3.1 The Portable Calibration Tool

In this subsection, we design a portable calibration tool, as shown in Fig. 2. The portable calibration tool consists of three folded rods and an angle indicator plate with a scale. Each folded rod includes three parts, where the length of the first part is 1.23 m, the length of the second part is 1.2 m and the length of the third part is 1.15 m. Therefore, the total length of the unfolded rod is 3.6 m, which can meet the calibration precision for most of the surveillance cameras. What's more, due to the angle indicator plate, a mapping relationship can be established in some special cases, such as the calibration tool cannot be opened completely.

Fig. 2. The portable calibration tool.

In order to obtain the corresponding relationships between the space points and their image points, we design circular discs with radial pattern on the calibration tool, as shown in Fig. 3. The circular disc can help to identify the calibration points easily and to determine their positions in the image, thereby we can establish the correspondence relationships between the space points and their image points can be established, and then estimate the camera parameters. Therefore, the circular discs can not only meet the need for accurate identification of the calibration points, but also can solve the low-resolution images due to the higher mounting position for the surveillance camera. That is, the designed calibration tool can improve the accuracy of the correspondence relationships and increase the adaptability to calibrate the different surveillance cameras.

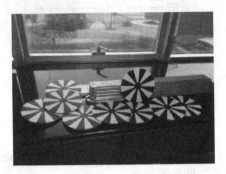

Fig. 3. A set of calibration circular discs.

3.2   Surveillance Camera Calibration

In this subsection, we estimate the surveillance camera parameters based on the designed calibration tool. Firstly, the calibration tool is opened and put on the ground. Then, set up a word coordinate system $\{O - X, Y, Z\}$ on the calibration tool. As shown in Fig. 4, O as the origin; the poles in the edge of the calibration tool as the $X-$ axis and $Y-$ axis respectively; the line through the origin O and orthogonal $X-$ axis and $Y-$ axis as $Z-$ axis.

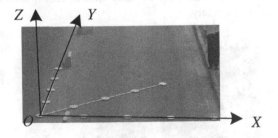

Fig. 4.   The state of the calibration tool when it is unfolded.

With the development of the camera manufacturing technology, some parameters of the surveillance camera can be known in advance: the principal point p can be estimated through the image center and the skew factor s is 0.

Based on the above assumptions, we only need to estimate the aspect ratio r_c, the focal length f_c, the rotation matrix R and the translation vector t.

In the literature [1], based on the known space points M and their image points m, we can obtain the homography matrix H between the ground and the image plane. Denote $H = (h_1 \ h_2 \ h_3) = K_c(r_1 \ r_2 \ t)$, according to the properties of the rotation matrix, we have

$$
\begin{aligned}
h_1^T K_c^{-T} K_c^{-1} h_2 &= 0, \\
h_1^T K_c^{-T} K_c^{-1} h_1 &= h_2^T K_c^{-T} K_c^{-1} h_2
\end{aligned}
\tag{2}
$$

By Eq. (2), we obtain two constrains about the camera intrinsic parameters. Thus, the focal length and aspect ratio can be calibrated by the homography matrix H. Then, the rotation matrix and the translation vector are as follows respectively:

$$
\begin{cases}
r_1 = \lambda K_c^{-1} h_1 \\
r_2 = \lambda K_c^{-1} h_2 \\
r_3 = r_1 \times r_2 \\
t = \lambda K_c^{-1} h_3
\end{cases}
\tag{3}
$$

where $\lambda = 1 / \left\| K_c^{-1} h_1 \right\| = 1 / \left\| K_c^{-1} h_2 \right\|$.

4   Object Measurement System

In this section, we develop an object measurement system, which can estimate the physical quantities of objects on the surveillance video-images, as long as the surveillance camera is one-time calibrated based on the portable calibration tool.

As shown in Fig. 5, the object measurement system consists of four modules: the acquisition module of calibration data, the computation module of camera parameters, the selection module of measurement endpoint and the computation module of object. Firstly, based on the calibration tool, obtain the calibration data through acquisition module of calibration data, and then establish the corresponding relationships between space points and image points; secondly, according to the corresponding relationships, estimate intrinsic and extrinsic parameters of the surveillance camera through the computation module of camera parameters; then, get the coordinates of the object endpoint through the selection module of measurement endpoints; finally, using the estimated camera parameters and the endpoints coordinates, compute the physical quantities of object through the computation module of object.

1. The acquisition module of calibration data

As shown in Fig. 4, these points with known relative positions are selected as the calibration points. On the portable calibration tool, the distance from each calibration point to the origin is known, thus the coordinates of calibration points in the world coordinate system $\{O - X, Y, Z\}$ can be obtained.

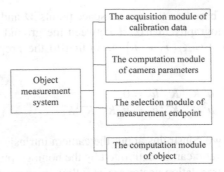

Fig. 5. Object measurement system based on surveillance video-images.

2. The computation module of camera parameters

Based on the surveillance camera calibration method in Sect. 3.2, the homography matrix H can be estimated between the ground and the image plane from the obtained calibration points. Then, by Eqs. (2) and (3), the intrinsic and extrinsic parameters of the surveillance camera K_c, R, t can be calibrated, which are necessary to compute the physic quantities of objects in the surveillance video-images.

3. The selection module of measurement endpoint

Through the human-computer interaction interface, select the measurement endpoints of object in the surveillance video-images and record their image coordinates as the input data of the object measurement. What's more, for this module, we design an open interface to meet its scalability to detect object endpoints automatically. That is, when the object endpoints can be extracted accurately in the video-images, the object measurement system can be linked directly to other system with the object extraction to realize the automatic object measurement.

4. The computation module of object

The computation module of object can not only estimate the vehicle speed, but also can measure the body height.

Vehicle speed estimation. In order to estimate vehicle speed, extract different frames from one vehicle video, and select the corners on the license plate as measurement endpoints. Then, compute the real distance between measurement endpoints. Finally, vehicle speed can be estimated using dividing the distance by the interval time.

Denote the camera transformation matrix as $P = K_c[Rt]$. Let the image coordinates of one corner on the license plate in the previous frame image be m_1, then the spatial coordinates of this corner is $M_1 = P^+m_1$, where P^+ is the generalized inverse matrix of the transformation matrix P. In the current frame image, let the image coordinates of the corresponding corner on the license plate be m_2, then the spatial coordinates of the this corresponding corner is $M_2 = P^+m_2$. According to the interval time Δt, the vehicle speed v can be estimated:

$$v = \|M_2 - M_1\| / \Delta t.$$

If the same vehicle appears in K frames continuously, we can obtain $K - 1$ vehicle speeds: $v_1, v_2, \ldots, v_{K-1}$. In theory, the $K - 1$ vehicle speeds are equal. However, due to the errors, the estimated $K - 1$ vehicle speeds are unequal. Therefore, the average vehicle speed is

$$\bar{v} = \sum_{i=1}^{K-1} v_i / K - 1.$$

Body height measurement. Denote the image coordinate of the foot as $m_p = \left(u_p \; v_p \; 1 \right)^T$ and the image coordinate of the head as $m_h = \left(u_h \; v_h \; 1 \right)^T$. In essence, the body height measurement is to obtain the distance from the head to its vertical projection point on the ground. Under the world coordinate system $\{O - X, Y, Z\}$, let the coordinate of the head be $M_h = \left(x_h \; y_h \; z_h \; 1 \right)^T$, then the coordinate of the foot is $M_p = \left(x_p \; y_p \; 0 \; 1 \right)^T$. Because the foot is on the ground, the relationship between the foot and its image point satisfy the following equation:

$$\alpha_p m_p = H \begin{pmatrix} x_p \\ y_p \\ 1 \end{pmatrix} = H \tilde{M}_p \Rightarrow 1/\alpha_p \tilde{M}_p = H^{-1} m_p. \tag{4}$$

By Eq. (4), the coordinate of the foot M_p can be estimated through the homography matrix H. Therefore, for the coordinate of the head M_h, only z_h is unknown.

By the image formation under a pinhole camera Eq. (1), we have

$$\alpha_h m_h = P_{3\times4} M_h,$$

where $P_{3\times4} = \left(p_{ij} \right)_{3\times4}$. Due to K_c, R and t have been estimated, then

$$z_h = \frac{\left(p_{11} - p_{31}u_h\right)x_p + \left(p_{12} - p_{32}u_h\right)y_p + p_{14} - p_{34}u_h}{p_{33}u_h - p_{13}}$$
$$= \frac{\left(p_{21} - p_{31}v_h\right)x_h + \left(p_{22} - p_{32}v_h\right)y_h + p_{24} - p_{34}v_h}{p_{33}v_h - p_{23}}.$$

Under the world coordinate system, z_h is the coordinate of the head in the $Z-$ axis direction. Therefore, we compute the body height.

5 Experiments

In this section, we perform experiments to evaluate the performance of our calibration tool and object measurement system. Firstly, through the method described in Sect. 3.2, the parameters of the surveillance camera can be calibrated based on the portable calibration tool. Then, by the object measurement system developed in

Sect. 4, the body height and vehicle speed are measured based on the surveillance video-images. The hardware environment: Intel Core I3 2.93 GHz, 4 core CPU, 2.0 GB RAM; soft-ware development environment: Microsoft Windows 7 Home operating system, Visual C++ 2010, OpenCV 2.4.5 [14].

5.1 Surveillance Camera Calibration Based on the Portable Calibration Tool

About the surveillance camera intrinsic parameters, because the skew factor is 0 and the principal point can be estimated through the image center, we can only need to calibrate the focal length and the aspect ratio based on the portable calibration tool. Firstly, open the calibration tool and put it on the ground to capture the video-images, as shown in Fig. 6.

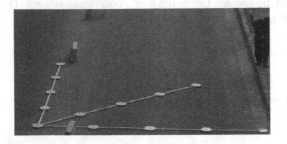

Fig. 6. The image of the portable calibration tool on the ground

Next, as shown in Fig. 7, the calibration points are manually extracted on the video-images and the correspondence relationships between the spatial points and their image points are established. Finally, the aspect ratio and the focal length of the surveillance camera are estimated. The calibration results are shown in Table 1. It can be seen that, the calibration results are close to the true value.

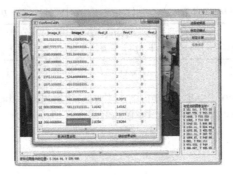

Fig. 7. The corresponding relationship on the portable calibration tools.

Table 1. The real and estimated intrinsic parameters of the surveillance cameras

Intrinsic parameter	Real	Estimated
f_c (Pixel)	6100	6116
r_c	1	1.013

5.2 Object Measurement System

In the traffic checkpoint, experiments to estimate vehicle speed and measure the body height are carried out to verify the effectiveness of the object measurement system.

Vehicle speed estimation. For the surveillance camera installed at junction, when the vehicle moves at 40 km/h and 50 km/h, we capture 16 frame images to estimate the vehicle speed respectively, as shown in Fig. 8. Frame dimensions are 1920×1080 pixels and the frequency is 16 frames/s. Based on the object measurement system, extract the corners on the license plate manually by the selection module of measurement endpoints, and then estimate the vehicle speeds, as shown in Table 2.

Fig. 8. The surveillance images when the vehicle moves at 40 km/h (left) and 50 km/h (right).

Table 2. The measurement results

	Real	Estimated
Vehicle speed (km/h)	40	40.9557
Vehicle speed (km/h)	50	49.1142
Body height (m)	1.75	1.734

Body height measurement. For the surveillance camera installed at junction, we capture one image as shown in Fig. 9. Based on the object measurement system, label the head and foot endpoints with different color cross cursor manually by the selection module of measurement endpoints, and then measure the body height, as shown in Table 2.

Fig. 9. Label the head and foot endpoints.

From Table 2, It can be seen that the error of the estimated vehicle speed is within the acceptable range (±3 km/h) and the error of the measured body height is within the acceptable range (±3 cm). That is, the developed object measurement system is very effective.

6   Conclusion

In this paper, we design a portable calibration tool to estimate the surveillance camera parameters and develop an object measurement system. This system can measure physical quantities of objects in surveillance video-images, as long as the surveillance camera is one-time calibrated. In addition, experimental results about vehicle speed estimation and body height measurement have shown the effectiveness of the object measurement system. In the following work, it is necessary to study the method to extract measurement endpoints of object automatically and accurately.

Acknowledgement. The authors of this paper are members of Shanghai Engineering Research Center of Intelligent Video Surveillance. This work is sponsored by the National Natural Science Foundation of China (61403084, 61402116); by the Project of the Key Laboratory of Embedded System and Service Computing, Ministry of Education, Tongji University (ESSCKF 2015-03); and by the Shanghai Rising-Star Program (17QB1401000).

References

1. Zhang, Z.Y.: A flexible new technique for camera calibration. Trans. Pattern Anal. Mach. Intell. **22**, 1330–1334 (2000)
2. Hartley, R., Zisseman, A.: Multiple View Geometry in Computer Vision. Cambridge University Press, Cambridge (2003)
3. He, B.W., Li, Y.F.: Camera calibration from vanishing points in a vision system. Opt. Laser Technol. **40**, 555–561 (2008)
4. Wu, Y.H., Li, Y.F., Hu, Z.Y.: Detecting and handling unreliable points for camera parameter estimation. Int. J. Comput. Vision **79**(2), 209–223 (2008)

5. Peng, E., Li, L.: Camera calibration using one-dimensional information and its applications in both controlled and uncontrolled environments. Pattern Recogn. **43**, 1188–1198 (2010)
6. Krügera, L., Wöhler, C.: Accurate chequerboard corner localisation for camera calibration. Pattern Recogn. Lett. **32**(10), 1428–1435 (2011)
7. Lee, S.C., Nevatia, R.: Robust camera calibration tool for video surveillance camera in urban environment. In: Computer Vision and Pattern Recognition, pp. 62–67 (2011)
8. Liu, J.C., Collins, R.T., Liu, Y.X.: Robust auto calibration for a surveillance camera network. In: Applications of Computer Vision (2013)
9. Peter, G., Branislav, M., Roman, P.: Calibration methodology for distance surveillance cameras. In: European Conference on Computer Vision, pp. 162–173 (2014)
10. Gupta, P., Purohit, G.N., Rathore, M.: Estimating speed of vehicle using centroid method in MATLAB. Int. J. Comput. Appl. **102**, 1–8 (2014)
11. Lai, W.K., Kuo, T.H., Chen, C.H.: Vehicle speed estimation and forecasting methods based on cellular floating vehicle data. Appl. Sci. **6**, 1–19 (2016)
12. Li, W.S., Wang, W.X., Gao, H.B., et al.: Video-based real-time measurement for human body height. Opt. Eng. **51**(8), 527–529 (2012)
13. Wang, J., Duan, H.X., Wang, J. Mei, L.: A height measure method based on surveillance video camera calibration. In: International Conference on Smart Sustainable City and Big Data, pp. 155–159 (2015)
14. http://opencv.org/. Accessed Apr 2015

Human Action Classification in Basketball: A Single Inertial Sensor Based Framework

Xiangyi Meng[1], Rui Xu[1], Xuantong Chen[1], Lingxiang Zheng[1(✉)], Ao Peng[1], Hai Lu[1], Haibin Shi[1], Biyu Tang[1], and Huiru Zheng[2]

[1] School of Information Science and Engineering, Xiamen University, Xiamen, China
lxzheng@xmu.edu.cn
[2] School of Computing and Mathematics, University of Ulster, Newtownabbey, UK
h.zheng@ulster.ac.uk

Abstract. Human Action Recognition is becoming more and more important in many fields, especially in sports. However, conventional algorithm are almost camera-based methods, which make it cumbersome and expensive. As the wearable inertial sensor has developed a lot, in this paper, we present a novel human action classification algorithm using in basketball, based on a single inertial sensor, which is a application of multi-label classification. We performed experiment on real world datasets. The AUPRC, AUROC and confusion matrix of our results demonstrated that our novel basketball motion recognizer have a great performance.

Keywords: Basketball motion · Human action recognition
Single inertial sensor · Feature extraction · Support vector machine
Multi-label classification

1 Introduction

In recent years, human action recognition (HAR) becomes more and more useful in many ares, including some human-computer interaction (HCI) applications like somatic game, human health monitoring, robotics [2,3,14]. Formally, the aim of HAR is to automatically detecting, analyzing and recording human actions from information and data obtained from many sources both on-line and off-line, for example, wearable inertial sensors, annotated video segments, etc. [3,11]. Consequently, in terms of sensors type used in HAR applications, there are two principal method for HAR: vision-based HAR and inertial-based HAR [3].

An ideal way to recognize human actions is to use the vision information. Shuiwang et al. proposed a 3D convolutional neural network for human action recognition on RGB video data, which can handle 3D inputs [5]. With the development of the techniques used for depth extraction from video [7], many deep learning based HAR approach using depth video data are proposed [8,12,13].

© Springer Nature Singapore Pte Ltd. 2018
J. C. Hung et al. (Eds.): FC 2017, LNEE 464, pp. 152–161, 2018.
https://doi.org/10.1007/978-981-10-7398-4_16

However, those methods just perform well on the existing datasets, not showing their strengths on real situations. Preliminarily, to deploy those approaches on real world stages must satisfy those prerequisites: (1) Applicable cameras to get a large amount of undimmed video segments; (2) Proper and interruption-free place to setup the cameras; (3) Powerful CPU/GPUs to run deep learning algorithm efficiently.

Moreover, consider the case of a non-professional basketball player. Peter, a skillful programmer, is an amateur basketball player who proposed to develop an application helping himself to train shooting skill. Intuitively, he made the computer capable to capture his action each time he shot. As mentioned above, the best solution seemed to be a video sensor HAR system. However, it turned out inconvenient to setup the cameras and supporting devices before his shooting training. Such case is common in many situations of HAR with the challenges of occlusion, camera position, computational complexity, etc. [3], though, it performs well on large scale datasets. These limitations constrained the applications of HAR, especially where too many noises exist. To address such problems, empirical researches studied the HAR based on wearable inertial sensors with accelerometers and gyroscopes, which is convenient enough for individuals to use in their daily routines.

An effective way to analyze the basketball action is to dig out the characteristics in the data obtained from the hand used during the action, which helps coaches and athletes to evaluate their performance better and optimize the training projects. This problem can be regarded as a multi-class classification problem with some inevitable issues [1], as shown below:

Intraclass Variability. This is the first challenge of HAR that a well-performed HAR framework must be robust to the intraclass variability. Those variabilities are common because the same action might be performed differently by different individuals. For example, Stephen and Shawn Marion (an 20-plus-point scorer in the early 2000s NBA) have definitely different shooting postures, whereas they can both get many scores during a match.

Interclass Similarity. Plus there is a inverse challenge, namely, interclass similarity, meaning that different actions are fundamentally different but they have similar numerical characteristics. For example, considering two common actions: **shooting** and **high lobbing pass**, both of them need players to lift their hand and force the ball out of their hands, consequently returning similar sensor data, respectively.

The NULL class problem. Typically, only limited parts of motion types are manually classified and can be recognized by the HAR system. It's an intuition that given this imbalance of relevant versus irrelevant data, activities of interest can easily be confused with activities that have similar patterns but that are irrelevant to the application in question–the so called NULL class. Of course, in some certain HAR applications, such as basketball shooting, golf, etc., the NULL class problem is not particularly serious, given that the types of the motion is not too complex.

To address these problems, we designed a new motion classifier deployed on basketball motion recognition. Being different with other motion recognition frameworks that need many sensors providing several kinds of data like RGB, depth, inertial data, etc., our recognizer is just based on the single inertial sensor, which returns accelerometers and gyroscopes data, attached on the user's wrist of the shooting hand. Then we proposed a novel feature extraction formula to.

The rest of this paper is organized as follows. Section 2 introduces the related work of the sports motion recognition using wearable sensors. The construction and processing of our dataset are stated in Sect. 5.1. Then we defined our problem mathematically in Sect. 3. Section 4 includes the feature extraction method and basketball motion classification method. Then we deployed our methods and the results are shown and analyzed in Sect. 5. Finally, Sect. 6 summarizes our paper.

2 Related Works

In recent years, wearable devices, for example, smart-watch, smart bracelet, etc., have gained unprecedented development. Due to their portability and low power consumption, wearable devices play an important role in the area of activity monitoring, performance evaluation and feedback providing. Andrea et al. presented an comprehensive survey on human activity recognition using body-worn inertial sensors [1]. In this survey, the authors limn the background and some state-of-the-art HAR frameworks, which can be characterized as an process, named Activity Recognition Chain (ARC), giving researchers a clear and understandable tutorial of HAR. The ARC is a process combining the method of signal processing, pattern recognition and machine learning techniques, which receives the raw data returned from the sensors as an input, and responds an output carrying the classification result of the action corresponding to the raw data. That is a common framework of HAR. Following, some relevant researches with in the field of sports motion recognition using wearable sensors are introduced.

Technical statistics are important in sport competitions. However, it is time-consuming and boring to do that manually. Now, with the help of the wearable devices, the technical actions can be recognized and recorded automatically. Taking rugby as an example, Kelly et al. addressed the problem of the automatic recognition of the tackles and collisions in rugby using a GPS receiver and an accelerometer placed between the shoulder blades overlying the upper thoracic spine of each player. In detail, they applied support vector machine (SVM) and hidden conditional random field (HCRF) to identify those actions above, resulting into an excellent performance, where the recall and the precision were 93.3% and 95.8%, respectively.

As for swim, Bächlin et al. designed a wearable assistant, named SwimMaster. What make the SwimMaster helpful is its real-time performance evaluation system using the swimming parameters extracted from the data obtained from the sensors embedded in the assistant.

Le et al. studied the basketball activity recognition problem using wearable inertial measurement units (IMU) [9]. However, being different from our work, they deployed 5 IMUs on the body (two are on the foots, two are on the legs and the remaining one is on the back of the user), which is obviously uncomfortable and will constrain the actions of the user when playing basketball, intuitively.

3   Problem Definition

This section gives information about the basic idea of identifying the motions of playing basketball and the main problem we are facing.

3.1   Prerequisite

In order to effectively analyze performance of a basketball player, a precise identification of the entire motion is essential, namely, to distinguish shooting from other types of motion. Now the only thing we have is a sensor placed on the twist of the habitual basketball shooting hand to record the tri-axial accelerated and angular velocity of the chip, also, those of the wrist.

3.2   Basketball Shooting Recognition

The task of basketball motion recognition is to correctly recognize the type of a given motion from a number of motions belonging to various kinds. In a basketball match, shooting, pass and dribble are three kinds of motions of great importance. So our paper mainly investigated the classification of the three types of motions. In order to facilitate the description of our approach, here we defined the problem more mathematically.

Considering one of a whole process of the human motion, let's denote it as $\mathcal{S}_i = \left(\mathbf{d}_i^1, \mathbf{d}_i^2, \ldots, \mathbf{d}_i^{|\mathcal{S}_i|} \right)^{\mathrm{T}}$, representing the i-th motion in a test case. Formally, $\mathbf{d}_i^j = (a_x, a_y, a_z, g_x, g_y, g_z)$ consists of the data from the accelerometer and gyroscope. In addition, $\mathcal{S}$ represents the set of the motions in a test case. The task of basketball shooting recognition is to find a judging function $f : \mathcal{S} \rightarrow \mathcal{Y}$, where $\mathcal{Y} = \{y_1, y_2, \ldots, y_{|\mathcal{S}|}\}$, $y_i \in 0, 1, 2$ is a set of judging results for whether the motions in $\mathcal{S}$ is basketball shooting, pass or dribble.

Intuitively, this can be formalized as a multi-label classification problem. To build the classifier, for each motion, a feature vector $\mathbf{x}_i$ should be derived using the information of $\mathcal{S}_i$. Then, our task reduces to build a model to estimate the probability $P(y_i|\mathbf{x}_i)$. However, it is challenging to accurately define and compute $\mathbf{x}_i$ given that each motion $\mathcal{S}_i$ has its own length and the law of the data is hard to mine. In the next section, a novel feature extraction method were explained, then we deployed a multi-label SVM to classify the motions.

4   Methods

This section illustrates the methods we use to overcome the complicated analyzing process and to obtain credible identification of basketball motion. To give a clear illustration about our approach, here we give the flowchart of our proposed method, shown in Fig. 1.

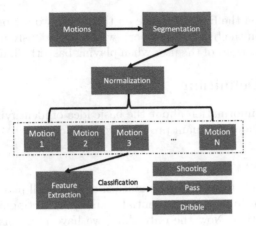

Fig. 1. This figure shows the flowchart of our basketball motion recognition method.

4.1   Feature Extraction

For each motion S_i, it can be denoted as a matrix,

$$S_i = \begin{pmatrix} d_i^1 \\ d_i^2 \\ \vdots \\ d_i^{|S_i|} \end{pmatrix} = \begin{pmatrix} a_x^1 & a_y^1 & a_z^1 & g_x^1 & g_y^1 & g_z^1 \\ a_x^2 & a_y^2 & a_z^2 & g_x^2 & g_y^2 & g_z^2 \\ \vdots & \vdots & \vdots & \vdots & \vdots & \vdots \\ a_x^{|S_i|} & a_y^{|S_i|} & a_z^{|S_i|} & g_x^{|S_i|} & g_y^{|S_i|} & g_z^{|S_i|} \end{pmatrix} \tag{1}$$

where each row in the matrix S_i represents a sample of the sensor data. Notice that different matrix S_i has its own size of row since motions are different in various of aspects. So we need to scale them into the same size which is set as the maximum size of each sample $max\{|S_i|\}, i = 1, 2, \ldots, |S|$. Therefore, before computing the feature vector, an extrapolation must be implemented to finish this preliminary.

Subsequently, we need to calculate the average vector of each row of the matrix, denoted as

$$\overline{d_i} = \sum_{j=1}^{|S_i|} d_i^j = (\overline{a_x}\ \overline{a_y}\ \overline{a_z}\ \overline{g_x}\ \overline{g_y}\ \overline{g_z}). \tag{2}$$

The average vector of the sensor data characterizes the comprehensive numerical feature, whereas we need to ensure our feature vector $\mathbf{x}_i$ is able to carry all of the information useful to describe an motion. An intuition is to calculate the distances between each row vector and the average vector, as defined below:

$$\mathbf{x}_i = \begin{pmatrix} dist(d_i^1, \overline{d}_i, \Sigma) \\ dist(d_i^2, \overline{d}_i, \Sigma) \\ \vdots \\ dist(d_i^{|\mathcal{S}_i|}, \overline{d}_i, \Sigma) \end{pmatrix} \tag{3}$$

where $dist(\mathbf{x}, \boldsymbol{\mu}, \Sigma)$ is the distance function. In this paper, we chose Mahalanobis distance as our metric, denoted as

$$dist(\mathbf{x}, \boldsymbol{\mu}, \Sigma) = \sqrt{(\mathbf{x} - \boldsymbol{\mu})^T \Sigma^{-1} (\mathbf{x} - \boldsymbol{\mu})} \tag{4}$$

where $\mathbf{x} = (x_1, ..., x_n) \in \mathbb{R}^n$, $\boldsymbol{\mu} = (\mu_1, ..., \mu_n) \in \mathbb{R}^n$ and

$$\Sigma = \begin{pmatrix} E[(X_1 - \mu_1)(X_1 - \mu_1)] & \cdots & E[(X_1 - \mu_1)(X_n - \mu_n)] \\ \vdots & \ddots & \vdots \\ E[(X_n - \mu_n)(X_1 - \mu_1)] & \cdots & E[(X_n - \mu_n)(X_n - \mu_n)] \end{pmatrix}.$$

Finally, the feature vector $\mathbf{x}_i$ of an entire specific motion $\mathcal{S}_i$ has been computed. Considering that the dimensionality of the feature vector $\mathbf{x}_i$ is too high, before deploying the classification algorithm, we performed Principal Component Analysis (PCA) to reduce the dimensionality.

4.2 Multi-label Support Vector Machine

Generally, there are two strategy, **one-vs-all** anb **one-vs-one**, for multi-label classification. In this paper, considering the potential inter-class similarities, we chose the **one-vs-one** strategy. Based on the selected strategy, we built a scalable linear support vector machine to classify the motions.

5 Experiments and Results

5.1 Dataset

In this section, we will introduce the composition of our dataset. Briefly, our dataset used in this paper is a combination of two sources: one is obtained from our own data collection process, the other is the UTD Multimodal Human Action Dataset (UTD-MHAD) [4].

Data Collection Process. In order to get the inertial data during an basketball motion, we deployed an *Arduino 101* board worn on the wrist of the experiment candidates.

Fig. 2. The mini Arduino 101 board we used in this paper.

Arduino 101 board has a 6-axis accelerometer/gyro and onboard Bluetooth LE (BLE) capabilities, shown in Fig. 2. Then we developed an android application to receive and store the data transferred from the board via BLE, moreover, this application can control when the board should stop collecting data and send the data back. Finally, we got 157 shooting motions, 70 pass motions and 80 dribble motions.

UTD-MHAD. UTD Multi-modal Human Action Dataset is a part of a research on human action recognition in The University of Texas at Dallas, US. It is a fusion of depth and inertial sensor data. In its collecting process, only one Kinect camera and inertial sensor were used. There are 27 kinds of human actions in UTD-MHAD, including basketball shooting, tennis serve, pickup and throw, etc. In this paper, we only used the single inertial sensor to recognize the basketball motion, consequently we only picked up the inertial data of basketball shooting, which is a subset of the UTD-MHAD, to supplement our experiments. The subset consists of 32 packages of sensor data of basketball shooting generated from 8 experiment candidates.

In order to accurately verify the performance of our proposed model, here we design a leave-one-out cross validation (LOOCV) for the training and prediction process. LOOCV is a strategy to increase the accuracy of the evaluation of the classifier, for which every sample in the sample set have the chance to be the test sample, and the other samples are regarded as the training set. LOOCV was selected as the method to construct the training and test set because the training set generated by the LOOCV accommodates almost all of the samples, which makes sure that the training set is quite similar to the original distribution of the samples.

Based on the dataset introduced in Sect. 5.1, we implemented our algorithms described in Sect. 4 using *Python* and its popular scientific calculation package *SciPy* [6] and a machine learning package *scikit-learn* [10].

Here we evaluated the classification performances using the area under the curve for Receiver Operating Characteristic Curves (AUROC) and Precision Recall Curves (AUPRC).

5.2 Results

The average AUROC and AUPRC of our experiment using the method described in Sect. 4 are summarized in Table 1. Moreover, as the Confusion Matrix showed in Fig. 3, the dribble recognition shows the best performance because of the dribble motion is significantly different from the other two motions. In addition, our classifier may sometimes confused shooting motions and pass motions for the reason that these two motions are somewhere similar with each other. That intuitively makes sense.

Table 1. The performances of the basketball motion classification using features we extracted.

Class	AUROC	AUPRC
Shooting	0.731	0.756
Pass	0.804	0.839
Dribble	0.967	0.982

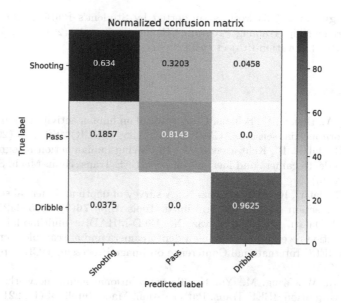

Fig. 3. The normalized confusion matrix of the multi-label SVM.

6 Conclusion

In this paper, we presented a novel human action recognition framework for basketball. Being different with conventional study on the human action recognition, our proposed framework is based on single inertial sensor, which is convenient to be deployed in real world situation and moreover, energy-saving. Our framework consists mainly of two parts: feature extraction and classification. Naive binary classifier is not enough to solve our multi-motion recognition task. To address problems in our task, we deployed a multi-label SVM, which fits the basketball motion recognition problem well. Moreover, the experiment based on the real world datasets demonstrated that our framework performed well in the task of basketball motion recognition.

However, some problems are still worth being studied. First, in our current framework, we need the whole sensor data, from the beginning to the end, to identify whether a motion is a motion. However, this makes it hard to deploy our framework in real time applications. Second, there still exists some singular motion posture, which is totally different with common style, but they are still effective. Current classifier has none knowledge about that "new" motion, which will lead to ridiculous mistakes. Moreover, the computation of the multi-label SVM is time-consuming and under the current calculation ability of portable devices, we need to send data to high performance servers to perform our algorithm. Unfortunately, the time wasted during the communication further decrease the possibility of deploying our framework in real time applications. Thus those aspects are what we should next focus on.

Acknowledgments. This work is supported by Student's Platform for Innovation and Entrepreneurship Training Program, Xiamen University (2016Y1123), and 2016 Google Student Innovation Project (64008066).

References

1. Bulling, A., Blanke, U., Schiele, B.: A tutorial on human activity recognition using body-worn inertial sensors. ACM Comput. Surv. (CSUR) **46**(3), 33 (2014)
2. Chen, C., Jafari, R., Kehtarnavaz, N.: Improving human action recognition using fusion of depth camera and inertial sensors. IEEE Trans. Hum.-Mach. Syst. **45**(1), 51–61 (2015)
3. Chen, C., Jafari, R., Kehtarnavaz, N.: A survey of depth and inertial sensor fusion for human action recognition. Multimed. Tools Appl. **76**(3), 4405–4425 (2015)
4. Chen, C., Jafari, R., Kehtarnavaz, N.: UTD-MHAD: a multimodal dataset for human action recognition utilizing a depth camera and a wearable inertial sensor. In: 2015 IEEE International Conference on Image Processing (ICIP), pp. 168–172. IEEE (2015)
5. Ji, S., Xu, W., Yang, M., Yu, K.: 3D convolutional neural networks for human action recognition. IEEE Trans. Pattern Anal. Mach. Intell. **35**(1), 221–231 (2013)
6. Jones, E., Oliphant, T., Peterson, P., et al.: SciPy: Open source scientific tools for Python (2001). http://www.scipy.org/
7. Karsch, K., Liu, C., Kang, S.B.: Depth extraction from video using non-parametric sampling. In: European Conference on Computer Vision, pp. 775–788. Springer (2012)
8. Kuo, W.Y., Kuo, C.H., Sun, S.W., Chang, P.C., Chen, Y.T., Cheng, W.H.: Machine learning-based behavior recognition system for a basketball player using multiple kinect cameras. In: 2016 IEEE International Conference on Multimedia & Expo Workshops (ICMEW), p. 1. IEEE (2016)
9. Nguyen, L.N.N., Rodríguez-Martín, D., Català, A., Pérez-López, C., Samà, A., Cavallaro, A.: Basketball activity recognition using wearable inertial measurement units. In: Proceedings of the XVI International Conference on Human Computer Interaction, p. 60. ACM (2015)
10. Pedregosa, F., Varoquaux, G., Gramfort, A., Michel, V., Thirion, B., Grisel, O., Blondel, M., Prettenhofer, P., Weiss, R., Dubourg, V., Vanderplas, J., Passos, A., Cournapeau, D., Brucher, M., Perrot, M., Duchesnay, E.: Scikit-learn: machine learning in Python. J. Mach. Learn. Res. **12**, 2825–2830 (2011)
11. Poppe, R.: A survey on vision-based human action recognition. Image Vis. Comput. **28**(6), 976–990 (2010)

12. Shotton, J., Sharp, T., Kipman, A., Fitzgibbon, A., Finocchio, M., Blake, A., Cook, M., Moore, R.: Real-time human pose recognition in parts from single depth images. Commun. ACM **56**(1), 116–124 (2013)

13. Simonyan, K., Zisserman, A.: Two-stream convolutional networks for action recognition in videos. In: Advances in neural information processing systems, pp. 568–576 (2014)

14. Xu, Y., Shen, Z., Zhang, X., Gao, Y., Deng, S., Wang, Y., Fan, Y., Chang, E.I., et al.: Learning multi-level features for sensor-based human action recognition. arXiv preprint arXiv:1611.07143 (2016)

A Smartphone Inertial Sensor Based Recursive Zero-Velocity Detection Approach

Yizhen Wang[1], Xiangyi Meng[1], Rui Xu[1], Xuantong Chen[1], Lingxiang Zheng[1], Biyu Tang[1(✉)], Ao Peng[1], Lulu Yuan[1], Qi Yang[1], Haibin Shi[1], Xiaoyang Ruan[1], and Huiru Zheng[2]

[1] School of Information Science and Engineering, Xiamen University, Xiamen, China
{lxzheng,tby}@xmu.edu.cn
[2] School of Computing and Mathematics, University of Ulster, Newtownabbey, UK
h.zheng@ulster.ac.uk

Abstract. A reliable and robust zero velocity points (ZVPs) detection approach is important to restrain the accumulative error in the pedestrian inertial navigation systems. A novel recursive zero-velocity detection (RZVD) approach for smartphone based pedestrian dead reckoning systems is proposed in this paper. It combined the adaptive threshold and context information of the vertical velocity to verify the correctness of ZVP detection and fixed the incorrect ZVPs recursively. The test results show that the performance of the proposed approach is better than original method. It indicates that the proposed approach is helpful to eliminate the serious estimation error caused by false detection of ZVPs.

Keywords: ZUPT · Smartphone · Pedestrian dead reckoning system
Recursive zero-velocity detection

1 Introduction

Zero-velocity-update (ZUPT) is widely used in the inertial measurement unit (IMU) based pedestrian dead reckoning systems. The inertial sensors error is hard to eliminate, but the error growth of the IMU can be bounded during zero velocity period which reduces the accumulated error effectively. It means that a reliable ZUPT algorithm with few false detection of ZVPs is important for the error remove.

To carry out the zero-velocity detection, some threshold values (for sensor values) must be assigned. Accelerometer and gyroscope are used wildly in zero-velocity (ZV) detection. In paper [5], the ZV detection is only based on accelerometer output which is good for normal walking scenarios, but not for both walking and running. In order to improve the robustness of ZV detection, Paper [7] uses the threshold method that combines the module value of

© Springer Nature Singapore Pte Ltd. 2018
J. C. Hung et al. (Eds.): FC 2017, LNEE 464, pp. 162–168, 2018.
https://doi.org/10.1007/978-981-10-7398-4_17

acceleration and its variance for accurate detection. In paper [10], the ZVPs are determined based on accelerometer and gyroscope which obtain much better result than using one of them separately. Paper [6] proposed an adaptive threshold method instead of fixed threshold to ZV detection for detecting ZVPs of different walking speed. A more reliable ZV detector based on Bayesian network inference instead of threshold methods is proposed in [8]. Those methods proposed above have improved the robustness of ZUPT to some extent, but due to the unpredictability of noise interference, the accuracy of ZVPs still cant be guaranteed in all cases. And most methods aim at the ZUPT of foot-mount IMU, which can get more reliable data in the sampling phase and are good for ZV detection. However, this method is very inconvenient in real life. With the development and popularization of smartphones, smartphone based location system has a good prospect. But there will be more noise interference during the sampling process, which makes ZV detection more difficult and more prone to incorrect detection. The method of ZUPT is similar between smartphone and foot mounted IMU. [2,3] have adopted a threshold method for acceleration to do ZV detection. But they also did not make an analysis of the accuracy of ZVPs. A traditional method of ZUPT was used in our previous work [9]. In this paper, a novel recursive zero-velocity detection (RZVD) approach is proposed for making ZVPs detection more reliable. It tries to improve the robust of ZVPs detection by using a method of adaptive threshold [6] instead of fixed threshold. The vertical velocity after ZUPT is used to judge whether every ZVP detected is correct or not and correct those incorrect ZVPs accordingly.

2   Methods

As shown in Fig. 1, there are four steps in our proposed ZUPT system. The data of pedestrian's vertical acceleration are obtained by using 3-axis acceleration and gravity sensor from smartphone [9]. In step 1, an adaptive threshold is determined by calculating the data [6]. In step 2, the threshold is combined with data for ZV detection. In step 3, Kalman filter is employed in ZUPT to reduce

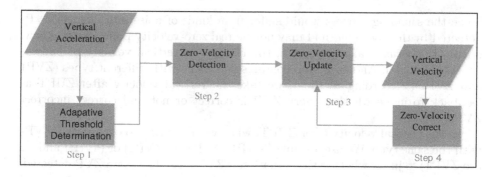

Fig. 1. System framework

the vertical velocity offset caused by accelerometer drift errors [1]. In step 4, the vertical velocity after ZUPT is used as a feedback to judge whether every ZVP detected in step 2 is right or not, and make appropriate correction to incorrect ZVPs. We repeat step 3–4 until no incorrect ZVP is detected.

2.1   Threshold Determination and Zero-Velocity Detection

According to [4], when the heel-touching-ground event happened, the pedestrian's center of gravity was at its lowest position and the vertical acceleration reached a maximum with downward direction. When the sole of one foot parallel to the ground, the pedestrian's center of gravity was at its highest position and the vertical acceleration reached a maximum with upward direction. The vertical velocity during those two moments is zero. Given that the ZV correction mechanism would be carried out in later step, we can use a method with high efficiency instead of high robustness in ZV detection. So we choose the adaptive threshold method [6] to detect ZVPs in every gait cycle. The algorithm is as follows.

Firstly, we use the formula (1) to identify local maximum that is greater than zero and the local minimum that is less than zero respectively.

$$\begin{cases} P_{\max} = \{a_i | a_i > a_{i+1} \cap a_i > a_{i-1} \cap a_i > 0\} \\ P_{\min} = \{a_i | a_i < a_{i+1} \cap a_i < a_{i-1} \cap a_i < 0\} \end{cases} \tag{1}$$

where a_i is i^{th} vertical acceleration. Then, T1 and T2 are defined in formula (2):

$$\begin{cases} T_1 = \mu\left(P_{max}\right) + k_1 \sigma\left(P_{max}\right) \\ T_2 = \mu\left(P_{min}\right) + k_1 \sigma\left(P_{min}\right) \end{cases} \tag{2}$$

where μ, θ are mean and standard deviation, and k_1 is user-defined constant. Threshold T1 and T2 are used to detect ZVPs. We consider the peaks which are greater than T1(ZVP1) and valleys which are less than T2 (ZVP2) as ZVPs.

2.2   Zero-Velocity Correct

Since the sampling process would suffer from kinds of noise interference, ZVPs detected by the above method may not be real zero-velocity point. As we know, during the actual walking process, the changes in vertical velocity of body is nearly sinusoidal. And each step contains two ZVPs with different types (ZVP1 and ZVP2). According to them, we take the vertical velocity after ZUPT as feedback to judge whether every ZVP is correct or not and correct incorrect ZVPs.

In the vertical velocity after ZUPT, when we detected two consecutive ZVPs with the same type. We use formula (3) (P1 and P2 are ZVP2) or (4) (P1 and P2 are ZVP1) to judge whether this situation is ZVP missed or extra ZVP detected.

$$if \max\left(a\left(p_1 : p_2\right)\right) > k_2{}^*T_1 \tag{3}$$

$$if \min\left(a\left(p_1 : p_2\right)\right) < k_2{}^*T_2 \tag{4}$$

where P_1 and P_2 are two consecutive ZVPs, and k_2 is a user-defined constant. When the result of formula (3) or (4) is true, we consider it as ZVP-missed happened between two consecutive ZVPs with the same type. It was caused by a too small stride. We set the point with max acceleration from P_1 to P_2 as a ZVP.

When the result of formula (3) or (4) is false, we consider the later one of those two ZVPs is an extra ZVP. When the heel-touching-ground event happened, it requires a certain buffer, a gait cycle might contain two or more peaks during acceleration phase which are greater than T1. The extra ZVP made the result of ZUPT abnormal. So we remove the extra ZVP directly.

Because the changes in vertical velocity of body is nearly sinusoidal during walking process, the vertical velocity of ZVP detected by T1 (ZVP1) should be greater than its previous one and less than its next one. In contrast, the vertical velocity of ZVP detected by T2 (ZVP2) should be less than its previous one and greater than its next one. If the ZVP we detected is inconsistent with the above conditions, we consider it as an incorrect ZVP. Depending on different circumstances, we moving incorrect ZVP to its previous one or next one each time until the ZVPs are all correct.

3   Result

We analysed data collected using the accelerometer and Gyroscope embedded in the Google Nexus 5 phone with 25 Hz clock rate. The phone was placed at the waist position. The pedestrian in the experiment is a 24 years old healthy

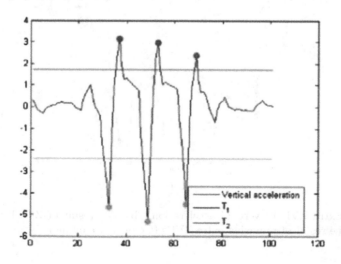

Fig. 2. The result of zero-velocity detection (Color figure online)

male. He walked on a predetermined route (total length: 37.0 m).The constants mentioned above were defined as k1 = 1/3 and k2 = 0.8 in our experiments. We used ZUPT without correction and ZUPT with correction in our indoor location system respectively. The results of experiments are presented in the following content.

Figure 2 shows the result of ZV detection with adaptive threshold method. Red points are ZVP_1 and green points are ZVP_2.

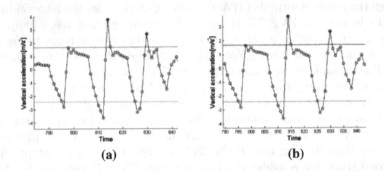

Fig. 3. (a) ZVP-missed (b) Supplement missing ZVP (Color figure online)

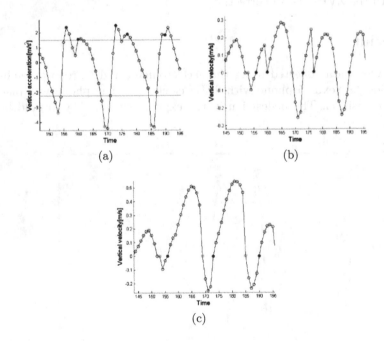

Fig. 4. (a) Extra ZVP in vertical acceleration. (b) The result of ZUPT with extra ZVPs. (c) The result of removing extra ZVPs (Color figure online)

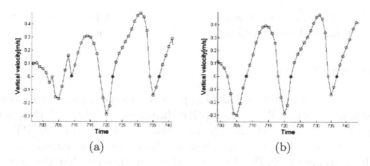

Fig. 5. (a) The result of ZUPT with incorrect ZVPs. (b) The result of ZUPT after correction (Color figure online)

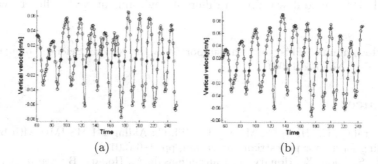

Fig. 6. (a) The result of ZUPT without correction (b) The result of ZUPT with correction

We propose three solutions for different error conditions (ZVP-missed, extra ZVP, ZVP of wrong location) respectively. The effects of those correction methods in our experiments are shown in Figs. 3, 4, and 5.

As shown in Fig. 6, the vertical velocity after original ZUPT contains many incorrect ZVPs, which makes the result of ZUPT is abnormal. Obviously, the velocity of ZUPT with correction is closer to the real changes of pedestrian.

To verify the performance of our proposed method in pedestrian dead reckoning systems. We carried out 4 groups of comparative experiments and the results

Table 1. The distance calculated and error rate with different methods.

Distance calculated/ Error rate	Exp1	Exp2	Exp3	Exp4
Without zero-velocity correct	32.16 m/13.1%	32.17 m/13.1%	34.13 m/7.9%	32.22 m/13.0%
Zero-velocity correct	36.39 m/1.7%	37.53 m/1.3%	39.1 m/5.5%	37.27 m/0.61%

are shown in Table 1. We can see that, the distance's precision of our method has greatly improved compared to original ZUPT method.

4   Conclusion

This paper presented an novel RZVD approach for smartphone based pedestrian dead reckoning systems. We use adaptive threshold to detect ZVPs quickly and implement the ZUPT by them. Then we use the result after ZUPT to judge whether every ZVP detected by adaptive threshold [4] is correct or not and correct those incorrect ZVPs. The test of result showed that the method can correct the incorrect ZVPs effectively and ensure the relative high accuracy of ZUPT. It can be effectively used in smartphone based pedestrian dead reckoning systems. Future work will be focused on the improvement of the robustness of this method to make it working in different environment with different walking speed.

Acknowledgments. This work was supported by the Natural Science Foundation of China (NSFC, No. 61201196).

References

1. Abdulrahim, K., Hide, C., Moore, T., Hill, C.: Aiding MEMS IMU with building heading for indoor pedestrian navigation, pp. 1–6 (2010)
2. Ayub, S., Zhou, X., Honary, S., Bahraminasab, A., Honary, B.: Sensor placement modes for smartphone based pedestrian dead reckoning. Springer, Netherlands (2012)
3. Lan, K.C., Shih, W.Y.: Early detection of neurological disease using a smartphone: a case study. In: International Conference on Sensing Technology, pp. 461–467 (2015)
4. Lan, K.: Using smart-phones and floor plans for indoor location tracking. IEEE Trans. Hum. Mach. Syst. **44**(2), 211–221 (2014)
5. Mi, L., Guo, M., Zhang, X., Zhang, Y.: An indoor pedestrian positioning system based on inertial measurement unit and wireless local area network. In: Chinese Control Conference, pp. 5419–5424 (2015)
6. Thang, H.M., Viet, V.Q., Thuc, N.D., Choi, D.: Gait identification using accelerometer on mobile phone. In: International Conference on Control, Automation and Information Sciences, pp. 344–348 (2012)
7. Xiao-Dong, Z., Ming-Rong, R., Pu, W., Kai, P.: A new zero velocity update algorithm for the shoe-mounted personal navigation system based on IMU. In: Control Conference, pp. 5297–5302 (2015)
8. Xu, Z., Wei, J., Zhang, B., Yang, W.: A robust method to detect zero velocity for improved 3D personal navigation using inertial sensors. Sensors **15**(4), 7708–7727 (2015)
9. Zheng, L., Wu, Z., Zhou, W., Weng, S., Zheng, H.: A smartphone based hand-held indoor positioning system (2016)
10. Zheng, L., Zhou, W., Tang, W., Zheng, X.: A foot-mounted sensor based 3D indoor positioning approach. In: IEEE Twelfth International Symposium on Autonomous Decentralized Systems, pp. 145–150 (2015)

A Cloud Platform for Compatibility Testing of Android Multimedia Applications

Chien-Hung Liu[✉]

Department of Computer Science and Information Engineering,
National Taipei University of Technology, Taipei 106, Taiwan
cliu@ntut.edu.tw

Abstract. Along with the widespread use of smartphones, Android has become one of the major platform for multimedia applications (apps). However, due to the fast evolution of Android operating system and the fragmentation of Android devices, it becomes important for an Android multimedia app to be tested on different devices to ensure that the app is compatible with and run well on any of the devices so as to provide consistent user experiences. This paper presents a cloud testing platform (CTP) that allows Android multimedia apps to be tested automatically against a scalable number of physical devices in parallel. Particularly, CTP provides four types of testing to ensure the compatibility of apps from different perspectives. Further, to facilitate identifying the bugs of apps, in addition to test results, CTP also provides the video, screenshots, and performance data corresponding to the tests. The case study shows that CTP can be effective in ensuring the compatibility of Android multimedia apps while saving test time and effort.

Keywords: Software testing · Android testing · Android compatibility testing

1 Introduction

As smartphones become widespread, the number of mobile apps has increased rapidly. According to statistics [1], the number of mobile apps in leading app stores as of March 2017 is more than 6.2 million. Among these mobile apps, many of them are multimedia apps, such as social network, entertainment, game, and voice and video communication; and the number of mobile multimedia apps is expected to grow continuously. Particularly, smartphones are equipped with camera, microphone, GPS, network, and many sensors, that makes smartphone a great platform for running multimedia apps. Thus, nowadays smartphone has become one of the major platform for users to access multimedia apps.

As more users can access more multimedia apps through smartphones (or devices), there is a growing concern about the quality and reliability of the mobile multimedia apps. In particular, Android is an open source software stack that allows manufacturers to adapt for developing a wide-array of devices with different display resolution, screen size, camera, CPU processor power, memory storage, and network compatibility. In addition to diverse hardware characteristics and combinations, the devices may also use

© Springer Nature Singapore Pte Ltd. 2018
J. C. Hung et al. (Eds.): FC 2017, LNEE 464, pp. 169–178, 2018.
https://doi.org/10.1007/978-981-10-7398-4_18

different versions of Android APIs. This creates the so-called *Android fragmentation* problem [2]. The fragmentation of Android devices can introduce inconsistent presentation and interaction of multimedia content among different devices for the same app and result in poor user experiences. Therefore, it is important for a multimedia app to be tested on different kinds of Android devices to ensure that the app is compatible with and run well on any of the devices and to discover any defects or problems related to using the apps on non-targeted devices.

One approach to testing the compatibility of an Android multimedia app is to test the app on a variety of Android devices one by one to verify if the presentation of multimedia content and interaction behaviors of the app are correct. However, this approach can be time-consuming, labor-intensive, and error-prone. Even though we can write a test script to automate the user interactions with the target app and verify if it behaves correctly, executing the test script on a set of Android devices one by one is still tedious and requires considerable human efforts to prepare test environment and record test results for each device. The cost of the efforts can become very significant as the number of Android devices used for the app compatibility testing increases.

To reduce the time and cost required for Android multimedia app compatibility testing, this paper presents a cloud testing platform (CTP) that allows Android multimedia apps to be tested automatically against a scalable number of physical devices in parallel. Specifically, users can upload a test script for the target app and select a set of Android devices to run the compatibility tests. CTP will automatically execute the test script on the selected devices concurrently. Further, to facilitate identifying the potential bugs related to content and presentation styles as well as to device hardware and network capabilities for the multimedia apps, CTP also provides video, screenshots, and performance data regarding the test execution so that users can look into these artifacts and uncover possible problems. A case study was conducted to show the effectiveness of CTP in compatibility testing of Android multimedia apps.

The rest of the paper is structured as follows. Section 2 briefly reviews existing work related to our approach. Section 3 presents the design and implementation of CTP. Section 4 provides a case study of CTP to illustrate its effectiveness. The concluding remarks and future work are described in Sect. 5.

2 Related Work

In this section, we briefly review existing researches related to our work. Several commercial testing services that support compatibility testing of Android apps are also listed and compared with CTP. Kaasila et al. [3] presented a testing platform for Android apps, called Testdroid. The platform allows users to upload an app and its test script. The script is then executed on a set of physical devices concurrently to test the app. The test results of the app on the devices then can be accessed through the platform. By analyzing the test results, the compatibility issues of the app can be identified.

Huang [4] presented the AppACTS, an automated compatibility testing service for Android mobile apps. The AppACTS is based on the master/slave architecture of Hadoop Distributed File System (HDFS) where the master node controls the process of testing and the slave node controls and executes tests on the Android devices. The AppACTS allows users to upload their mobile apps, select devices to perform tests, and view the test results. Moreover, AppACTS supports a set of commands that allow users to develop scripts to interact and test an Android app automatically.

Prathibhan et al. [5] proposed a cloud-based testing tool, called Mobile Application Testing (MAT), to support functional, performance, and compatibility testing of Android mobile apps. For the compatibility testing, they measured the response time, throughput, and network latency of an app for each device. These three metrics are normalized and averaged to yield the compatibility measure of the app. To test an app, users need to interact the app first and record the UI events using the tool and the recorded test seems not able to be executed on multiple devices or emulators in parallel.

Zhang et al. [6] presented a method to choose a small set of Android devices to execute compatibility testing for mobile apps in order to save test cost. Basically, the method aims to generate a test sequence of devices that covers most compatibility features, such as camera, screen resolution, and network connections, for a set of Android devices. The compatibility features are first represented using a tree model. Then, with the model, existing Android devices are classified into several clusters using the K-means algorithm. The clusters are ranked according to the market shares of the devices in the clusters. Finally, one popular device is selected form each cluster in sequence so that the number of devices used for compatibility testing can be reduced while covering most of the compatibility features.

Villanes et al. [7, 8] presented a framework, called Automated Mobile Testing as a Service (AM-TaaS), which supports automated tests for mobile apps. The framework allows users to upload an app and execute test scripts provided by users or run default test cases implemented based on a set of test criteria proposed by App Quality Alliance (AQuA) [9]. The criteria cover various features to be tested, such as install and launch, memory use, and connectivity, for ensuring the quality of mobile apps. The architecture and implementation of AM-TaaS are described in [8]. Currently, AM-TaaS is implemented on a local server instead of a cloud infrastructure. The testing of apps can be performed only on Android emulators instead of physical devices.

Ma et al. [10] presented a toolset for automated testing of Android apps, called BugRocket. The toolset is based a master/slave architecture where the master server interacts with users and controls the testing process and the slave servers connect to devices and perform the tests. Particularly, BugRocket can take an Android app and its configuration file as input and generate corresponding GUI test scripts automatically. It then distributes and performs the tests on a variety of devices and generates a test report containing several testing statistics, such as pass rate and installation/uninstallation success rate of the tests. A case study is provided to illustrate the feasibility of BugRocket.

In addition to aforementioned work, several commercial Android app testing services have been available on the market [11–14]. Basically, these testing services offer various automated tests, such as functional tests and stress tests, for mobile apps on multiple devices. Table 1 shows the comparison of these commercial test services with our CTP from different perspectives, including the supported testing tools, collected performance data, and recorded artifacts.

Table 1. Comparison of related Android cloud testing platforms

Platform	Supported testing tools	Performance data collection	Video and screen snapshot recording
Bitbar [11]	Appium, Calabash, Espresso, Robotium, Uiautomator	Support	Partially support
AWS Device Farm [12]	Appium, Calabash, Espresso, Robotium, Uiautomator	Support	Support
Firebase Test Lab [13]	Espresso, Robotium, Uiautomator	Support	Support
Xamarin Test Cloud [14]	Cucumber, Appium	Support	Partially support
CTP	Installer, Monkey, Monkeytalk, Robotium, Robot framework, Uiautomator, Espresso	Support	Support

3   The Design and Implementation of CTP

This section presents the design and implementation of CTP, including the system architecture of CTP, the compatibility testing process in CTP, the supported testing tools, and the video recording and performance data collection.

● **The System Architecture of CTP**

Figure 1 shows the system architecture of CTP which consists of two parts, the CTP Server and the Integration Station (IS). The CTP Server and IS communicate with each other using Rabbit Message Queue [15], network protocol, and file system. Particularly, the CTP Server consists of a frontend Web UI, Project Information Center (PIC), IS Monitor, Integrator, Job Scheduler, Device Manager, Test Report Generator, and Workspace subsystems. The Web UI is responsible for providing the user interface. The PIC is responsible for managing test projects. The IS Monitor takes charge of monitoring the status of IS. The Integrator controls the whole process of compatibility testing. The Job Scheduler is in charge of task scheduling. The Device Manager will handle the device management. The Test Report Generator takes charge of creating test report and the Workspace is responsible for environment setting and data processing.

Moreover, the IS consists of the Build Process, Builder, Executor, and Workspace. Each IS instance runs on an OpenStack [16] VM or a Docker [17] container. The Build Process is in charge of retrieving the test job from the Message Queue, creating a Builder instance for the test job, and returning the test result and related artifacts. The Builder takes charge of executing the test job and recoding Video by calling the Executor. The Executor will connect/disconnect the physical devices and execute the commands of

Builder. The Device Monitor will interact with the Executor and run on the device to collect test results and performance data.

Note that CTP supports both OpenStack and Docker. This enables us to deploy CTP to a cloud or non-cloud infrastructure so as to satisfy both needs of IT operators and developers. Moreover, CTP allows the Executor to connect devices using either wireless Wi-Fi or USB connection. The Wi-Fi connection allows CTP connecting to a scalable number of devices without concerning their physical locations. The disadvantage of Wi-Fi connection is that it may affect the performance of data transmission for the apps under test. On the other hand, the USB connection can have a better data transmission performance. Yet, it can limit the number of physical devices connected to the cloud platform. To overcome this problem, CTP has implemented a Master/Slave architecture for USB connection so that the number of devices connected to CTP using USB becomes scalable.

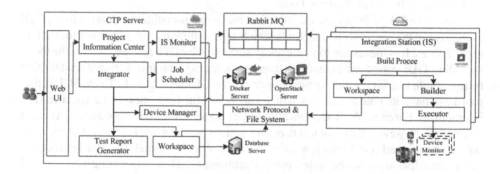

Fig. 1. The system architecture of CTP

• The CTP Compatibility Testing Process

Figure 2 shows the compatibility testing process in CTP. To start the testing, users need to specify the type of testing tool and select a set of devices for running the tests. The users then upload the apk file and test script of the app under test and submit the test request. The users may need to set the test configuration, such as test schedule and time out. Note that there are numerous Android devices connected to CTP. However, only the devices available for testing will be provided online and allow users to select. Each selected device is then assigned to a test job which will be scheduled into a job queue according to CTP task scheduling policy and retrieved later by an IS.

After obtaining a test job, each IS will create a corresponding test builder to handle the test and interacts with the device specified in the job. The test builders will download the apk file and script to their target devices and execute the tests in parallel automatically. When a device completes its test, the test result, video, and performance data of the device will be collected by its associated builder. The builder then sends these artifacts back to CTP server where a test report is generated. Finally, users can access the test report online for each selected device and look into the video and performance data so as to analyze possible compatibility issues related to the devices for the app.

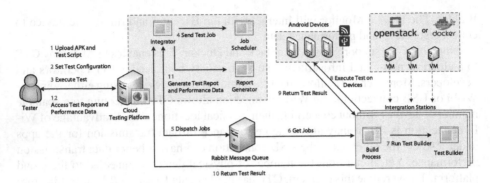

Fig. 2. The compatibility testing process in CTP

• Support of Different Testing Tools

To ensure the compatibility and quality of mobile apps from different perspectives, CTP currently supports various testing tools which offer different types of testing for mobile apps. Table 2 lists the testing tools currently supported by CTP and the types of the tools, including GUI testing, acceptance testing, stress testing, and installation/uninstallation testing. Note that CTP supports many open source GUI testing tools. Each GUI testing tool may have its unique features which can be useful for testing some specific GUI interactions. For example, Robotium allows users to test an app developed using Android WebView. However, it cannot test user interactions across apps. In contrast, there is no way to access and test WebView using Uiautomator. Nevertheless, GUI interactions across multiple apps can be tested using Uiautomator. Thus, supporting various GUI testing tools is indeed necessary and important.

Table 2. The testing tools supported in CTP

Testing tools	Testing types				Require user scripts
	GUI	Acceptance	Stress	Installation/ uninstallation	
Installer				X	
Monkey			X		
Monkeytalk	X	X			X
Robotium	X				X
Robot Framework		X			X
Uiautomator	X				X
Espresso	X				X

To support multiple testing tools, CTP employs the strategy pattern [18], a software design pattern that enables an algorithm's behavior to be selected by users at runtime. By using the strategy pattern, adding a new algorithm to the existing algorithms is straightforward and can be achieved simply by adding a subclass that implements the strategy interface. As mentioned in the testing process of CTP, users can specify a test tool for conducting the compatibility testing, and a test builder corresponding to the specified tool will be created at runtime. Each test builder implements the interface of Builder in Fig. 1. With this design, it would be easy to extend CTP for supporting more types of testing tools.

● **Video Recording and Performance Data Collection**

The performance of mobile multimedia apps can be affected significantly by the hardware characteristics and network capability of the devices. Moreover, the user experiences of mobile apps can also be affected considerably by the multimedia content and presentation styles. To help developers identifying the bugs related to content and presentation styles as well as to device hardware and network capabilities, CTP also provides video and performance data regarding the tests for each device. These artifacts allow users to visually verify the quality of app's multimedia content and presentation styles and to analyze app's performance related to the hardware and network of the devices in order to uncover any compatibility problems.

To allow users visualize the process of testing, an Android screen recorder is employed to capture the entire test process on each selected device. The recorded videos and captured screenshots will be compressed and sent back to CTP and users can access them online for analyzing and evaluation. Further, CTP deploys a monitor to each device to collect the app's performance metrics regarding hardware and network. The recoded metrics are retuned back to CTP and a summary report will be generated. Figure 3 shows the screenshots of performance data collected by CTP, including the usages of CPU and memory, battery power consumption, and network traffic.

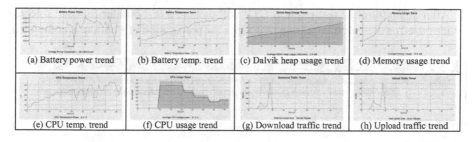

| (a) Battery power trend | (b) Battery temp. trend | (c) Dalvik heap usage trend | (d) Memory usage trend |
| (e) CPU temp. trend | (f) CPU usage trend | (g) Download traffic trend | (h) Upload traffic trend |

Fig. 3. Examples of performance data collected by CTP

4   The Case Study of CTP

To illustrate the usefulness of CTP, a case study was carried out. In the study, an in-house app called WhatHappen is used. This app allows users to track the GPS locations where they have visited. In addition, users can also add notes to places on google map and the app will remind users when they are nearby the locations. Figure 4 shows the screenshot for configuring the test. To set up the compatibility tests, users first need to select the testing tool, such as Robotium, and the target devices. To select the target devices, users can specify the selecting criteria which can be the versions of Android operating system and/or the brands of the smartphones. CTP will automatically selects the available devices that meet the specified criteria. On the other hand, users can select the devices directly from a list of available devices. As shown in Fig. 4, four devices are selected for the compatibility testing. After selecting a set of devices, users need to upload the apk file and test script. If the app is developed in-house, CTP can check out the app source code from a repository and compile and generate the apk automatically.

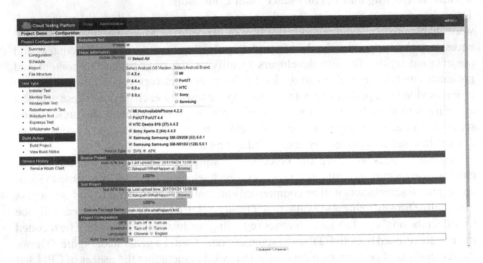

Fig. 4. A screenshot of CTP compatibility test configuration

After submitting the test request, CTP will prepare a test builder for each selected device based on the specified testing tool. Once the preparation of a builder is completed,

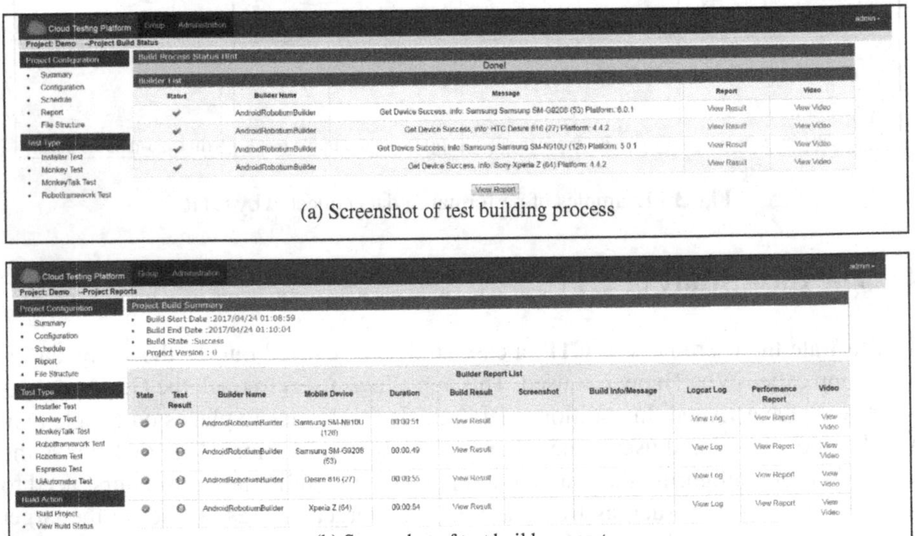

(a) Screenshot of test building process

(b) Screenshot of test builder report

Fig. 5. The screenshots of test building process and builder report

the builder will execute its compatibility test automatically. Figure 5(a) shows the screenshot of test building process in which four test builders are created and each builder has completed its test. When the tests are completed, users can view the builder report summary as shown in Fig. 5(b) which contains the information of test status, the duration of test execution, and the links to the test result, test log, performance data, and video for each device.

By clicking the links in Fig. 5(b), users can see the details of the tests for the selected devices, such as the test report, the recoded video, and the summary of performance data as shown in Fig. 6. By analyzing and evaluating these test result artifacts, users can easily identify if the app under test is compatible with the selected devices and if the multimedia presentation and app performance on the devices are acceptable. With CTP, the compatibility testing of Android multimedia apps can be performed automatically on a scalable number of devices in parallel and, hence, the test time and effort can be reduced significantly.

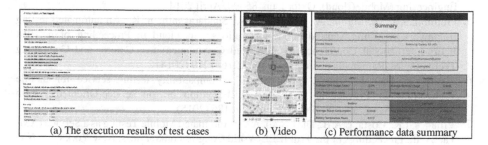

| (a) The execution results of test cases | (b) Video | (c) Performance data summary |

Fig. 6. The screenshots of some test result artifacts

5   Conclusions and Future Work

This paper has presented a cloud testing platform, called CTP, to support compatibility testing for Android multimedia apps. In particular, CTP allows Android multimedia apps to be tested automatically against a scalable number of physical devices in parallel. It supports various testing tools to facilitate different types of testing for Android apps. Further, CTP can also provide the video, screenshots, and performance data regarding the compatibility testing to ease the identification of possible defects in the apps. The design of CTP is described and a case study is presented to illustrate the usefulness of CTP. With CTP, the time and effort to ensure the compatibility of Android multimedia apps can be greatly reduced.

In the future, we plan to extend CTP to support more testing tools, such as crash testing, and provide more artifacts about the testing results, such as crash and code coverage reports, so as to facilitate the compatibility and quality assurance of Android multimedia apps. Moreover, we plan to conduct empirical studies to quantitatively evaluate the effectiveness of CTP in compatibility testing for Android multimedia apps.

Acknowledgement. This work was supported in part by the Ministry of Science and Technology, Taiwan, under the grant No. MOST 105-2221-E-027-086.

References

1. Number of apps available in leading app stores as of March 2017, Statista. https://www.statista.com/statistics/276623/number-of-apps-available-in-leading-app-stores/. Accessed Mar 2017
2. Android fragmentation. http://en.wikipedia.org/wiki/Fragmentation_%28programming%29. Accessed Mar 2017
3. Kaasila, J., Ferreira, D., Kostakos, V., Ojala, T.: Testdroid: automated remote UI testing on Android. In: Proceedings of the 11th International Conference on Mobile and Ubiquitous Multimedia, (MUM 2012), New York (2012)
4. Huang, J.-F.: AppACTS: mobile app automated compatibility testing service. In: Proceedings of the 2nd IEEE International Conference on Mobile Cloud Computing, Service, and Engineering (MobileCloud), Oxford, pp. 85–90, April 2014
5. Prathibhan, C.M., Malini, A., Venkatesh, N., Sundarakantham, K.: An automated testing framework for testing Android mobile applications in the cloud. In: Proceedings of the 2014 IEEE International Conference on Advanced Communications, Control and Computing Technologies, Ramanathapuram, India, pp. 1216–1219 (2014)
6. Zhang, T., Gao, J., Cheng, J., Uehara, T.: Compatibility testing service for mobile applications. In: Proceedings of the 9th IEEE Symposium on Service-Oriented System Engineering, San Francisco Bay, CA, pp. 179–186 (2015)
7. Villanes, I.K., Costa, E.A.B., Dias-Neto, A.C.: Automated Mobile Testing as a Service (AM-TaaS). In: Proceedings of the 11th IEEE World Congress on Services, New York City, NY, pp. 79–86 (2015)
8. Rojas, I.K.V., Meireles, S., Dias-Neto, A.C.: Cloud-based mobile app testing framework: architecture, implementation and execution. In: Proceedings of the 1st Brazilian Symposium on Systematic and Automated Software Testing (SAST), New York (2016)
9. App Quality Alliance. http://www.appqualityalliance.org/. Accessed Apr 2017
10. Ma, X., Wang, N., Xie, P., Zhou, J., Zhang, X., Fang, C.: An automated testing platform for mobile applications. In: Proceedings of the 2016 IEEE International Conference on Software Quality, Reliability and Security Companion (QRS-C), Vienna, pp. 159–162 (2016)
11. Bitbar. http://bitbar.com/. Accessed Mar 2017
12. AWS Device Farm. https://aws.amazon.com/tw/device-farm/. Accessed Mar 2017
13. Firebase Test Lab for Android. https://firebase.google.com/docs/test-lab/. Accessed Mar 2017
14. Xamarin Test Cloud. http://xamarin.com/. Accessed Mar 2017
15. RabbitMQ. https://www.rabbitmq.com/. Accessed Mar 2017
16. OpenStack. https://www.openstack.org/. Accessed Mar 2017
17. Docker. https://www.docker.com/. Accessed Mar 2017
18. Gamma, E., Helm, R., Johnson, R., Vlissides, J.: Design Patterns: Elements of Reusable Object-Oriented Software. Addison Wesley, Boston (1995)

Critical Task Scheduling for Data Stream Computing on Heterogeneous Clouds

Yen-Hsuan Kuo, Yi-Hsuan Lee, Kuo-Chan Huang,
and Kuan-Chou Lai[(✉)]

Department of Computer Science, National Taichung University of Education,
Taichung City, Taiwan, R.O.C.
kclai@mail.ntcu.edu.tw

Abstract. Internet of Things is an emerging paradigm to enable easy data collection and exchange among a wide variety of devices. When the scale of Internet of Things enlarges, the cloud computing system could be applied to mine these big data generated by Internet of Things. This paper proposes a task scheduling approach for time-critical data streaming applications on heterogeneous clouds. The proposed approach takes the tasks in critical stages into consideration, and re-schedules these tasks to appropriate resources to shorten their processing time. In general, selecting the time-critical task to give more resources may remove the execution bottleneck. A small-scale cloud system including 3 servers is built for experiments. The performance of the proposed approach is evaluated by three micro-benchmarks. Preliminary experimental results demonstrate the performance improvement of the critical task scheduling approach.

Keywords: Critical path · Task scheduling · Data stream computing
Spark · Mesos · Cloud computing

1 Introduction

As an emerging technology, Internet of Things enables easy data collection and exchange among a wide variety of devices across existing network infrastructure. When the quasi-continuous data stream is captured and collected by the Internet of Things, these collected data have to be analyzed and mined for identifying significant patterns. When the volume of these collected data grows rapidly with the popularization of new IoT devices, processing these huge-volume and high-velocity data is a time-costly work. As the scale of Internet of Things enlarges, the collected data sets may be too large resulting in that traditional computing system is inadequate to deal with them; therefore, the convergence of cloud computing and Internet of Things is an inevitable trend.

In order to meet the real-time requirement of time-critical data streaming applications, cloud computing systems have to provide adequate computing resources to prevent the violation of service level agreements. In the meantime, several new data stream processing engines have been developed recently, e.g., Apache Storm [6], Apache Spark [5], Apache Samza [4] and Apache Flink [3]. An important challenge in efficiently using cloud resources is to develop task scheduling algorithms for data

© Springer Nature Singapore Pte Ltd. 2018
J. C. Hung et al. (Eds.): FC 2017, LNEE 464, pp. 179–186, 2018.
https://doi.org/10.1007/978-981-10-7398-4_19

stream processing applications. In general, the problem of minimizing makespans by task scheduling been shown as the NP-complete problem. Therefore, solving such kind problems in a polynomial time complexity is very difficult; and then, most approaches adopt heuristic algorithms to solve the task scheduling problem. The object of minimizing the schedule length for task scheduling is one of major factors in determining system performance. However, multiple objects could be defined to satisfy clients' demands in cloud computing systems.

This paper proposes a task scheduling approach for time-critical data streaming applications on heterogeneous clouds. The proposed approach takes the tasks in critical stages into consideration, and re-schedules these tasks into appropriate resources to shorten their processing time. The proposed approach is applied to our previous developed system. The previous proposed system framework consists of the virtual cluster computing system, the resource management system, the profiling system and the monitoring system.

A small-scale cloud system including 3 servers is built for experiments. The cloud OS is OpenStack (Grizzly), and Apache Mesos is used to supply the virtual clusters in which virtual resources may be obtained from multiple physical servers. Apache Spark is adopted to be the computing platform. Three micro-benchmarks are used to evaluate the performance of the proposed approach. Preliminary experimental results demonstrate the performance improvement of the critical task scheduling approach.

This paper is organized as follows. Section 2 discusses related works. Section 3 presents the system framework. Section 4 proposes the critical task scheduling approach. Section 5 presents experiment results. Final section gives the conclusions and future works.

2 Related Works

The technology of Internet of Things [2] allows network devices to sense and collect data from the world around us; these sensed and collected data could be mined and applied for different applications. In general, Internet of Things (IoT) is a network of connected devices which communicate with each other to perform certain tasks. The application of Internet of Things includes location tracking, energy saving, transportation, safety control, ..., and so on.

However, the rapid growth of Internet of Things also increases the rate at which data is generated by IoT resulting in big data. In the big data age, time-critical data streaming [10, 12] applications have the time-limit requirement; therefore, the cloud computing system is a good candidate to handle these applications with sophisticated resource management.

Cloud computing [13] applies the virtualization technology to provide diverse services by the pay-as-you-go model. In cloud computing, task scheduling is a very important issue in improving system performance. Gupta et al. [1], presents a holistic viewpoint to the suitability of high performance computing applications running on clouds. The authors also propose optimizing approaches to improve the performance of HPC applications according to the execution patterns. Simulation results show that significant improvement in average turnaround time. Tasi et al. [8], proposes a hyper-heuristic scheduling algorithm to obtain schedules by diversity detection and improvement detection dynamically. Experimental results show that the schedule length could be reduced significantly. Rodriguez and Buyya [11] present a resource provisioning and scheduling approach for scientific workflows to meet users' QoS requirements on clouds. Their approach adopts the meta-heuristic optimization technique to minimize the workflow execution cost with satisfying deadline constraints. Simulation results show the performance improvement of their approach. Kanemitsu et al. [9], proposes a clustering-based task scheduling algorithm for effective execution in Directed Acyclic Graph (DAG) applications. The proposed approach adopts two-phase scheduling. In the first phase, the characteristics of the system and applications are considered to obtain the number of necessary processors. In the second phase, tasks are assigned and clustered to minimize the schedule length. Experimental results show the performance improvement in term of schedule length and efficiency. Zhang et al. [14] presents an online stack-centric scheduling algorithm for cloud brokers to schedule multiple customers' resource requests. The proposed approach tries to exploit the volume discount pricing strategy. Simulation results present the superiority of their approach. Chen [7] introduces a two-phase approach taking system reliability into consideration. The proposed approach adopts a linear programming strategy to obtain minimal makespan, and then, uses a task-duplication strategy to improve system efficiency. Experimental results show the improvement in terms of schedule length ratio, reliability and speedup.

3 System Framework

As shown in Fig. 1, the proposed system framework consists of the virtual cluster computing system, the resource management system, the profiling system and the monitoring system. The virtual cluster computing system provides the computing environment of virtual clusters. The computing node in the virtual cluster is the virtual machine provisioned by the cloud platform. Several Spark systems are executed above the virtual cluster computing system, which are integrated with our proactive task scheduling algorithm. The historical log of resource usage is generated by the profiling system, and the current resource usage is captured by the monitoring system. All virtual resources are efficiently managed by the resource management system, in conjunction with resource usage forecasting model and over-commit module. The proactive task scheduling algorithm running in Spark system is proposed aiming for shorter processing time.

When a Spark system starts to execute an application, the profiling system first checks whether the profiled log has existed. If the profiled log is not existed, the profiling system would be activated to generate the corresponding profiled log for this newly-arrived application. In the profiled log the Spark system could find the execution pattern of this application, such as the number of stages and data dependencies between stages. In the meantime, the resource management system would inform the Spark system about available resource information. The proactive task scheduling algorithm would decide an appropriate sequence to execute stages according to profiled execution pattern and resource information, and aims to shorten the overall processing time.

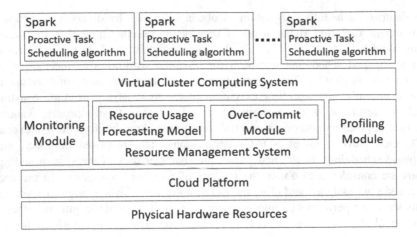

Fig. 1. The proposed system framework.

4   Critical Task Scheduling Algorithm

In the current Spark system, its default task scheduling algorithm is efficient but intuitive. If there is only one Spark system executing, all resources are allocated to this Spark system, and the default algorithm simply schedules stages using the FIFO mechanism. If there are multiple Spark systems executing in parallel, another mechanism is applied to make sure all Spark systems get fair resource usages. In other words, the default task scheduling algorithm treat all stages ready for execution equivalent. When a Spark system is informed that an available resource is allocated to it, the ready stage with smallest id would be picked for execution first. However, in order to shorten the overall processing time, a stage belonging to the critical path should be given higher priority for execution. Apparently, the default task scheduling algorithm in the Spark system cannot identify the stage belonging to the critical path, hence its overall processing time could probably be improved further.

In the proposed framework, a proactive critical task scheduling algorithm is designed to generate an appropriate sequence to execute stages without violating any data dependencies. The main idea is to execute stages belonging to critical path early, and the goal is to achieve shorter processing time. Figure 2 shows the pseudo-code of proposed proactive critical task scheduling algorithm. When a Spark system starts to execute an application, the execution pattern could be found in the profiled log. Based on data dependencies between stages, each stage would be designated a Rank and all stages belonging to critical path would be marked. The Rank of a stage with no successors is designated to 0. For a stage with successors, its Rank is designated to the maximum Rank of successors +1. After above process the stage has no predecessors and belongs to critical path will be designated to maximum Rank. Other stages also belong to critical path could then be easily identified and marked according to data dependencies.

Input: A DAG with N stages
Output: $Assign(S_i)$ /* The sequence to execute stages */

1.   **if** the profiling log is not existed
2.   Activate the profiling system
3.   **end**
4.   Determine $Rank(S_i)$
5.   Determine $CP(S_i)$ /* Mark stages belong to critical path */
6. $\forall\ i, Assign(S_i) \leftarrow 0$
7. $Ready \leftarrow$ Stages S_i with no predecessors
8.   **for** $k = 1$ to N
9. $T_1 \leftarrow arg\ \max Rank(S_i)\ \forall\ S_i \in Ready$
10.   **if** $|T_1| > 1$
11. $T_2 \leftarrow \forall\ S_i \in T_1$ and $CP(S_i) ==$ True
12.   **if** $|T_2| > 1, T_{select} \leftarrow S_i \in T_1$ with minimum i
13.   **else** $T_{select} \leftarrow T_2$
14.   **end**
15.   **else** $T_{select} \leftarrow T_1$
16.   **end**
17. $Assign(T_{select}) = k$
18. $Ready \leftarrow Ready - T_{select}$ /* Remove T_{select} from Ready */
19.   Update $Ready$
20.   **end**

Fig. 2. Pseudo-code of proposed proactive critical task scheduling algorithm.

In Fig. 2, stages whose predecessors have all been assigned were collected in a Ready set. At first the Ready set would contain stages without any predecessors in the given DAG (Line 7). A loop (Line 8-20) would then repeat N times, while in each iteration one stage was selected. In each iteration, this task scheduling algorithm found stages with maximum Rank in the Ready set (Line 9). If there are more than one stages all with maximum Rank, only stages belonging to critical path would be retained (Line 11). If there are still more than one stages remained, the stage with smallest id would be selected in this iteration (Line 12). After N iterations, an appropriate sequence to execute all stages in the given DAG would be generated.

5 Experimental Results

A small-scale cloud system including 3 servers is built for experiments. Table 1 lists the hardware specification of 3 servers. The OS of physical machine is Ubuntu 13.04, and the hypervisor is KVM 3.5.0-23. The cloud OS is OpenStack (Grizzly), and Apache Mesos is used to supply the virtual clusters in which virtual resources may be obtained from multiple physical servers. Each virtual machine occupies 4 VCPU, 4 GB RAM and 40 GB disk. In each virtual machine the Spark executor occupies one VCPU and others are used to execute tasks of applications.

Table 1. Hardware specification.

IBM blade center model	CPU	Memory	Hard disk
HS22	Intel Xeon E5520*16	26 GB	146 GB
HS22	Intel Xeon E5620*16	30 GB	300 GB
HS22	Intel Xeon E5620*16	30 GB	300 GB

The proposed proactive critical task scheduling algorithm is implemented in Spark. When a Spark system received an available resource, it would pick a stage which was ready for execution according to the sequence generated from the proactive task scheduling algorithm. Three synthesized DAGs with various stage parallelisms, as shown in Fig. 3, were used as benchmarks for evaluation. Each stage consists of

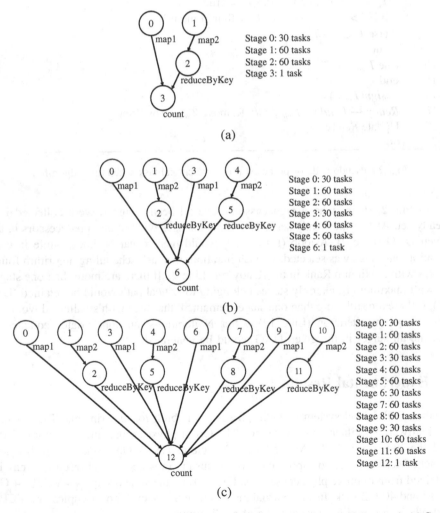

Fig. 3. Three synthesized DAGs. (a) Stage parallelism = 2; (b) Stage parallelism = 4; (c) Stage parallelism = 8.

several tasks which could be executed in parallel. When a stage is selected for execution, all tasks it contained must be finished before executing the next stage.

Figure 4 shows the processing time of three synthesized DAGs with various parallelisms. Each experiment is executed 5 times and shows the average processing time. Blue bars are the average processing time applying default task scheduling algorithm based on FIFO in the Spark system. Orange bars are the average processing time applying proposed proactive critical task scheduling algorithm. These preliminary experimental results show that using proposed algorithm could improve processing time around 4.7%, 4.3% and 1.3% for various stage parallelisms. The improvement when stage parallelisms equal to 2 and 4 are stable, but is slight when stage parallelism increases to 8. This is mainly because although the number of stages could be executed in parallel increases, the amount of available resources is still the same. When the number of resources is relatively insufficient, the affect caused by task scheduling algorithm would become small.

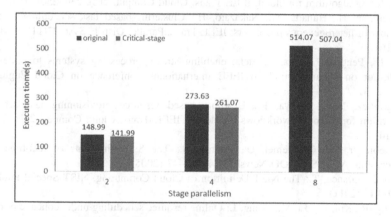

Fig. 4. Execution time of DAGs in Fig. 3.

6 Conclusions and Future Works

This work addresses the time-critical task scheduling problem for data streaming applications on heterogeneous clouds. In this study, we propose a task scheduling algorithm to execute tasks in the stages belonging to critical path as soon as possible in order to shorten the overall execution time. Experimental results show that the proposed algorithm indeed reduces the schedule length. Although, due to the scale of our cloud system, the preliminary experimental results don't show significant improvement. We will extend the experiment with more factors, such as resource heterogeneity, application parallelism, and system scale in the future work.

Acknowledgement. This study was sponsored by the Ministry of Science and Technology, Taiwan, R.O.C., under contract numbers: MOST 103-2218-E-007-021 and MOST 103-2221-E-142-001-MY3.

References

1. Gupta, A., Faraboschi, P., Gioachin, F., Kale, L.V., Kaufmann, R., Lee, B.-S., March, V., Milojicic, D., Suen, C.H.: Evaluating and improving the performance and scheduling of HPC applications in cloud. IEEE Trans. Cloud Comput. **4**(3), 307–321 (2016)
2. Al-Fuqaha, A., Guizani, M., Mohammadi, M., Aledhari, M., Ayyash, M.: Internet of Things: a survey on enabling technologies, protocols, and applications. IEEE Commun. Surv. Tutorials **17**(4), 2347–2376 (2015)
3. Apache Flink. https://flink.apache.org
4. Apache Samza. http://samza.apache.org/
5. Apache Spark. https://spark.apache.org
6. Apache Storm. http://storm.apache.org/
7. Chen, C.-Y.: Task scheduling for maximizing performance and reliability considering fault recovery in heterogeneous distributed systems. IEEE Trans. Parallel Distrib. Syst. **27**(2), 521–532 (2016)
8. Tsai, C.-W., Huang, W.-C., Chiang, M.-H., Chiang, M.-C., Yang, C.-S.: A hyper-heuristic scheduling algorithm for cloud. IEEE Trans. Cloud Comput. **2**(2), 236–250 (2014)
9. Kanemitsu, H., Hanada, M., Nakazato, H.: Clustering-based task scheduling in a large number of heterogeneous processors. IEEE Trans. Parallel Distrib. Syst. **27**(11), 3144–3157 (2016)
10. Xu, L., Peng, B., Gupta, I.: Stela: enabling stream processing systems to scale-in and scale-out on-demand. In: 2016 IEEE International Conference on Cloud Engineering, pp. 22–31
11. Rodriguez, M.A., Buyya, R.: Deadline based resource provisioning and scheduling algorithm for scientific workflows on clouds. IEEE Trans. Cloud Comput. **2**(2), 222–235 (2014)
12. Stonebraker, M., Çetintemel, U., Zdonik, S.: The 8 requirements of real-time stream processing. ACM SIGMOD Newsl. **34**(4), 42–47 (2005)
13. Mell, P., Grance, T.: The NIST Definition of Cloud Computing. NIST Special Publication 800-145 (2011)
14. Zhang, R., Kui, W., Li, M., Wang, J.: Online resource scheduling under concave pricing for cloud computing. IEEE Trans. Parallel Distrib. Syst. **27**(4), 1131–1145 (2016)

An Exploratory Study of Multimodal Perception for Affective Computing System Design

Chih-Hung Wu[(⊠)] and Bor-Chen Kuo

National Taichung University of Education,
Taichung, Taiwan, R.O.C.
chwu@mail.ntcu.edu.tw, kbc@gmail.com

Abstract. Affective computing (AC) is an emerging research direction to deal with the great challenge of creating emotional intelligence for a machine. Affective computing is a cross-disciplinary research knowledge that integrates recognition, interpretation, and simulation of human emotion into a system. This article describes the design of multimodal perception of affective computing system. Our multimodal physiological channels include facial expression recognition, heart rate monitoring, blood oxygen level (SpO2), skin conductance response (SCR), and electroencephalogram (EEG) signals for building our affective computing system. To solve the various data sampling problem, we developed a concurrent control integration mechanism to automatically average all the sensors' data into the same data sampling rate (one record per second) and rearrange all the data into the same time. We believed the proposed system design is benefit for helping researchers in collecting and integrating experiment data in affective computing area.

Keywords: Affective computing · Physiological response
Multimodal affective computing system · Electroencephalogram (EEG)
Skin conductance response (SCR) · Blood oxygen level (SpO2)

1 Introduction

Nowadays, the benefits of affective computing are widely acknowledged in cognitive procedure such as attention, learning, and memory since Picard proposed the concept of affective computing [1]. Regarding to the importance of emotion in learning, the affectivity has an accelerating or perturbing role in learning is incontestable [2] and inextricable bound to learning. [3]. It is now widely accepted that intelligent learning environments are expected to care about both learners and tutors, and to have a good understanding of the variety of learning contexts. A facial expression recognition system, which can get, recognize and analyze facial expression and emotion state when students learning. The studies for physiology signals recognition in learning have used eye tracking technology to observing the visual attention [4–7], also have used EEG and Heart rate variability in ECG to measure the learning emotion [8, 9] and mental workload [10, 11].

© Springer Nature Singapore Pte Ltd. 2018
J. C. Hung et al. (Eds.): FC 2017, LNEE 464, pp. 187–193, 2018.
https://doi.org/10.1007/978-981-10-7398-4_20

Therefore, the purpose of this study is to present a system design of multimodal perception for affective computing system. The system design integrates multi-channel physiological signals that includes facial expression recognition, heart rate monitoring, electroencephalogram (EEG) signals, blood oxygen level (SpO2), and skin conductance response (SCR) for building our affective computing system.

2   Multimodal of Affective Computing System

Multimodal of affective computing can enhance the quality and efficacy of e-learning by including the learner's emotional states [1, 12, 13]. Bahreini et al. [12] proposed a framework for improving learning through webcam and microphones (FILTWAM) for real-time emotion recognition in e-learning by using webcams. The system provides timely and emotion feedback based on learner's facial expression and verbalizations. Wu et al. [1] proposed a comprehensive multimodal-based affective recognition system as shown in Fig. 1.

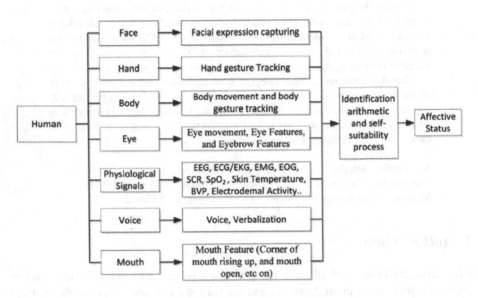

Fig. 1. A comprehensive multimodal-based affective recognition system [1]

In the system, human's emotion can be observed and recorded by 7 different channels that included facial expression, hand gesture tracking, body movement and gesture, eye movement, physiological signals, verbalization, and mouth feature to develop an effective affective computing recognition system. In physiological signals, some of the main physiological signals highly adopted for human emotion assessment are: Electrocardiogram (ECG), Electromyogram (EMG), Skin Conductive Resistance (SCR), and Blood Volume Pressure (BVP). Several approaches have been found that the correlation between the emotional changes and EEG signals [14]. Facial expression recognition is one of the most important emotion expression channels with powerful channel of nonverbal communication and plenty of affective information [15, 16]. Facial expressions are able to provide important clues about emotions [17].

Multimodal of affective computing system can be used in development of affective tutoring system (ATS) [1, 18]. Thompson and McGill [19] presents the design and evaluation of an affective tutoring system (ATS) that detected and real-time emotional states responded to the learner. Lin, Wu et al. [18] designed a personalized e-learning with an ATS to enable assisted teaching and strengthen students' learning effectiveness. The novel ATS which includes four modules: affective recognition module (combines facial emotion recognition and semantic emotion recognition), tutor agent module, content module, and instruction strategies module. The novel ATS demonstrates the good usability of system with high learning performance.

3 Multimodal Perception for Affective Computing System Design

Scientific evidences from academic studies about understanding human emotions are still very limited [15]. The need for interdisciplinary knowledge as well as technological solutions to integrate measurement data from a diversity of physiological sensor equipment is probably the main motivation for the current lack of multimodal databases of observations dedicated to human emotional experiences. This study designs an effective multimodal of affective computing system then use it for assisting in recording and identifying learners' emotion status as shown in Fig. 2.

Our multimodal affective computing data integration system used PHP program to develop the data integration system. We developed a program that transmitted physiological sensors (heart rate, SpO2, SCR) to the XLS format file. The brainwave information included attention, mediation, alpha, beta, theta, and delta information were real-time recorded in the XLS format file via our developed FLASH and PHP program with Neurosky's SDK. Emwave sensor can identify and store participants' emotion status every 5 s as JSON format via its building software. Human facial expressions can be classified into six archetypal emotions: surprise, fear, disgust, anger, happiness, and sadness that are widely accepted from psychological theory. Our facial expression system used Microsoft's EMOTION API in cognitive services that can real-time classified 7 facial expressions that include angry, contempt, disgust, fear, happy neutral, sad, and surprise.

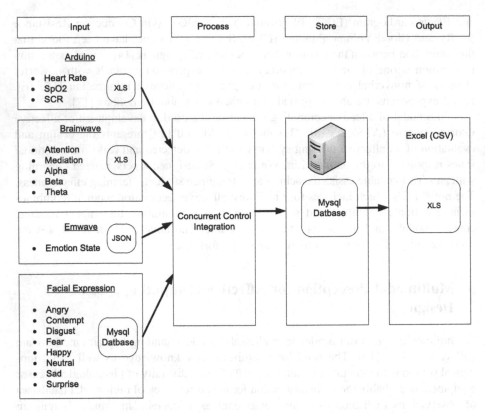

Fig. 2. Multimodal affective computing data integration system design

4 System Development

Our affective computing data integration system has four steps (input, process, store, and output). In the input step, each sensor (Arduino, brainwave, Emwave, and Facial expression system) have to upload data into our main database in our server. The major problem of a multimodal emotion recognition system is that each physiological signal sensor has a different data sampling rate [1, 20].

In our system, heart rate and heart rate variances information from Emwave sensor has slowest sampling rate by 5 s per data. The other physiological sensors—Arduino has the data sampling rate by 2 records per second. The Neurosky's brainwave sensor and our facial expression system has the data sampling rate by 1 records per second as shown in Fig. 3.

Second	Time	Arduino			Brainwave				Emwave	Facial Expression								
		HR	SPo2	SCR	Attention	Mediation	alpha	beta	Emotion	AN	CO	DI	FE	HA	NE	SA	SU	EM
1	07:14:53	83	97	1.83	78	74	11872	24347	-	0	0	0	0	0.8	0.1	0	0	Happiness
	07:14:53	83	97	2.02														
	07:14:53																	
	07:14:53																	
2	07:14:54	82	97	1.14	74	75	11034	9136	-	0	0	0	0.1	0.8	0	0	0.1	Neutral
	07:14:54	82	97	1.05														
	07:14:54																	
	07:14:54																	
3	07:14:55	83	97	2.51	75	57	23836	14782	-	0	0	0	0	0.7	0.1	0	0	Happiness
	07:14:55	83	97	2.71														
	07:14:55																	
	07:14:55																	
4	07:14:56	84	97	3.1	77	60	4000	35124	-	0	0	0	0	0.6	0	0.1	0	Happiness
	07:14:56	84	97	2.51														
	07:14:56																	
	07:14:56																	
5	07:14:57	83	97	2.41	91	57	26017	3292	1	0	0	0	0	0.8	0	0	0	Happiness
	07:14:57	83	97	2.9														
	07:14:57																	
	08:14:57																	

Fig. 3. Data example of raw data from various physiological sensors

Thus, the biggest problem with multimodal emotion recognition is how to integrate the emotional states of various physiological signals. Based on a prior study [1, 20], we designed a concurrent control mechanism for data integration to solve the problem in process step. Because each sensor's data has various data sampling rate, we developed a concurrent control integration mechanism to automatically average all the sensors' data into the same data sampling rate (one record per second) and rearrange all the data into the same time in the process step. After that, system will store all the data into the main database in the store step. After that, the researcher can view the organized experiment data in the website and choose to download their data into a XLS format file. The detail user interface of abovementioned functions is shown in Fig. 4.

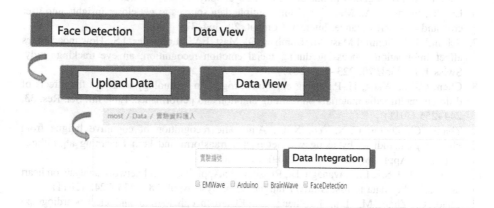

Fig. 4. The design of user interface multimodal affective computing system

5 Conclusion

Our multi-channel affective computing system integrates data that include facial expression recognition, heart rate monitoring, brain wave (attention, mediation, alpha, beta, gamma, theta, delta), SpO2, and SCR for building our affective computing system. To solve the various data sampling problem, we developed a concurrent control integration mechanism to automatically average all the sensors' data into the same data sampling rate (one record per second) and rearrange all the data into the same time. This study presents a system design of multimodal perception for affective computing system. We believed the proposed system design is benefit for helping researchers in collecting and integrating experiment data in affective computing area.

Acknowledgements. This work was financially supported by Ministry of Science and Technology for support (Grant No. MOST 104-2410-H-142-017-MY2).

References

1. Wu, C.-H., Huang, Y.-M., Hwang, J.-P.: Review of affective computing in education/learning: trends and challenges. Br. J. Educ. Technol. **47**, 1304–1323 (2016)
2. Piaget, J.: Les Relations entre l'affectivité et l'intelligence dans le développement mental de l'enfant. Centre de documentation universitaire, Paris (1964)
3. Ben Ammar, M., Neji, M., Alimi, A.M., Gouardères, G.: The affective tutoring system. Expert Syst. Appl. **37**, 3013–3023 (2010)
4. Dimigen, O., Sommer, W., Hohlfeld, A., Jacobs, A.M., Kliegl, R.: Coregistration of eye movements and EEG in natural reading: analyses and review. J. Exp. Psychol. Gen. **140**, 552–572 (2011)
5. Latanov, A.V., Konovalova, N.S., Yermachenko, A.A.: EEG and EYE tracking for visual search task investigation in humans. Int. J. Psychophysiol. **69**, 140 (2008)
6. Lin, T., Imamiya, A., Mao, X.: Using multiple data sources to get closer insights into user cost and task performance. Interact. Comput. **20**, 364–374 (2008)
7. Schmid, P.C., Schmid Mast, M., Bombari, D., Mast, F.W., Lobmaier, J.S.: How mood states affect information processing during facial emotion recognition: an eye tracking study. Swiss J. Psychol. **70**, 223–231 (2011)
8. Chen, C.-M., Wang, H.-P.: Using emotion recognition technology to assess the effects of different multimedia materials on learning emotion and performance. Libr. Inf. Sci. Res. **33**, 244–255 (2011)
9. Zhang, C., Zheng, C.-X., Yu, X.-L.: Automatic recognition of cognitive fatigue from physiological indices by using wavelet packet transform and kernel learning algorithms. Exp. Syst. Appl. **36**, 4664–4671 (2009)
10. Patel, M., Lal, S.K.L., Kavanagh, D., Rossiter, P.: Applying neural network analysis on heart rate variability data to assess driver fatigue. Exp. Syst. Appl. **38**, 7235–7242 (2011)
11. Zhao, C., Zhao, M., Liu, J., Zheng, C.: Electroencephalogram and electrocardiograph assessment of mental fatigue in a driving simulator. Accid. Anal. Prev. **45**, 83–90 (2012)
12. Bahreini, K., Nadolski, R., Westera, W.: Towards multimodal emotion recognition in e-learning environments. Interact. Learn. Env. **24**, 590–605 (2016)

13. Wu, C.H.: New technology for developing facial expression recognition in e-learning. In: 2016 Portland International Conference on Management of Engineering and Technology (PICMET), pp. 1719–1722 (2016)
14. Chanel, G., Ansari-Asl, K., Pun, T.: Valence-arousal evaluation using physiological signals in an emotion recall paradigm. Lecturer Notes in Computer Science, vol. 1, pp. 530–537 (2007)
15. Soleymani, M., Lichtenauer, J., Pun, T., Pantic, M.: A multimodal database for affect recognition and implicit tagging. IEEE Trans. Affect. Comput. 3, 42–55 (2012)
16. Chen, L., Zhou, C., Shen, L.: Facial expression recognition based on SVM in E-learning. IERI Procedia 2, 781–787 (2012)
17. Busso, C., Deng, Z., Yildirim, S., Bulut, M., Lee, C.M., Kazemzadeh, A., Lee, S., Neumann, U., Narayanan, S.: Analysis of emotion recognition using facial expressions, speech and multimodal information. In: Sixth International Conference on Multimodal Interfaces ICMI 2004 (2004)
18. Lin, H.-C.K., Wu, C.-H., Hsueh, Y.-P.: The influence of using affective tutoring system in accounting remedial instruction on learning performance and usability. Comput. Hum. Behav. 41, 514–522 (2014)
19. Thompson, N., McGill, T.J.: Genetics with jean: the design, development and evaluation of an affective tutoring system. Educ. Tech. Res. Dev. 65, 279–299 (2017)
20. Gonzalez-Sanchez, J., Chavez-Echeagaray, M.E., Atkinson, R., Burleson, W.: ABE: an agent-based software architecture for a multimodal emotion recognition framework. In: 2011 9th Working IEEE/IFIP Conference on Software Architecture (WICSA), pp. 187–193 (2011)

Heading Judgement for Indoor Position Based on the Gait Pattern

Lulu Yuan[1], Weiwei Tang[1], Tian Tan[1], Lingxiang Zheng[1], Biyu Tang[1], Haibin Shi[1], Hai Lu[1(✉)], Ao Peng[1], and Huiru Zheng[2]

[1] School of Information Science and Engineering, Xiamen University, Xiamen, China
{lxzheng,luhai}@xmu.edu.cn
[2] School of Computing and Mathematics, University of Ulster, Jordanstown Campus, Shore Road, Newtownabbey, UK
h.zheng@ulster.ac.uk

Abstract. In the inertial sensing unit based indoor positioning systems, the gyroscope drift is the primary source of heading error. To reduce this error, we proposed that the heading drift and the real heading change can be distinguished by the similarity of the gait pattern in the same movement model. It use the curve fitting method to find out the gait pattern in walking straightly. The Frechet distance is used to discriminate the gait of walking in turn and walking straightly. Experiments show that this method can recognize the walking in turn successfully with no mistake and the rate of mismatch walking in straight to walking in turn is less than 17.39%. Although there are some mistakes of match the walking in straight to walking in turn model, it will have few impact because heading drift is little in a short time. The result of test two shows that it get the best result compared with the other two methods when doing heading correction. It indicates that the proposal can promote the performance of heading correction and reduce the effect of sensor drift.

Keywords: Heading estimation · Frechet distance · Curve fitting

1 Introduction

In general, MEMS gyroscope is used for attitude calculation. Due to the gyroscope drift, it is hard to maintain the correct heading in a long period of time when using an IMU based indoor positioning system. In order to achieve precise indoor positioning, scholars have proposed a variety of heading correction algorithms. The Heuristic Drift Reduction (HDR) method, proposed by the University of Michigan Johann Borenstein, corrects gyroscope drift by specifying a number of main directions [1, 2]. However, the HDR is based on the ideal straight line walking. If the possibility of straight line walking is low, the HDR has no effect; if this possibility is high, the HDR will correct the output of the gyroscope. In 2010, Johann Borenstein proposed the Heuristic Drift Elimination method (HDE) [3]. Instead of modifying the gyro measurement, the angular velocity between the two steps is modified. The HDE method reduces the drift, in some cases, can even eliminate the error to almost zero. However, the HDE algorithm is also based on straight

© Springer Nature Singapore Pte Ltd. 2018
J. C. Hung et al. (Eds.): FC 2017, LNEE 464, pp. 194–199, 2018.
https://doi.org/10.1007/978-981-10-7398-4_21

line walking, which will cause an inaccuracy in tracking. Jimenez, A.R. proposed a further improvement of the HDE algorithm, improved heuristic drift elimination (iHDE), which introduced the motion pattern analysis to improve the HDE correction algorithm [4]. However, there are still some problems such as linear jitter yet to be solved.

The previous work of HDR or HDE and their variant can reduce the error. However, it is based on the hypothesis that people walk according the dominant direction of the building and it is not suitable for tracking walkings in open space or in the environment without dominant direction. In this paper, we aim to eliminate this hypothesis for HDE. We proposed a gait pattern recognition algorithm to identify whether a person is walking straightly or not. The hypothesis is that gait pattern will be similar and repetitive when people is walking straightly and the gait pattern will be changed when people is making turns. It can be used to classify the straight walking and turning. We use the curve of the accelerometer data in a gait cycle to present the gait pattern. The pattern can be obtained by a curve fitting algorithm. The Frechet distance is used to evaluate the similarity of gait pattern and classify the straight walkings and turnings.

The Sect. 2 of this paper will introduce the algorithm and method used to discriminate the walking model. Section 3 presents the experiments and results. The paper is concluded by summary and future work in Sect. 4.

2   Method

As shown in Fig. 1, we first select the appropriate features from the accelerometer sensor data, then remove the noise using a low-pass filter. The data is then normalized to facilitate the later curve fitting and similarity matching. The degree of fitting are evaluated after curve fitting. Finally, the state of the walk is classify to a straight line or a curve by pattern discrimination.

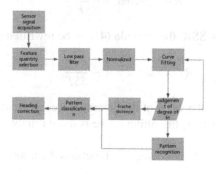

Fig. 1. Walking pattern classification

(1) Selection of feature: For the preprocessing phase of the data, the appropriate feature is extracted from inertia data, and get trained data by the normalization process. (2) Low-pass filter: due to the noise of the sensor, signal fluctuations are

large. It can reduce the high-frequency components, that is, the noise, using a low-pass filter. (3) Normalization of the data: in this paper, the normalization includes two aspects, one is the normalization of the amplitude. The other one is the normalization of the time period of the gait cycle. Based on the above considerations, the Fourier linear normalization is chosen as the normalization method under the premise that the data do not conform to the approximate normal distribution [5].

$$A_n \sin(nx + \varphi_n) = A_n \sin \varphi_n \cos nx + A_n \cos \varphi_n \sin nx \qquad (1)$$

Calculates the parameters of (1), Making the error minimum.

After curve fitting, we use R - squared method [6] to judge the fitting degree. In statistics, the coefficient of determination, denoted R2 or r2 and pronounced "R squared", is a number that indicates the proportion of the variance in the dependent variable that is predictable from the independent variable(s).

Before the introduction of coefficient of determination (R-squared), it is necessary to introduce two parameters Sum of squares of the regression (SSR) and Total sum of squares (SST). SSR computation formula is as follows:

$$SSR = \sum_{i=0}^{m} (\hat{y}_i - \bar{y})^2 \qquad (2)$$

SST computation formula is as follows:

$$SST = \sum_{i=0}^{m} (y_i - \bar{y})^2 \qquad (3)$$

The determination of the coefficient is defined as the ratio of SSR to SST, there are:

$$R - squared = \frac{SSR}{SST} \qquad (4)$$

Due to SST = SSE + SSR, the formula (4) can be rewritten as:

$$R - squared = 1 - \frac{SSE}{SST} \qquad (5)$$

According to Eq. (5), the value of the determined coefficient is in the range of [0,1]. The coefficient is closer to 1, the fitting curve is more reliable, and the model is more accurate.

For the comparison of the objective functions Gauss8, Fourier6 and Fourier8, Fourier6 achieved the best objective function, which can reduce the system load due to the complexity of the fitting curve. Therefore, we choose Fourier6 for curve fitting.

Finally, the curve is matched by the Frechet distance. After obtaining the standard straight curve, the further work is to select the appropriate curve similarity matching algorithm to classify the action according to the matching result. The simplest algorithm for curve similarity is the least squares method. The method obtains the parameters of the curve similarity by calculating the square of the error of the corresponding points.

If we define two functions:

$$\begin{cases} f(x) & X \in [0, 1] \\ g(x) & X \in [0, 1] \end{cases} \tag{6}$$

The Frechet distance between them is:

$$\delta F(f, g) = \inf_{\alpha, \beta \in [0,1]} \max(d(f(\alpha), g(\beta))) \tag{7}$$

After calculating the corresponding Frechet distance, the pattern can be identified by the following formula:

$$M_i = \begin{cases} 1, & \delta F_i(f, g) < T_d \\ 0, & \text{otherwise} \end{cases} \tag{8}$$

Among them, $\delta F_i(f, g)$ is the Frechet distance of step i and T_d is the set Frechet distance threshold.

3 Experiment

Two groups of experiments were carried out on the classification of the movement. The Group1 was tested at the Software Park. The walking distance was about 240 m and the time was 207 s. The total number of steps was 184, the actual turn was 8 times and the straight line was 176 times. Group 2 received about 10 min of rectangle walking test, including 3-step straight and 1-step turn rectangular test, for a total of 264 steps, of which 66 steps turn, 198 steps straight.

3.1 Frechet Distance in Different Modes of Motion

The data collection contains normal speed straight walking and 90 degrees turn.
    The testers tested several groups of straight lines and turned for about 60 steps per group. The Frechet distance in the different motion modes is shown in Table 1, and the sliding window algorithm is added by default.

Table 1. Frechet distance

Motion mode	Mean	Maximum value	Minimum value	Standard deviation
Non-sliding window	1.2213	2.7242	0.5272	1.3188
Straight line1	0.7761	1.7048	0.3764	0.8371
Straight line2	0.6498	1.3991	0.3279	0.6962
90 degrees turn	4.1478	5.3065	2.4425	4.2045

Compared with the non-sliding window straight line 1 and the sliding window straight line 1, it can be found that the Frechet distance error due to the phase difference of the step landing can be effectively reduced after opening the sliding window.

Compared with straight line 1 and straight line 2, it can be found that that the standard curve is consistent in the normal walking process, and the distribution of inertial data is basically the same, the mean of the two line is closer.

In addition, it can be found in the turning experiments that the Frechet distance of direction-changed turn behavior is larger, which can meet the requirements of distinction between straight and turning.

According to the above analysis, it can be found that under different motion modes, the data of the sensor will be distributed in different periods, and then we can use the similarity matching by curve fitting to achieve the purpose of moving pattern classification.

3.2   Gait Pattern Based Heading Correction

In these test, we apply the gait pattern recognition algorithm to the HDE and evaluate the performance when using it to do the heading correction. In the Fig. 2. The red line represents the results of the EKF [7–9]; the blue line represents the result of the original HDE [3]; and the green line represents the gait pattern based heading correction with HDE (gait based HDE) we proposed. In this algorithm, we use the gait curve of the previous step as the standard curve, that is, the standard curve is dynamically changing. This is mainly because a person's gait will change slightly in the course of walking. If a person walks two steps in a straight line, the changes will be slight, but if turning happened, the changes will be big. From Fig. 2. It can be found that the Gait Based HDE got the best result in the experiment.

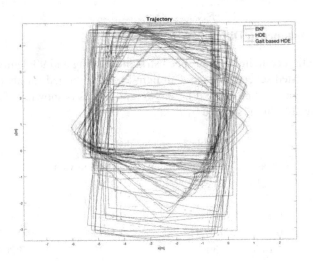

Fig. 2. Trajectory (Color figure online)

4 Conclusion

Experiments show that the proposed heading error elimination algorithm based on extended Kalman filter has good applicability in the field of inertial navigation system, and can effectively eliminate the problem of heading drift caused by gyro error. It can satisfy long time accuracy requirements for strap down inertial navigation systems.

With the development of computer science and information technology, the demand for the indoor positioning is getting bigger and bigger, and the requirements for accuracy are getting higher and higher. However, the existing technology cannot measure accurately because of the limitation of gyroscope's acceleration, which cannot accurately measure and judge the actions when people walking with the acceleration, turn or other acts. In this paper, data analysis is carried out by means of normalization and curve fitting, and then the Frechet distance and pattern recognition method are used to detect the similarity between the training set and the test set, and we get the recognition rate of 91.97%. The above results show in this paper the information of the feet-tied sensor, acceleration and gyroscope can effectively reflect the behavioral characteristics of the user, so as to judge the behavior of the user and improve the accuracy of the indoor positioning based on the inertial sensing unit to a certain extent.

References

1. Borenstein, J., Ojeda, L., Kwanmuang, S.: Heuristic reduction of gyro drift in a personal dead-reckoning system. J. Navig. **62**(1), 41–58 (2009)
2. Borenstein, J., Ojeda, L.: Heuristic reduction of gyro drift in gyro-based vehicle tracking. In: SPIE Defense, Security, and Sensing, pp. 730507–730507-11. International Society for Optics and Photonics (2009)
3. Borenstein, J., Ojeda, L.: Heuristic drift elimination for personnel tracking systems. J. Navig. **63**(4), 591–606 (2010)
4. Jiménez, A.R., Seco, F., Zampella, F., et al.: Improved heuristic drift elimination (iHDE) for pedestrian navigation in complex buildings. In: 2011 International Conference on Indoor Positioning and Indoor Navigation (IPIN), pp. 1–8. IEEE (2011)
5. Hu, F., Zhu, Z., Zhang, J.: Mobile panoramic vision for assisting the blind via indexing and localization. In: European Conference on Computer Vision, pp. 600–614. Springer International Publishing (2014)
6. Fages, A., Ferrari, P., Monni, S., et al.: Investigating sources of variability in metabolomic data in the EPIC study: the Principal Component Partial R-square (PC-PR2) method. Metabolomics **10**(6), 1074–1083 (2014)
7. Zheng, L., Zhou, W., Tang, W., Zheng, X., Peng, A., Zheng, H.: A 3D indoor positioning system based on low-cost MEMS sensors. Simul. Model. Pract. Theory **65**, 45–56 (2016)
8. Zheng, L., Zhou, W., Tang, W., Zheng, X., Yang, H., Pu, S., Li, C., Tang, B., Chen, Y.: A foot-mounted sensor based 3D indoor positioning approach. In: 2015 IEEE Twelfth International Symposium on Autonomous Decentralized Systems (ISADS), pp. 145–150. IEEE (2015)
9. Zheng, X., Yang, H., Tang, W., Pu, S., Zheng, L., Zheng, H., Liao, B., Wang, J.: Indoor pedestrian navigation with shoe-mounted inertial sensors. In: Multimedia and Ubiquitous Engineering, pp. 67–73. Springer, Berlin, Heidelberg (2014)

Multilayer Perceptron Application
for Diabetes Mellitus Prediction
in Pregnancy Care

Mário W. L. Moreira[1,2], Joel J. P. C. Rodrigues[1,3,4(✉)], Neeraj Kumar[5],
Jianwei Niu[6], and Arun Kumar Sangaiah[7]

[1] Instituto de Telecomunicações, Universidade da Beira Interior, Covilhã, Portugal
[2] Instituto Federal de Educação, Ciência e Tecnologia do Ceará (IFCE),
Aracati, CE, Brazil
[3] National Institute of Telecommunications (Inatel),
Santa Rita do Sapucaí, MG, Brazil
joeljr@ieee.org
[4] University of Fortaleza (UNIFOR), Fortaleza, CE, Brazil
[5] Computer Science and Engineering Department,
Thapar University, Patiala, Punjab, India
[6] School of Computer Science and Engineering, Beihang University, Beijing, China
[7] School of Computing Science and Engineering,
Vellore Institute of Technology (VIT), Vellore, India

Abstract. The human intelligence modeling by brain components simulation, such as neurons and their connections, is part of leading smart decision computing paradigms. In Health, artificial neural networks (ANN) have the capacity to adapt to uncertainty situations and learn even with inaccurate data. This paper presents the modeling and performance evaluation of an ANN-based technique, named multilayer perceptron (MLP), for gestational diabetes mellitus (GDM) prediction that is responsible for several severe complications and affects 3 to 7% of pregnancies worldwide. Results show that this approach reached a precision of 0.74, Recall 0.741, F-measure 0.741, and ROC area 0.779. These indicators show that this method is an excellent predictor of this disease. This contribution offers a computational intelligence (CI) tool capable of identifying risk cases during pregnancy and, thus, reduce possible sequels for both pregnant woman and fetus.

Keywords: Intelligent decision computing
Artificial neural networks · Multilayer perceptron
Gestational diabetes mellitus · Pregnancy

1 Introduction

Diabetes is considered a disease of current ways of life and represents one of the major economic problems in health management systems. This metabolic disease

© Springer Nature Singapore Pte Ltd. 2018
J. C. Hung et al. (Eds.): FC 2017, LNEE 464, pp. 200–209, 2018.
https://doi.org/10.1007/978-981-10-7398-4_22

is responsible for several direct and indirect costs, including an adverse impact on gross domestic product (GDP) around the world. The systematic review of World Health Organization (WHO) estimates the direct annual cost of diabetes tripled between 2003 and 2013 and it represents more than US$ 827 billion actually [16]. Jiwani *et al.* assess gestational diabetes mellitus (GDM) prevalence in 173 countries, presenting the practices related to screening and management of this metabolic disease [6]. This study shows that treatment of short- and long-term complications caused by GDM can be very costly. Treatment costs for perinatal complications can reach US$ 9,000 (United States dollar) during the first year of life, and treatment costs for type 2 diabetes mellitus can reach up to US$ 3,500 per year in the United States. Guariguata *et al.* analyze the global prevalence of DMG [5]. This study estimates that more than 90% of GDM cases in pregnancy occurs in low- and middle-income countries. The major number of occurrence is in the South-East Asia with prevalence about 25.0% and the North America and the Caribbean with about 10.4%. Hyperglycemia in pregnancy affects about 21.4 million of live births worldwide. This condition represents a threat to global maternal health. The report of WHO provides a criterion for the diagnostic and classification of hyperglycemia first detected during pregnancy [11]. This study shows that DMG represents a significant risk of adverse perinatal outcomes. Chronic cases require management and evaluation of complications and need pharmacological intervention, especially when detected earlier in the pregnancy. It represents high costs and several consequences of the disease, requiring for more robust health systems to ensure better surveillance, prevention, and more efficient diabetes management [3].

Smart decision support systems (DSSs) represent an essential solution for health management in predicting, identifying, and managing chronic diseases [8]. These models are a combination of inference mechanisms and datasets, designed to support decision-making process in complex problems that involve a large amount of information. Therefore, DSSs offers an important support in the strategic decision-making of a hospital health system, improving the effectiveness of the health management systems [10,12]. The applicability of artificial neural networks (ANNs) as inference mechanism for DSSs in healthcare can predict early gestational diabetes by the risk factors and symptoms presented by the patient. The paper presents the modeling and performance evaluation of an ANN-based technique, named multilayer perceptron (MLP), for gestational diabetes mellitus (GDM) prediction. Thus, this method can reduce the high amount of unnecessary consultations and hospitalizations caused by this disease [13].

The rest of the paper is organized as follows. Section 2 addresses the related work discussing the ANN-based method, named MLP in healthcare. Section 3 shows the modeling proposal capable to identify gestational diabetes mellitus based on risk factors and symptoms presented by patients. Performance evaluation and the result analysis considering the proposed approach are shown in Sect. 4. Finally, Sect. 5 provides the conclusion and suggestions for further works.

2   The ANN-Based MLP Method and Its Application in Healthcare

The multilayer perceptron (MLP) method is based on a network strongly connected with feed-forward connections. In other words, it is a network in which layers are organized a certain order. Neurons of a layer stimulate all the next layer neurons. In this approach, a neuron cannot stimulate a neuron from the same layer or previous layers. An MLP network has a layer with input signals that stimulate the network. This layer has no neurons. Neurons that generate the network output forms the output layer, *i.e.*, the network response to a stimulus. Intermediate layers are those that lie between the input and output layers. There is no mandatory or limit for the number of intermediate layers. Figure 1 shows the architecture of the proposed neural approach in a generalized way. It includes a MLP network with several intermediate layers.

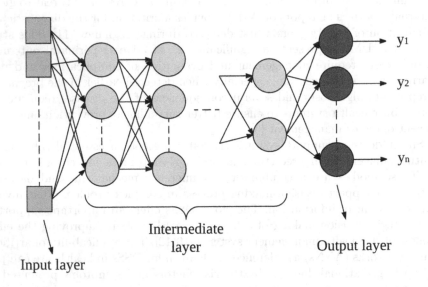

Fig. 1. Neural architecture of the MLP model.

In health, Naraei *et al.* discuss the use of data mining (DM) and machine learning (ML) approaches as effective solutions for analysis and knowledge discovery in healthcare [13]. This work seeks to the accuracy improvement of DSSs in the medical area. For the performance comparison between the MLP and support vector machine (SVM) algorithms, the authors performed analyzes in a heart diseases dataset using several indicators. Results show that SVM algorithm is more effective that the MLP technique regarding to accuracy. Ahmad *et al.* improve the genetic algorithm approach applying automatic and simultaneous parameter optimization and the attribute selection of MLP technique for medical diagnosis [2]. This method improves the accuracy of the proposed algorithm in several databases, such as diabetes, heart diseases, and cancer. Zang *et al.* develop a pathological detection system to assist physicians in the decision-making process on magnetic resonance brain images [18]. This method uses the fractional Fourier entropy features to submit these attributes for a MLP classifier reaching good results. In [1], Aguiar *et al.* use the MLP and self-organizing map methods for pulmonary tuberculosis assess in a high complexity patients, in Brazil. This study uses clinical and radiologic data collected for performance evaluation of this classifiers. Both of them obtained excellent results. These studies show that the MLP approach based on ANNs has high applicability in health care. For the GDM risk cases prediction, an adaptation of this approach can also generate exceptional results.

The MLP algorithm has been applied successfully to solve several complex problems. This method uses a supervised training manner. The most widely used algorithm is the back-propagation error, which is based on the error-correction learning rule. This is a generalization of the least-mean-square (LMS) adaptive filter algorithm [17]. This approach presents two stages, namely, a step forward and a step backward. In the first step, for the sensory nodes of the network, the algorithm applies an activity pattern (input vector), and its effect is propagated along the network layers. Finally, the network response produces a set of outputs. During the step forward, the algorithm fixes the synaptic weights, while to the step backward, the synaptic weights are adjusted according to a rule of error correction. Specifically, the real response is subtracted from the desired response to produce an error signal. This error propagates back through the network in the opposite direction to the stimulus. To make the real response approach close to the desired response, the weights are adjusted through a statistical method.

An MLP trained with back-propagation can perform a non-linear input-output mapping. Let m_0 being the amount of input nodes of an MLP and $M = m_L$ the amount of neurons in the output layer. The network input-output relationship defines a map of a m_0-dimensional Euclidean input space to a M-dimensional space, which is infinite, continuous, and differentiable when the activation function also present these equal characteristics. To evaluate the ability of an MLP from the input-output mapping point of view, it is necessary to determine the minimum number of intermediate layers that provides an approximation of any continuous mapping.

Let $\phi(\cdot)$ being a non-constant, limited, monotonic-increasing, and continuous function. Let I_{m_0} being the unitary hypercube m_0-dimensional $[0,1]^{m_0}$. The space of continuous functions in I_{m_0} is denoted by $C(I_{m_0})$. Then, given any function $f \ni C(I_{m_0})$ and $\epsilon > 0$, there is an integer M, sets of real constants α_i, b_i, and w_{ij}, where $i = 1, \cdots, m_1$ and $j = 1, \cdots, m_0$ to define the Eq. 1 as an approximate realization of the function $f(\cdot)$.

$$F(x_1, \cdots, x_{m_0}) = \sum_{i=1}^{m_1} \alpha_i \phi \left(\sum_{j=1}^{m_0} w_{ij} x_j + b_i \right) \tag{1}$$

In this sense, $|F(x_1, \cdots, x_{m_0}) - f(x_1, \cdots, x_{m_0})| < \epsilon$ for all $x_1, x_2, \cdots, x_{m_0}$ that are in the input space. The theory named universal approximation theorem [4] is directly applicable to MLP algorithm. Equation 2 represents the logistic input function used as the nonlinearity function of MLP. This equation is non-constant, limited, and monotonic-increasing.

$$\phi_j(v_j(n)) = \frac{1}{1 + exp(-av_j(n))} \tag{2}$$

Where $a > 0$ and $-\infty < v_j(n) < \infty$. Equation 1 represents the output of an MLP described as a network that has m_0 input, nodes $x_1, \cdots, x_{m_0}$, and a single intermediate layer with m_1 neurons. The intermediate neuron i has synaptic weights $w_i, \cdots, w_{m_0}$, and bias b_i. The network output is a linear combination of the intermediate neuron outputs, with $\alpha_1, \cdots, \alpha_{m_1}$ defining the synaptic weights of the output layer. This equation also generalizes approximations by finite Fourier series. This work establishes that a single intermediate layer is sufficient for a MLP to compute a uniform approximation for a given training set represented by the set of inputs $x_1, \cdots, x_{m_0}$ and the desired output $f(x_1, \cdots, x_{m_0})$.

The essence of back-propagation learning is encoding the input-output mapping into the synaptic weights and thresholds of an MLP. It is expected that network is well trained.

3   Identifying Gestational Diabetes Mellitus Using the MLP Algorithm

This research adopts the well-known evaluation technique, called k-fold cross-validation, for performance assessment of the ANN-based proposal. In this method, k subsets split the diabetes database. On them, a subset is retained and used as test set. The remaining k-1 subsets constitute the training set. Then, the cross-validation rule repeats this process k times using each subsets exactly once as test set for model validation.

Gestational diabetes mellitus is a condition characterized by increased blood glucose levels (hyperglycemia) and appears for the first time during pregnancy. The condition occurs in approximately 4% of all pregnancies. Usually, this metabolic disease disappears after childbirth. However, women who have suffered from this disease presents a greater risk of contracting type 2 diabetes. Therefore, it is important to continue medical accompaniment after gestation. Any woman can develop diabetes mellitus in pregnancy. Some women groups present higher risk, such as black, Hispanic, Indian, or Asian race. The main risk factors for this disease are the following: age greater than 25 years old, familiar history of diabetes, previous gestational diabetes, previous pregnancies with babies born over 4 kg of weight and with unexplained stillbirth, reduced glucose tolerance or Increased fasting blood glucose, increased amniotic fluid, excess weight before pregnancy, and excessive weight gain in pregnancy.

This study enrolled 394 women with, at least, 21 years old. This dataset belongs to the National Institute of Diabetes and Digestive and Kidney Diseases (NIDDK) [9]. This research disregarded data with missing attributes. The women participating in this research were native people of USA, who live on the banks of Gila and Salt rivers, in the southern part of Arizona state. The Pima Indians have the highest rate of diabetes worldwide. This group also has elevated levels of obesity and hypertension. Table 1 shows the attributes average selected for this study. They represent patients' risk factors and symptoms.

Table 1. Main risk factors and symptoms for gestational diabetes mellitus prediction.

Risk factor or symptoms presented by pregnant woman	Gestational diabetes mellitus
Number of pregnancies	3.6 ± 3.3
Plasma glucose concentration	122 ± 33
Diastolic blood pressure	69.1 ± 18.1 mmHg
Triceps skin fold thickness	23.1 ± 14.7 mm
2-hour serum insulin	$102.1 \pm 126.6\,\mu U/ml$
Body mass index	$33.3 \pm 7.6\,kg/(\text{height in } m^2)$
Diabetes pedigree function	0.5 ± 0.3
Age	32.6 ± 11.5 years

Several intervals divide each attribute to allow the performance evaluation study of the proposed method. Next section discusses the 10-fold cross-validation method and other important indicators to validate this proposal.

4 Performance Evaluation and Results Analysis

This study adopts the well-known assess technique, named k-fold cross-validation, to evaluate the ANN-based proposal. Under this method, the diabetes database is split in k. Among them, a subset is retained and used as test set. The remaining $k-1$ subsets compose the training set. Then, the cross-validation rule repeats this process k times using each subsets exactly once as test set for the model validation. The final result of this process is the average performance of the classifier in the k tests (experiments). Repeating multiple tests increases the reliability of the classifier accuracy estimate. Figure 2 shows an example for $k = 4$ subsets. For this work, the performance evaluation of the predictive accuracy for the MLP classifier uses $k = 10$.

Table 2 shows the experimental results for the diabetes database. The true positive rate (TPR) indicates women with diabetes in both medical evaluation and classification tests. The false positive rate (FPR) represents women who is

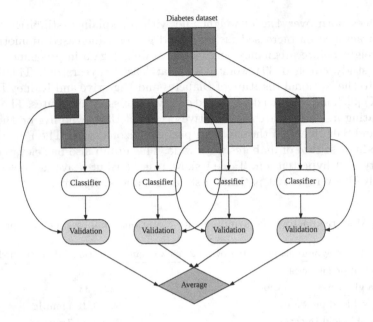

Fig. 2. Cross-validation method schema for $k = 4$ subsets.

not considered with the disease in the medical evaluation but they are considered sick by the classifier. These indicators allow obtaining the Recall, which is equal to the percentage of positive samples classified correctly divided by the total of positive data, and precision, that represents the proportion of positive samples correctly classified divided by the total of cases classified as positive. F-measure is a weighted average between precision and Recall. Class value "1" is interpreted as "tested positive for GDM risk".

Table 2. Performance evaluation of the MLP algorithm by the 10-fold cross-validation method.

	TPR	FPR	Precision	Recall	F-measure	Class
	0.811	0.400	0.805	0.811	0.808	0
	0.600	0.189	0.609	0.600	0.605	1
Weighted Avg.	**0.741**	**0.331**	**0.740**	**0.741**	**0.741**	

For TP Rate, the best possible values are closer to 1 while, for the FP Rate, the best values are close to 0, *i.e.*, this rate refers to the false alarm. As lower is its value, the lowest is the probability that the classifier predicts positive risk of GDM in patients who are not at risk. The presented results show an excellent performance of the MLP algorithm. Other related works show similar performance in the diabetes mellitus prediction. Table 3 shows these results,

Table 3. Performance comparison of the MLP algorithm with the ANN-based algorithm RBF Network, Naive Bayes classifier, based on Bayesian networks, and the decision tree-based C4.5 algorithm.

	Approach	TPR	FPR	Prec.	Rec.	F-measure
Moreira *et al.*	MLP	0.741	0.331	0.740	0.741	0.741
Sa'di *et al.* [15]	RBF Network	0.743	0.381	0.735	0.743	0.737
Kahramanli and Allahverdi [7]	Naïve Bayes	–	–	0.738	–	–
	C4.5	–	–	0.730	–	–

presenting a performance comparison of the MLP algorithm with the ANN-based algorithm RBF Network, Naive Bayes classifier, based on Bayesian networks, and the decision tree-based C4.5 algorithm.

The receiver operating characteristic (ROC) analysis has been used in several areas related to medicine. More recently, this indicator has been introduced to areas such as ML and DM. This curve is obtained by the relation between the TP and FP rates. Figure 3 shows this curve for the class "1" that represents "tested positive for GDM risk" and class "0".

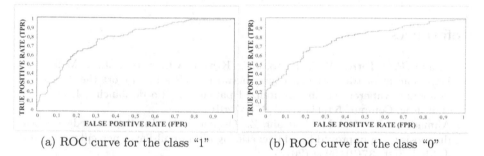

(a) ROC curve for the class "1"   (b) ROC curve for the class "0"

Fig. 3. ROC curve for both classes "1" and "0"; Curves close to the point $(0, 1)$ represent excellent predictive classifiers.

Panchal *et al.* use a method for classifying the accuracy of a classification model by the area under ROC [14]. The MLP algorithm presented ROC area 0.779. According to the authors, it represents a fair performance.

5   Conclusion and Future work

In this paper, an important paradigm of computational intelligence known as ANN was investigated to classify cases of women with risk of developing GDM based on the symptoms and risk factors presented by expectant mothers. The MLP algorithm, which is based on ANNs, was applied and modeled to predict possible pregnancy disorders due to this metabolic disease, presenting an excellent performance in several performance evaluation indicators. This work also introduced the back-propagation algorithm. Using real data and the cross-validation method, the proposed algorithm was evaluated, presenting an excellent performance in all studied indicators. The performance of the ANN model was also compared to other papers in the literature that wrap conventional classifiers based on Bayesian networks, decision trees, and ANNs. The results revealed the model proposed in this study is a suitable solution for the identified problem. Further works suggest the use of other ANN techniques, such as minimal sequential optimization, simple logistic, and radial basis function network. The use of methods to evaluate other disorders that can represent risk for pregnant are also strongly recommended. Reducing diseases that offers risks for both mothers and fetus during pregnancy are part of the United Nations Millennium Development Goals. This work tries to contribute to achieve this goal.

Acknowledgments. This work was supported by the by National Funding from the FCT - Fundação para a Ciência e a Tecnologia through the $UID/EEA/50008/2013$ Project, by Brazilian National Council for Research and Development (CNPq) via Grant No. 309335/2017-5, and by Finep, with resources from Funttel, Grant No. 01.14.0231.00, under the Centro de Referência em Radiocomunicações - CRR project of the Instituto Nacional de Telecomunicações (Inatel), Brazil, and by Ciência sem Fronteiras of CNPq, Brazil, through the process number $207706/2014 - 0$.

References

1. Aguiar, F.S., Torres, R.C., Pinto, J.V., Kritski, A.L., Seixas, J.M., Mello, F.C.: Development of two artificial neural network models to support the diagnosis of pulmonary tuberculosis in hospitalized patients in Rio de Janeiro, Brazil. Med. Biol. Eng. Comput. **54**(11), 1751–1759 (2016)
2. Ahmad, F., Isa, N.A.M., Hussain, Z., Osman, M.K.: Intelligent medical disease diagnosis using improved hybrid genetic algorithm-multilayer perceptron network. J. Med. Syst. **37**(2), 9934 (2013)
3. Claudi, A., Sernani, P., Dragoni, A.F.: Towards multi-agent health information systems. Int. J. E-Health Med. Commun. (IJEHMC) **6**(4), 20–38 (2015)
4. Fard, S.P., Zainuddin, Z.: The universal approximation capability of double flexible approximate identity neural networks. In: Computer Engineering and Networking, pp. 125–133. Springer (2014)
5. Guariguata, L., Linnenkamp, U., Beagley, J., Whiting, D., Cho, N.: Global estimates of the prevalence of hyperglycaemia in pregnancy. Diab. Res. Clin. Pract. **103**(2), 176–185 (2014)
6. Jiwani, A., Marseille, E., Lohse, N., Damm, P., Hod, M., Kahn, J.G.: Gestational diabetes mellitus: results from a survey of country prevalence and practices. J. Matern.-Fetal Neonatal Med. **25**(6), 600–610 (2012)
7. Kahramanli, H., Allahverdi, N.: Design of a hybrid system for the diabetes and heart diseases. Expert Syst. Appl. **35**(1), 82–89 (2008)
8. Li, J., Ray, P., Bakshi, A., Seale, H., MacIntyre, R.: Tool for e-health preparedness assessment in the context of an influenza pandemic. Int. J. E-Health Med. Commun. (IJEHMC) **4**(2), 18–33 (2013)

9. Lichman, M.: UCI machine learning repository. School of Information and Computer Sciences, University of California, Irvine (2013). http://archive.ics.uci.edu/ml. Accessed 14 Apr 2017
10. Liu, V., Caelli, W., Chen, Y.N.M.: Using a public key registry for improved trust and scalability in national e-health systems. Int. J. E-Health Med. Commun. (IJEHMC) 4(4), 66–83 (2013)
11. López Stewart, G.: Diagnostic criteria and classification of hyperglycaemia first detected in pregnancy: a world health organization guideline. Diab. Res. Clin. Pract. **103**, 341–363 (2014)
12. Misra, S.C., Bisui, S., Fantazy, K.: Identifying critical changes in adoption of personalized medicine (PM) in healthcare management. Int. J. E-Health Med. Commun. (IJEHMC) **7**(3), 1–15 (2016)
13. Naraei, P., Abhari, A., Sadeghian, A.: Application of multilayer perceptron neural networks and support vector machines in classification of healthcare data. In: Future Technologies Conference (FTC), December 6–7, San Francisco, CA, USA, pp. 848–852. IEEE (2016)
14. Panchal, I., Sawhney, I., Sharma, A., Dang, A.: Classification of healthy and mastitis murrah buffaloes by application of neural network models using yield and milk quality parameters. Comput. Electron. Agric. **127**, 242–248 (2016)
15. Sa'di, S., Maleki, A., Hashemi, R., Panbechi, Z., Chalabi, K.: Comparison of data mining algorithms in the diagnosis of type II diabetes. Int. J. Comput. Sci. Appl. (IJCSA) **5**(5) (2015). https://doi.org/10.5121/ijcsa.2015.5501
16. WHO: Global report on diabetes. Technical report, World Health Organization, 20 Avenue Appia, 1211 Geneva 27, Switzerland (2016)
17. Widrow, B.: Hebbian learning and the LMS algorithm. In: IEEE 15th International Conference on Cognitive Informatics and Cognitive Computing (ICCI*CC), August 22 23, Palo Alto, CA, USA, p. 2. IEEE (2016)
18. Zhang, Y., Sun, Y., Phillips, P., Liu, G., Zhou, X., Wang, S.: A multilayer perceptron based smart pathological brain detection system by fractional fourier entropy. J. Med. Syst. **40**(7), 1–11 (2016)

Analysis of GLV/GLS Method for Elliptic Curve Scalar Multiplication

Yunqi Dou[1(✉)], Jiang Weng[2], Chuangui Ma[3], and Fushan Wei[1]

[1] State key Laboratory of Mathematical Engineering and Advanced Computing,
Zhengzhou, China
douyunqi@126.com
[2] Information and Navigation College, Air Force Engineering University,
Xi'an, China
[3] Department of Basic, Army Aviation Institution, Beijing, China

Abstract. GLV method is an important research direction to accelerate the scalar multiplication on classes of elliptic curves with efficiently computable endomorphisms, which can reduce the number of doublings by using Straus-Shamir simultaneous multi-scalar multiplication technique. Researchers explore to generalize the method to higher dimension, and then evaluate the effect of accelerating the scalar multiplication. In this paper, we consider various multi-scalar multiplication algorithms, and analyze the computational cost of scalar multiplication under different dimensions to select the optimal multi-scalar multiplication algorithm and parameters. On this basis, the multi-scalar multiplication algorithm is applied to the GLV method, and the computational cost of scalar multiplication is analyzed. Higher dimension usually means fewer doublings, but more precomputation, there is a trade-off. The analysis results show that the limit of GLV method to accelerate the scalar multiplication is dimension 8, and the GLV method will lose its effect of speedup for higher dimension. In particular, dimension 3 or 4 may be the optimal choice for the case that resource constrained or the cost of endomorphism is large.

Keywords: Elliptic curve · Scalar multiplication · GLV method
Multi-scalar multiplication

1 Introduction

The fundamental operation in elliptic curve cryptosystems is scalar multiplication, which dominates the execution time of ECC algorithms. In order to accelerate elliptic curve cryptosystems, various scalar multiplication methods have been developed. Especially, the rapid development of Internet of things and wireless sensor network puts forward higher requirement for the efficiency of the scalar multiplication algorithm. There are two main research hotspots to

Y. Dou—This work is supported by the National Natural Science Foundation of China (No. 61379150, 61309016, 61602512).

accelerate the scalar multiplication: 1. Using multiple base chain to reduce the Hamming weight of scalar [1,2]; 2. Using the efficiently computable endomorphism to reduce the number of doublings by using Straus-Shamir simultaneous multi-scalar multiplication technique [3–5].

In 2001, Gallant et al. [3] proposed a general method (a.k.a. GLV method) to accelerate the scalar multiplication on a class of elliptic curves over prime field with efficiently computable endomorphisms. Given an elliptic curve E over finite field F_p and a point $P \in E(F_p)$ of large prime order n such that $\phi(P) \in \langle P \rangle$, the computation of kP can be decomposed as $k_1 P + k_2 \phi(P)$ where $|k_1|, |k_2| \approx \sqrt{n}$. This immediately enables the elimination of half the doublings by using the Straus-Shamir trick for simultaneous scalar multiplication. Suppose that ϕ satisfies the characteristic polynomial $X^2 + rX + s$, there exists $\lambda \in [0, n-1]$ such that $\phi(P) = \lambda P$, where λ is a root of $X^2 + rX + s \bmod n$. The integers k_1 and k_2 can be computed by solving a closest vector problem in a lattice $L = \{(x, y) \in \mathbb{Z}^2 : x + y\lambda \equiv 0 \bmod n\}$.

There is a vast literature on GLV method to speed up scalar multiplication. On the one hand, one expects to find more efficiently computable endomorphisms to extend the applicable range of the GLV method. On the other hand, one explores to generalize the GLV method to higher dimension for further reducing the number of doublings. At present, the efficiently computable endomorphisms are constructed only on a few kinds of elliptic curves [5], and it is a challenging task to seek such endomorphisms. In 2009, Galbraith et al. [4] presented a method to construct an efficiently computable endomorphism on elliptic curve $E(F_{p^2})$ by exploiting the action of Frobenius map, where $E(F_{p^2})$ is a quadratic twist of elliptic curve defined over F_p. Combing the result with the GLV method, they generalized the GLV method to a large class of elliptic curves over F_{p^2} (a.k.a GLS curves).

In 2010, Zhou et al. [6] discovered two distinct efficiently computable endomorphisms on a subclass of GLS elliptic curves, by which they generalized the GLV method to dimension 3 on GLS elliptic curves with j-invariant 0. In 2012, Hu et al. [7] presented a 4-dimensional GLV method for faster scalar multiplication on some GLS curves with j-invariant 0. Their implementation result shows the 4-dimensional GLV method is roughly 22% faster than the 2-dimensional GLV method on the same curve. In 2014, Longa and Sica [5] generalized the GLS method to all GLV curves. By using endomorphisms Φ and Ψ over F_{p^2}, they constructed a 4-dimensional decomposition

$$kP = k_1 P + k_2 \Phi(P) + k_3 \Psi(P) + k_4 \Phi\Psi(P), \text{with } \max_i\{|k_i|\} \leq Cn^{1/4},$$

for some explicitly computable $C > 0$. They also presented an efficient algorithm to compute 4-dimensional decomposition of k, consisting in two applications of the extended Euclidean algorithm in $\mathbb{Z}$ and $\mathbb{Z}[i]$.

The motivation of this paper is not to seek more efficiently computable endomorphisms or to generalize the GLV method to higher dimension. But from another point of view, we expect to evaluate the performance of GLV method under different dimensions. While implementing the scalar multiplication using

GLV method, one expects to select the optimal multi-scalar multiplication algorithm and parameters, which is related to many factors, such as the dimension of multi-scalar multiplication, length of scalar, memory constraints and so on. Firstly, we evaluate the efficiency of some main multi-scalar multiplication algorithms. The result shows that the w-NAF-based interleaving method is optimal in most cases. Considering the precomputation process of multi-scalar multiplication algorithm, we evaluate the influence of GLV dimension on the performance of scalar multiplication. With the increase of dimension of multi-scalar multiplication, its speedup is less and less obvious. The GLV method above dimension 8 almost fails to accelerate the scalar multiplication. Finally, for the cases that resource constrained or the cost of endomorphism is large, dimension 3 or 4 may be the optimal choice.

The sequel of the paper is organized as follows: In Sect. 2, we review some background about elliptic curve cryptography and provide a brief description for some multi-scalar multiplication algorithms. Section 3 compares the multi-scalar multiplication algorithms from a theoretical point of view. Section 4 analyzes the limit of GLV method for elliptic curve scalar multiplication. Finally, we conclude the paper in Sect. 5.

2 Preliminaries

2.1 Elliptic Curve Cryptography

This section presents a brief introduction to the basic knowledge of elliptic curve. The reader can refer to [8,9] for details. Let $p > 3$ be a prime, elliptic curve $E : y^2 = x^3 + ax + b$ is defined over the finite field F_p, where $a, b \in F_p$. The rational points on the elliptic curve E form an abelian group, in which the point at infinity $\mathcal{O}$ plays the role of identity. The two basic group operations: addition $(P + Q)$ and doubling $(2P)$ are defined by the well-known law of chord and tangent.

Let M and S denote the cost of field multiplication and squaring respectively, and the costs of field addition, field subtraction and multiplication by small constants are ignored. This article mainly focuses on the Weierstrass curves using Jacobian coordinates with $a = -3$ (Jacobian-3). We consider the following operations: doubling $(2P)$, tripling $(3P)$, addition $(P + Q)$ and mixed addition which are denoted by DBL, TPL, ADD and mADD, respectively. The costs of state-of-the-art point formulae can refer to [10].

2.2 Multiple Scalar Multiplication

Algorithm 1 presents an extended version of simultaneous multi-scalar multiplication algorithm, which corresponds to the joint binary representation while $w = 1$.

Yen et al. [11] improved simultaneous 2^w-ary method to propose the simultaneous sliding window method (Algorithm 2). Due to the use of sliding window, at least one of i_j is odd in $(i_1, i_2, \cdots, i_n)$, which reduces the number of points in

---

Algorithm 1. Simultaneous 2^w-ary method

Input: integer n, scalar k_i, point P_i, window width w, $1 \le i \le n$

Output: $\sum_{i=1}^{n} k_i P_i$

1. precompute $i_1 P_1 + \cdots + i_n P_n$, for all $i_1, \cdots, i_n \in \{0, 1, \cdots, 2^w - 1\}$.
2. compute the w-NAF representation of k_i: $\text{NAF}_w(k_i) = \sum_{j=0}^{b-1} k_{i,j} 2^j$, for $1 \le i \le n$.
3. $R \leftarrow \mathcal{O}$
4. For i from $b - 1$ downto 0 do
5. $\quad R \leftarrow 2^w R$
6. $\quad R \leftarrow R + (k_{1,i} P_1 + \cdots + k_{n,i} P_n)$
7. return R

---

precomputation stage and additions in evaluation stage. However, the precomputation of above two methods involves combinations of the points. With the increase of dimension of multi-scalar multiplication, the number of precomputed points increases exponentially and the performance of scalar multiplication will decrease.

Möller [12] proposed w-NAF-based interleaving method (Algorithm 3), in which the precomputation involves only a single point. In the calculation of $k_1 P_1 + k_2 P_2 + \cdots + k_n P_n$ ($n \ge 2$), the interleaving algorithm allows different widths w_i to be used for each scalar k_i.

Solinas [13] proposed a joint sparse form (JSF) of two integers

$$\binom{k}{l} = \binom{k_{d-1} \cdots k_0}{l_{d-1} \cdots l_0}_{\text{JSF}}$$

such that k_i and l_i satisfy certain properties. The author also presents an algorithm to compute JSF representation of integers k and l, and demonstrates that its joint Hamming weight is minimal among all signed binary representations of the same pair of integers. Its average density is $1/2$. Using the precomputed points $P + Q$ and $P - Q$, the algorithm requires $d + 1$ doublings and $\frac{d}{2}$ additions on average to calculate $kP + lQ$. In 2009, Doche et al. [14] proposed joint double-base chain based on the concept of double-base number system. Since the method focuses on the double scalar multiplication, we only present its computational cost in Table 1.

3   Efficiency Estimation on Multi-scalar Multiplication

The multi-scalar multiplication algorithms considered in this section includes two phases: the precomputation stage and the evaluation stage. Firstly, we compare the multi-scalar multiplication algorithms above from several aspects, including the number of precomputed points, the computational costs of precomputation stage and evaluation stage. Considering the calculation of $k_1 P_1 + k_2 P_2 + \cdots + k_n P_n$, we suppose the bit length of scalar k_i is similar and d is the maximum. Algorithms 4 and 5 represent the scalar multiplication algorithms based on joint sparse representation and joint double base chain representation, respectively.

Table 1 compares the multi-scalar multiplication algorithms (window width $w >$ 1), where #S denotes the number of precomputed points, the precomputed points include $P_1, P_2, \cdots, P_n$.

Table 1. Comparison of multi-scalar multiplication algorithms

Algorithm	#S	Precomputation stage	Evaluation stage
Algorithm 1	$2^{nw} - 1$	$(2^{n(w-1)} - 1)\text{DBL} +$ $(2^{nw} - 2^{n(w-1)} - n)\text{ADD}$	$\lceil d/w \rceil \cdot w\text{DBL} +$ $\frac{d}{w} \cdot \left(1 - \frac{1}{2^{nw}}\right)\text{ADD}$
Algorithm 2	$2^{nw} - 2^{n(w-1)}$	$n\text{DBL} +$ $(2^{nw} - 2^{n(w-1)} - n)\text{ADD}$	d DBL + $d/\left(w + \frac{1}{2^n - 1}\right)\text{ADD}$
Algorithm 3	$n \cdot 2^{w-1}$	$n\text{DBL} +$ $n(2^{w-1} - 1)\text{ADD}$	d DBL + $\frac{nd}{w+2}\text{ADD}$
Algorithm 4	4	2ADD	$(d+1)\text{DBL} +$ $\left(\frac{1}{2}d + 1\right)\text{ADD}$
Algorithm 5	4	2ADD	$0.55d$ DBL + $0.28d$ TPL + $0.3945d$ ADD

Note: Algorithms 4 and 5 are only for the case $n = 2$; In Algorithm 3, the same window width w is used.

According to the random elliptic curves recommended by NIST in the FIPS 186-2 standard, we consider the scalar k to be 160 bits to 512 bits. Firstly, Table 2 compares the costs of Algorithms 2 and 3 for the case $n = 2$, in which the optimal parameter d is given in parentheses.

Table 2. Comparison of Algorithms 2 and 3 for $n = 2$

w	Algorithm 2		Algorithm 3	
	#S	Cost	#S	Cost
1	3	$14.65d + 10.2$	2	$13.8d$
2	12	$11.37d + 116$ (80–256 bits)	4	$12.1d + 34.4$
3	–	–	8	$11.08d + 75.2$ (80–120 bits)
4	–	–	16	$10.4d + 156.8$ (121–256 bits)

Assume $n = 2$, d is about 80–256 bits. For Algorithm 2, the optimal choice of w is 2. For Algorithm 3, the optimal choice of w is $w = 3$ for $80 \leq d < 120$, and $w = 4$ for $120 \leq d \leq 256$. For Algorithm 4 and 5, the number of precomputed points is 4, the computational cost is $(12.2d + 37.6)$M and $(11.4d + 20.4)$M, respectively. Therefore, the Algorithm 3 is more efficient than Algorithms 2, 4 and 5 in general. However, the Algorithm 5 may be more efficient than Algorithm 3 for resource constrained case. Figure 1 compares the costs of multi-scalar multiplication algorithms, (a) shows the case that storage points are limited to

4, (b) shows the case that there is no storage limit, where the abscissa represents the bit length of scalar k, and the ordinate represents the computational cost.

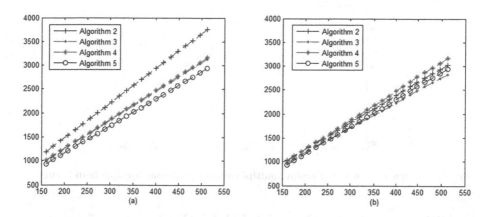

Fig. 1. Comparison of multi-scalar multiplication algorithms for Jacobian-3 curves

When $n > 2$, the Algorithm 3 is more efficient than Algorithm 2. Suppose the single scalar multiplication can be converted into multi-scalar multiplication of different dimensions by the GLV method, Table 3 compares the costs of Algorithm 3 under different dimensions.

Table 3. Comparison of costs of Algorithm 3 under different dimensions

w	$n = 3$		$n = 4$		$n = 6$		$n = 8$	
	#S	Cost	#S	Cost	#S	Cost	#S	Cost
1	3	$17.2d$	4	$20.6d$	6	$27.4d$	8	$34.2d$
2	6	$14.65d + 51.6$	8	$17.2d + 68.8$	12	$22.3d + 103.2$ (27–40 bits)	16	$27.4d + 137.6$ (20–45 bits)
3	12	$13.12d + 112.8$ (54–120 bits)	16	$15.16d + 150.4$ (40–120 bits)	24	$19.24d + 225.6$ (41–86 bits)	32	$23.32d + 300.8$ (46–64 bits)
4	24	$12.1d + 235.2$ (121–170 bits)	32	$13.8d + 313.6$ (121–128 bits)	–	–	–	–

The theoretical analysis shows that the efficiency of multi-scalar multiplication is related to the dimension, length of scalar and window width. In general, the Algorithm 3 is optimal. The efficiency of the Algorithm 3 can be improved with the increase of dimension, but the growth rate becomes smaller. When the bit length of scalar is 160–256 bits, the 8-dimensional multi-scalar multiplication algorithm is about 2% faster than the 6-dimensional algorithm. Even if the bit length is 512 bits, speedup is only about 4%, as shown in Fig. 2.

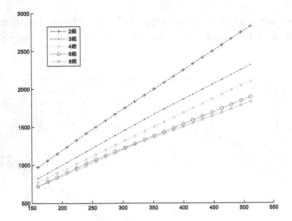

Fig. 2. Comparison of multi-scalar multiplication algorithms for Jacobian-3 curves

4   Efficiency Evaluation of GLV Method

From the analysis of Sect. 3, we have known that/ the Algorithm 3 is optimal while implementing multi-scalar multiplication. In this section, we only consider Algorithm 3 and carry out more detailed analysis. We expect to analyze the cost of Algorithm 3 and the optimal parameter choices under different dimensions. Finally, we analyze the limit of GLV method for elliptic curve scalar multiplication.

To compute $kP = \sum_{i=1}^{n} k_i \psi^{i-1}(P)$, we require three stages: precomputation stage, evaluation stage and coordinate conversion, whose computational costs are denoted as C_{precomp}, $C_{\text{evaluation}}$ and C_{Affine} respectively. Therefore, the total cost is

$$\text{Cost} = C_{\text{precomp}} + C_{\text{evaluation}} + C_{\text{Affine}}.$$

Just as Sect. 3, we need to store the same number of precomputed points in the precomputation stage. However, the efficiently computable endomorphism can be used to reduce the computational cost of precomputation if the endomorphism is more efficient than point addition. In this case we need to precompute $P_j = jP, j\psi^i(P) = \psi^i(P_j)$ for $j \in \{1, 3, \cdots, 2^w - 1\}, i \in \{1, \cdots, n - 1\}$. In order to use the mixed addition formulas in evaluation stage, the precomputed points need to be converted to affine coordinates. Following the precomputation scheme in [15], only one inversion operation is needed. Denote by H the cost of performing the endomorphism ψ, the cost of precomputation stage for Jacobian-3 curves is $1I + 9 \cdot (2^{w-1} - 1)M + (2^w + 4))S + 2^{w-1}(n - 1)H$.

In general, I/M is considered to be 50–100 for prime filed [9]. We suppose I/M = 80 in this paper. Finally, at the end of scalar multiplication the result needs to be converted to affine. The cost is $1I + 3M + 1S$ for the Jacobian-3 curve. The total cost of scalar multiplication for the Jacobian-3 curves is related to the parameters n, w, H, I and d.

According to the costs of point operations on the Jacobian-3 curve, the total cost to calculate kP is

$$\text{Cost} = \begin{cases} \text{I} + (n-1)\text{H} + (3.4nd + 4.08n + 7d + 3.8)\text{M}, w = 1 \\ 2\text{I} + 2^{w-1}(n-1)\text{H} + \left(5.3 \cdot 2^w + 10.2 \cdot \frac{nd}{w+2} + 4.08n + 7d - 2\right)\text{M}, w > 1 \end{cases}$$

Figure 3 compare the costs of GLV scalar multiplication of different dimensions, which corresponds to the cases that H = 1M, H = 4M, H = 8M and H = 12M respectively. With the increase of computational cost of endomorphism, the speedup of high-dimensional GLV method is less and less obvious with respect to that of low dimension. When $4\text{M} \le \text{H} \le 12\text{M}$, the GLV method above 8-dimensional cannot accelerate the scalar multiplication any more. Therefore, the limit of GLV method to accelerate the scalar multiplication is 8-dimensional. As far as the current multi-scalar multiplication algorithm is concerned, there is no need to seek the higher dimensional GLV method.

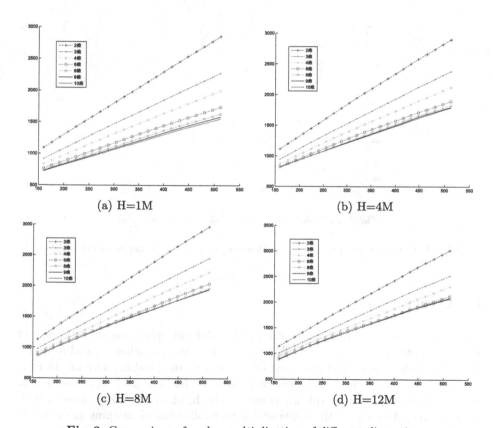

(a) H=1M   (b) H=4M

(c) H=8M   (d) H=12M

Fig. 3. Comparison of scalar multiplication of different dimensions

For some elliptic curves ([5, Example 3, 4, 5, 6]), the efficiently computable endomorphisms may be complex, which involve the inversion operation.

Assuming $H = I = 80M$, we have to change the precomputation scheme: first precompute $Q_i = \psi^i(P)$ for $i \in \{1, \cdots, n-1\}$, and then compute the jP, jQ_i for $i \in \{1, \cdots, n-1\}, j \in \{3, 5, \cdots, 2^w - 1\}$ using the scheme in [15]. In this case, the precomputation cost is

$$C_{\text{precomp}} = nI + 9n \cdot (2^{w-1} - 1)M + n(2^w + 4)S + (n - 1)H.$$

The total cost is

$$\text{Cost} = \begin{cases} I + (n-1)H + (3.4nd + 4.08n + 7d + 3.8)M, w = 1 \\ (n+1)I + (n-1)H + \left(5.3 \cdot 2^w + 10.2 \cdot \frac{nd}{w+2} - 1.72n + 7d + 3.8\right)M, w > 1 \end{cases}$$

As shown in Fig. 4, the costs of scalar multiplication under different dimensions are estimated. In this case, the GLV method does not require high dimension: $n = 4$ is optimal for scalar of 160–350 bit length, $n = 3$ for 350–512 bit length.

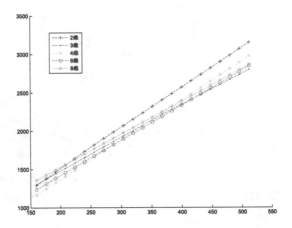

Fig. 4. Comparison of scalar multiplication of different dimensions ($H = 80M$)

5   Conclusion

In this paper, we first analyze several multi-scalar multiplication algorithms, and then estimate the efficiency of GLV based scalar multiplication algorithm under different dimensions. When the cost of efficiently computable endomorphism is large, the 3-dimensional or 4-dimensional GLV method has the best performance. When the cost of endomorphism is small, the limit of the GLV method is 8-dimensional. At present, the multi-scalar multiplication algorithms are efficient for the case of low dimension. With the increase of dimension, the efficiency is greatly affected. How to improve the efficiency of multi-scalar multiplication algorithm with higher dimension is still a problem to be studied.

References

1. Dimitrov, V., Imbert, L., Mishra, P.K.: Efficient and secure elliptic curve point multiplication using double-base chains. In: Roy, B. (ed.) ASIACRYPT 2005. LNCS 3788, pp. 59–78. Springer-Verlag (2005)
2. Longa, P., Gebotys, C.: Fast multibase methods and other several optimization for elliptic curve scalar multiplication. In: Jarecki, S., Tsudik, G. (eds.) PKC 2009. LNCS, vol. 5443, pp. 443–462. Springer, Heidelberg (2009)
3. Gallant, R.P., Lambert, J.L., Vanstone, S.A.: Faster point multiplication on elliptic curves with efficient endomorphisms. In: Kilian, J. (ed.) Advances in Cryptology - Proceedings of CRYPTO 2001, LNCS 2139, pp. 190–200. Springer-Verlag (2001)
4. Galbraith, S.D., Lin, X., Scott, M.: Endomorphisms for faster elliptic curve cryptography on a large class of curves. J. Cryptol. $24(3)$, 446–469 (2011)
5. Longa, P., Sica, F.: Four-dimensional Gallant-Lambert-Vanstone scalar multiplication. J. Cryptol. $27(2)$, 248–283 (2014)
6. Zhou, Z., Hu, Z., Xu, M., Song, W.: Efficient 3-dimensional GLV method for faster point multiplication on some GLS elliptic curves. Inf. Process. Lett. $77(262)$, 1003–1106 (2010)
7. Hu, Z., Longa, P., Xu, M.: Implementing 4-dimensional GLV method on GLS elliptic curves with j-invariant 0. Des. Codes Crypt. $63(3)$, 331–343 (2012)
8. Blake, I.F., Seroussi, G., Smart, N.P.: Advances in Elliptic Curve Cryptography. Cambridge University Press, Cambridge (2005)
9. Hankerson, D., Menezes, A., Vanstone, S.: Guide to Elliptic Curve Cryptography. Springer-Verlag (2004)
10. Bernstein, D.J., Lange, T.: Explicit-formulas database. http://hyperelliptic.org/EFD
11. Yen, S.-M., Laih, C.-S., Lenstra, A.K.: Multi-exponentiation. IEE Proc. Comput. Digit. Tech. $141(6)$, 325–326 (1994)
12. Möller, B.: Algorithms for multi-exponentiation. In: Vaudenay, S., Youssef, A.M. (eds.) Selected Areas in Cryptography - SAC 2001, LNCS 2259, pp. 165–180. Springer, Heidelberg (2001)
13. Solinas, J.A.: Low-weight binary representations for pairs of integers. Combinatorics and Optimization Research Report CORR 2001-41. Centre for Applied Cryptographic Research, University of Waterloo (2001)
14. Doche, C., Kohel, D.R., Sica, F.: Double-base number system for multi-scalar multiplications. In: Joux, A. (ed.) EUROCRYPT 2009. LNCS 5479, pp. 502–517. Springer-Verlag, Berlin (2009)
15. Longa, P., Miri, A.: New composite operations and precomputation scheme for elliptic curve cryptosystems over prime fields. In: Cramer R. (ed.) PKC 2008. LNCS 4939, pp. 229–247. Springer, Heidelberg (2008)

An Improved Algorithm for Facial Feature Location by Multi-template ASM

Li Benfu[⊠]

Southern Medical University, Guangzhou, China
da3831182neixi@163.com

Abstract. In order to improve the accuracy of the Shape Model Active method, we propose a new method to improve the accuracy of ASM (ASM) algorithm in face detection, and propose a new method to construct the local template. In the process of local localization, the paper uses form Closed-algorithm to segment the texture segmentation. Information is effectively improved the performance of the ASM method. The results show that the proposed algorithm can extract the feature points of most forward faces correctly. The proposed algorithm has a wide range of applications in image understanding of face tracking, recognition and facial expression analysis.

Keywords: Active shape model (ASM) · Multi-template · Narrow strip-map Closed-form algorithm

The active shape model (active shape model, ASM) algorithm [1] is one of the active vision algorithms. Because of its flexible adaptation to target contour, and to some extent, it can be used to detect the flexible object. It is the main face feature detection algorithm. The ASM algorithm [2] is based on the original ASM algorithm, and then uses the local template to achieve global and local optimization. The point Prof (ILE) is used to match the pixel value gradient information and training template. Therefore, it is difficult to make accurate segmentation of the region of the edge of the feature, such as the fuzzy edge region of the jaw and neck region or other features. For this problem, the Liuaiping et al. [3] uses the discrete cosine transform to model the feature points of the training image, and makes full use of the 2D texture information near the feature points; Yuhua et al. [4] use Gabor transform to model the local texture around the feature points and improve the robustness of the algorithm to illumination and noise; Li [5] et al. In the training process of multi template ASM algorithm, the Gabor feature information is used in the same direction as the feature edge, and the different states of the eye and mouth are set up respectively; Cristinacce et al. [6] The improved ASM algorithm is used to improve the search efficiency of the model. The improved model can improve the accuracy of the feature point location and reduce the influence of illumination and noise. However, the texture feature information is not rich, so they have little effect on the detection accuracy. Toth et al. [7] is used to implement the segmentation of magnetic resonance image(magnetic resonance image, MRI) first, and then use the ASM algorithm to extract the feature contour, which is suitable for the segmentation of the texture features of the smooth region, but only for the segmentation of MRI.

© Springer Nature Singapore Pte Ltd. 2018
J. C. Hung et al. (Eds.): FC 2017, LNEE 464, pp. 220–230, 2018.
https://doi.org/10.1007/978-981-10-7398-4_24

In this paper, we improve the traditional ASM algorithm for locating face image texture smoothing region. The 1 global templates and 7 local templates are established. The feature points are located near the feature region. Then, the -form Closed algorithm is used to segment the Profile vertices. Matching Because the -form Closed algorithm has the characteristics of smooth region segmentation, the feature edge of the smooth region is highlighted, which can improve the efficiency and accuracy of the algorithm.

1 Closed-Form Algorithm Analysis

Closed-form algorithm proposed by [8] and other Levin can be used to solve the problem of image segmentation. The algorithm is described as the I of any image, the Ii of any of its pixels are composed of foreground and background, and the proportion of the foreground pixels is α_i, that is $I_i = \alpha_i F_i + (1 - \alpha_i)B_i$. Thus available $\alpha_i \approx a_i I_i + b_i, \forall i \in W$; Among them $a_i = 1/(F_i - B_i), b_i = -B_i/(F_i - B_i)$, W for the image window. By minimizing

$$J(\alpha, a, b) = \sum_{j \in I} \left(\sum_{i \in W_j} (\alpha_i - a_j I_i - b_j)^2 + \varepsilon a_j^2 \right) \tag{1}$$

Can be obtained from the ratio of the foreground pixels A, Wj for the first j pixels of the image window.

By the segmentation results and the analysis of the [8], the Closed-form algorithm can be seen in the segmentation of the image smooth part. By the segmentation results and the analysis of the [8], the Closed-form algorithm can be seen in the segmentation of the image smooth part, (1) before the segmentation, the user interaction is needed to draw the background and foreground of the graph Fig. 1b (white as the foreground, black background), and the automation can not be completed; (2) Because of each pixel are required in to the center of the image window W formula (1), so the computing complexity is high, the high resolution image computation for a long time; (3) the sensitivity of the foreground and background edge noise is high, except for the special setting parameters, it is easy to produce burr or hole in the edge.

a Original   b User mark   c Segmentation result

Fig. 1. Form-closed algorithm partition diagram

To solve the first problems, this paper applies the ASM feature extraction to find the 2 point (Profile) as the initial marker, which is not required to be used as a result of the second questions, which have little effect on the computation time. Because the ASM can maintain a certain topology, it can better resist the segmentation edge. Based on the above 3 points, we can effectively integrate the -form Closed algorithm into ASM.

2   The Establishment of Multi-template ASM

In this paper, a multi-template is made up of 1 global templates and 7 local templates. The global template is used to locate all the feature points of the face image, and the local template is used to locate the local features after global positioning.

2.1   Selection of Feature Points of Multi-template ASM

The feature points are selected from the-3 CANDIDE model [9], which is used to establish the global template, which is based on the global template, which is based on the global template, which is based on the global template, which can be used to establish the local template, which is based on Fig. 2b. A feature point is formed between the 2 feature points on the line, and the local template is formed by 100 C, which can maintain the information of the feature points between the global template and the local template, and enhance the prior information of the feature points.

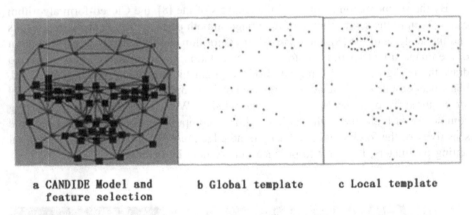

| a CANDIDE Model and feature selection | b Global template | c Local template |

Fig. 2. ASM template

2.2   Feature Points of the ASM Feature Points

According to Fig. 2b, the template of the Fig. 2c to the corresponding position of the training image is manual calibration, and the training image after the following standard processing: (1) Eliminate the bad quality of the image; (2) Select the face relative to the lens deflection is not more than 10b of the image; (3) Select the face image without wearing glasses, earrings and other accessories; (4) segmentation process, using the image of the training only to retain the face, neck and other background are set to 0, so as to better achieve the model and narrow band segmentation area on the Profile pixel matching.

After manual calibration of the image as shown in Fig. 3c, d, the global template feature points are obtained $S_i = (x_{i1}, y_{i1}, x_{i2}, y_{i2}, \cdots, x_{iN}, y_{iN})^T$ and local template feature points $S_{LOCAL_i} = (x_{i1}, y_{i1}, x_{i2}, y_{i2}, \cdots, x_{i2N}, y_{i2N})^T, i = 1, 2, \cdots, M$. Among them M, where I represents the first I image, N indicates the number of feature points. N indicates the number of feature points. Divide the feature points of the 7 into 7 regions $S_1, S_2, \cdots, S_7$ according to the feature points of the face in $S_{LOCAL_i}, i = 1, 2, \cdots, M$ regions. Respectively represent the left and right eyebrow, right and left eyes, nose, lips and facial contour. Through principal component analysis, to obtain the shape of the global template and seven local template description (feature sub space). Unified mind $M = \bar{S} + Pb$, among them, $\bar{S}$ is the average shape of the template, $P = (p_1, p_2, \cdots, p_t)$ is a set of template shape feature subspace, $b = (b_1, b_2, \cdots, b_t)$ is a shape parameter, different B corresponding to different shapes.

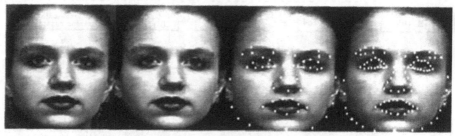

a Original image b Training image   c Global template   d Localtemplate
                                        feature point   feature point

Fig. 3. Original images and training images

Because the shape description is not always consistent with the true shape, it is required to obtain the local gray level information near the training process in order to adjust the shape during the matching process. Take the $2K + 1$ gray level sample $G_{ij} = [g_{ij0}, g_{ij1}, \cdots, g_{ij(2K)}]$ on the j Profile of the image i. The number of pixels selected from the K (K = 3) to the normal or outward direction of the normal, G_{ij} for differential, $dg_{ij} = (g_{ij2} - g_{ij1}, g_{ij4} - g_{ij3}, \cdots, g_{ij(2K+1)} - g_{ij(2K)})^T$, and its standardization $y_{ij} = dg_{ij} / \left(\sum_{l=1}^{2K} |dg_{ijl}| \right)$.

Calculate the gray difference mean and covariance matrix of the j feature points $\bar{g}_j = \frac{1}{N} \sum_{i=1}^{N} y_{ij}, \sum_j = \frac{1}{N} \sum_{i=1}^{N} (y_{ij} - \bar{y}_j)(y_{ij} - \bar{y}_j)^T$, the mean $\bar{g}_j$ and covariance matrix $\sum_j$ are preserved as the training results of this point.

3 Improved Multi-template ASM Algorithm

Specific steps of the algorithm as shown in Fig. 4. First of all to face feature using global template gives an overall positioning, again on the feature regions using template partial local location. In the process of localization, first find out where the profile vertex of the template feature points of a narrow zone; then will all feature points according to the order of the head and tail connected form a polygon, point to the inside the polygon normal to normal with profile, the profile vertex as a seed point initialization position; then with profile pointing outside the polygon normal to normal, the p As the initial position of the background seed point, the Closed-form algorithm is used to segment the foreground pixels in the narrow strip region and the background pixels are 0. Finally, the training template and the segmented images are matched. If the convergence condition is satisfied, the narrow strip is constructed by the new feature points Profile.

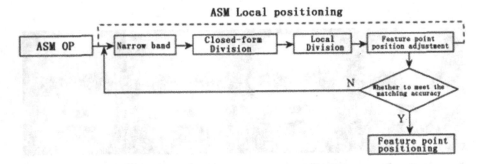

Fig. 4. Algorithm implementation block diagram

3.1 Overall Positioning

First overall positioning, so that local positioning stage of regional features of foreground and background markers initialization is more accurate and rational. First the ASM to initialize the location, according to the center of the left eye, right eye and lip center 3 respectively form model of triangle positioning and image positioning triangles, as shown in Fig. 5. The triangles of the model and the target image of the triangle, using triangular geometric features will align the model triangle and the triangle image, model is placed in the most close to the position of the target image to complete the initialization steps, then ASM overall positioning.

ASM overall positioning process is shown in Fig. 6. In Fig. 6, square as feature points, dots to the midpoint of the characteristic line, dotted lines feature points on the line for the profile of the i feature points on the $2K+1$ gray sampling $G_{ij} = [g_{ij0}, g_{ij1}, \cdots, g_{ij(2K)}]$. K is the number of pixels selected from the normal or the outward direction of the normal. The 3–4 is usually the one that affects the topological properties of the model. The triangle at the top of the dotted line is a Profile vertex, which is used to construct the narrow band and is used as a marker of the foreground and background. The I of the Profile of the feature points on the gray level sampling $G_{ij} = [g_{ij0}, g_{ij1}, \cdots, g_{ij(2K)}]$ for differential and normalized by $c_i = [c_{i0}, c_{i1}, \cdots, c_{i(2K)}]$,

By minimizing the objective function $f(c_i) = (c_i - \bar{g}_i)^T - \sum_i^{-1}(c_i - \bar{g}_i)$, the best matching points are obtained by minimizing the feature points on the Profile. In Fig. 6, the cross point is searched to the best matching point, where $\sum_i^{-1}$ is the inverse of the gray covariance matrix of the I feature points obtained from the training set, and GI is the I of the Profile feature points.

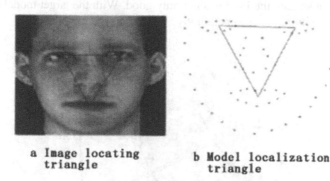

a Image locating triangle

b Model localization triangle

Fig. 5. Triangle registration

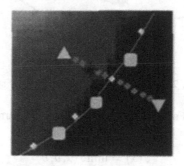

Fig. 6. Positioning process map

3.2 Local Positioning

Overall positioning is a global optimization, when the facial images in the presence of multiple feature details, some local characteristics of regional feature points may due to the role of global optimization and to the lack of accurate positioning, so after a global template search and location can be localized on the basis of global localization, localization process such as dashed lines in Fig. 4 of the box part.

3.2.1 Narrow Band Structure and -Form Closed Segmentation

For Fig. 4 in the narrow strip belt construction such as shown in Fig. 7. Figure 7 cloud form graphics image into foreground, dot in the foreground edge feature points dotted lines for feature points of attachment, and dotted vertical arrow said ASM feature points in the midpoint method to that profile. Profile (figure of hexagon and hexagon) vertex as a narrow strip of edge point, is connected to all the edge points, get surrounded by two solid narrow band region. In addition, in the foreground and background markers, because there have been overall positioning as a basis, feature point basic positioning in The assumption is that all normal directions point to feature points (Closed) in the -form Profile algorithm, which is based on the assumption that the foreground markers are in the foreground. Even though the above assumption is based on the feature points matching and adjustment, the Profile is not only good. With the target model matching, the narrow band will continue to be iterative tending to the hypothesis.

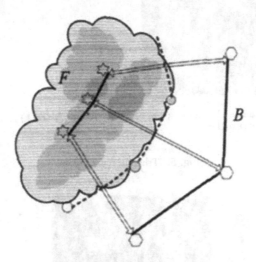

Fig. 7. The composition of the narrow strip

Then the -form Closed segmentation is done for the narrow strip region, pass type (1) solving

$$J(\alpha) = \min_{a,b} J(\alpha, a, b) \qquad (2)$$

By type (2), the A, rewrite formula (1) is obtained for the least squares form of a, b that is

$$J(a, b) = \sum_{j \in I} \Pi^{G_j} \begin{bmatrix} a_j \\ b_j \end{bmatrix} - \bar{\alpha}_{j\Pi}^2 \qquad (3)$$

One of the first j pixels of the image window W, there is

$$G_j = \begin{bmatrix} I_1 & I_2 & \cdots & I_i & \cdots & \sqrt{\varepsilon} \\ 1 & 1 & \cdots & 1 & \cdots & 0 \end{bmatrix}^T$$

$$\bar{\alpha}_j = [\alpha_1 \quad \alpha_2 \quad \cdots \quad \alpha_i \quad \cdots \quad 0]^T, i \in W_j$$

By minimizing (3)a, b. See literature [8].

3.2.2   Local Matching Search

Local positioning algorithm using the local template serial introduction, and through the local matching search, adjust the overall positioning results, steps are as follows:

Step 1. For $j = 1$, the initial shape description of a local template is introduced into the J.

Step 2. Adjust the position, size and angle of the local template, so that it is consistent with the global template corresponding to the global template.

Step 3. The feature points are constructed from the local feature points, and the Closed-form algorithm is used to segment the feature region.

Step 4. Local matching of all the features of a local template for the first j, that is, the first j feature points of the template I, read into the training data (mean $\bar{g}_{ij}$ and covariance matrix $\sum_{ij}$). Get the feature points corresponding to the Profile of the image gray level sampling, and then the difference is obtained $c_{ji} = [c_{ji0}, c_{ji1}, \cdots, c_{ji(2K)}]$. By minimizing the objective function $f(c_{ji}) = (c_{ji} - \bar{g}_{ji})^{T} - \sum_{ji}^{-1}(c_{ji} - \bar{g}_{ji})$. Get the best matching point on the image, and then adjust the feature points to the best matching point.

Step 5. If the convergence condition is satisfied, the positioning result of the feature region is adjusted to the best matching point, and the next step is to the Step 3. If the convergence condition is satisfied, the positioning result of the feature region is adjusted to the best matching point, and the next step is to the Step 3.

Step 6. Checks if all local templates have been introduced, and if, in the next step, the initial shape of the $j + 1$ local template is introduced, otherwise, the Step 2.

Step 7. Will get the results in the overall positioning of the global template shape feature subspace M to find the optimal shape.

Step 8. Output of the new positioning results.

4   Experiment and Result Analysis

In this paper, the training images are derived from the ORL database. We select 200 face images as the training sample set, which is used to detect the remaining images and some standard video sequences from the ORL database. The size of the unified 100×100 experiments are carried out on the Matlab platform, Some simulation results and experimental data analysis are as follows.

4.1   Simulation Result Analysis

Figure 8 shows the algorithm simulation results, Fig. 8a shows the results of the overall positioning, can be seen, the point of the chin is not convergence in place; Fig. 8b, c is shown as a partial localization process results from Fig. 8c can be seen that the chin point has been converging to the edge of the face. Figure 8 shows the algorithm simulation results, Fig. 8a shows the results of the overall positioning, can be seen, the point of the chin is not convergence in place; Fig. 8b, c is shown as a partial local-ization process results from Fig. 8c can be seen that the chin point has been converging to the edge of the face.

a Overall positioning   b Feature points   c Final positioning
  results   location after   result
                      narrow strip
                      segmentation

Fig. 8. Simulation results

a Contrast results 1   b Contrast results 2

c Contrast results 3   d Contrast results 4

e Contrast results 5   f Contrast results 6

Fig. 9. Simulation results comparison chart

Figure 9 shows for the comparison of the results of this algorithm with the traditional ASM algorithm detection map. In the comparison results, first for the results of this algorithm and the second for traditional ASM algorithm results, in obvious error detection area are painted lines highlighted. By Fig. 9 shows that traditional ASM algorithm of feature point location in the image smooth regions (such as frowning face, chin) exist certain errors, and feature point location algorithm in this paper is the positioning of the closer to real characteristics of the region.

4.2 Comparison of Positioning Accuracy

This paper presents the algorithm to improve the accuracy of the positioning accuracy of the data, the accuracy of the measurement of the accuracy of the Euclidean distance

of the average error $E = \frac{1}{N} \sum_{i=1}^{N} \left(\frac{1}{n} \sum_{j=1}^{n} \sqrt{\left(x_{ij} - x'_{ij}\right)^2 + \left(y_{ij} - y'_{ij}\right)^2} \right)$; Among them, N is

the total number of test images, n is the number of feature points in a face image, $\left(x_{ij}, y_{ij}\right)$ said that the coordinates of the j manual calibration points in the first I test image, $\left(x'_{ij}, y'_{ij}\right)$ said that after the convergence of the algorithm.

In order to have a more direct comparison of the 2 algorithms, we further calculate the overall improvement of the algorithm in this paper with respect to the traditional ASM algorithm $I = \frac{E_{ASM} - E'_{ASM}}{E_{ASM}} \times 100\%$. Among them, E_{ASM} represents the average error of the traditional ASM algorithm, and the average error of the E'_{ASM} represents the algorithm.

Table 1 shows that the results of this paper are compared with the traditional ASM algorithm for the remaining 200 images of the ORL database. The average detection error of the traditional ASM algorithm is 15.12 pixels, and the average error of this algorithm is only 7.08 pixels, the improvement of the positioning accuracy is 53.17%.

Table 1. Accuracy comparison based on multi template localization algorithm

Mean error/Pixel		Improvement degree/%
The algorithm of this paper	Traditional ASM algorithm	
7.08	15.12	53.17

5 Concluding Remarks

This paper presents a face feature point localization algorithm based on narrow band Closed-form. The advantages of this algorithm are: (1) Closed-form algorithm is more accurate, so it can be used to segment the training image and to detect the smooth region, while the background pixels are 0, which is advantageous for matching calculation; (2) local template is built on the basis of global template. The construction of the narrow strip can reduce the computation time of the Closed-form algorithm. The experimental results show that the convergence accuracy and the feature points of the image smoothing region can be improved by incorporating the Closed-form algorithm into the algorithm.

However, there are still some shortcomings in this paper. Firstly, if the overall positioning error is relatively large, it may lead to a narrow strip of Profile vertices, which can converge to the local optimal "false edge". Secondly, the initial position of the foreground and background marker is not reasonable, and it is easy to lead to segmentation.

In addition, how to apply this algorithm to 3D specific human face model, and improve the similarity between the 3D model and the real human face, is also the subject of further research.

References

1. Cootes, T.F., Taylor, C.J., Lanitis, A.: Multi-resolution search with active shape models. In: Proceedings of the 12th International Conference on Pattern Recognition, Manchester, vol. 1, pp. 610–612 (1994)
2. Jin, W.: Study on Video-Based Face Expression Modeling. Zhejiang University, Hangzhou (2003). (in Chinese)
3. Aiping, L., Yan, Z., Xinpu, G.: Application of improved active shape model in ace positioning. Comput. Eng. 33(18), 227–229 (2007). (in Chinese)
4. Yuhua, F., Jianwei, Ma.: ASM and improved algorithm for facial feature location. J. Comput. Aided Des. Comput. Graph. 19(11), 1411–1415 (2007). (in Chinese)
5. Li, Y., Lai, J.H., Yuen, P.C.: Multi-template ASM method for feature points detection of facial image with diverse expressions. In: Proceedings of the 7th International Conference on Automatic Face and Gesture Recognition, pp. 435–440. IEEE Computer Society Press, Washington D.C (2006)
6. Cristinacce, D., Cootes, T.: Boosted regression active shape models. In: Proceedings of British Machine Vision Conference, Warwick, vol. 2, pp. 880–889 (2007)
7. Toth, R., Tiwari, P., Rosen, M., et al.: A multi-modal prostate segmentation scheme by combining spectral clustering and active shape models. In: Proceedings of the SPIE, Bellingham: Society of Photo-Optical Instrumentation Engineers Press, vol. 6914, pp. 69144S.1–69144S.12 (2008)
8. Levin, A., Lischinski, D., Weiss, Y.: A closed form solution t o natural image matting. In: Proceedings of IEEE Computer Society Conference on Computer Vision and Pattern Recognition, New York, vol. 1, pp. 61–68 (2006)
9. Ahlberg, J.: CANDIDE-3—an updated parameterized f ace. Linkping: Linkping University. Image Coding Group, Department of Electrical Engineering (2001)

Correlation Analysis of Climate Indices and Precipitation Using Wavelet Image Processing Approach

Mingdong Sun[1], Xuyong Li[1(✉)], and Gwangseob Kim[2]

[1] Research Center for Eco-Environmental Sciences,
Chinese Academy of Sciences, Beijing, China
xyli@rcees.ac.cn
[2] Kyungpook National University, Daegu, South Korea

Abstract. For extensive research on the connection and the phase relationships between climate indices and the precipitation time series in Korea, three wavelet image processing are applied to examine the relation between the climate indices (NAO, SOI, NOI, PDO, WP and NP) and precipitation in Korea. The continuous wavelet, cross-wavelet analysis and wavelet coherence is utilized to expand and present regions with common high significant frequency and phase features of the precipitation and these climate indices time series. It is found that the all wavelet frequency spectrum of these climate indices time series have some similar significant frequency features as the spectrum of precipitation time series in Korea have. There are significant variations of around 4–6 years of periodicities in all spectrum analysis. It is illustrated that there is strong underlying connection between precipitation variability and climate indices that implied by the cross wavelet and wavelet coherence analysis. The results clearly demonstrate that the climate indices have the influential consistent correlation relationship with the precipitation variation in Korea.

Keywords: Climate indices · Precipitation · Wavelet image processing
Correlation analysis

1 Introduction

Climate change is a long-term, significant changes in a region or the earth as a whole, which may include temperature, precipitation and wind patterns change. Climate change strongly affects numerous aspects of hydrological systems associated with water resources, coastal zones and oceans. In recent years, climate changes and environmental changes are widely reaching effects especially in the atmosphere and the slowly changing oceans. A lot of meteorologists have paid great attention to some periodical meteorological variations as the hydrological management which impacted by the climate and environmental change.

As the desire for practical research on the climate changes and the expected impacts, climate indices are developed to describe the state and the variations in the climate system. It is defined as a simplified way to represent the complex incorporate information and relations of climate change system. Climate indices are constructed

© Springer Nature Singapore Pte Ltd. 2018
J. C. Hung et al. (Eds.): FC 2017, LNEE 464, pp. 231–240, 2018.
https://doi.org/10.1007/978-981-10-7398-4_25

with certain parameters and descriptions of particular aspects of the climate. Numerous climate indices that have been defined and examined in numerous researches (Hurrell 1995; Trenberth 1997; Stenseth et al. 2003). Trenberth (2001) described the natural characteristic and evolution of ENSO events and suggested two indices are necessary to capture the features of ENSO. Stenseth et al. (2002) provided a description of the interaction effect of two climate phenomena the North Atlantic Oscillation and the El Niño-Southern Oscillation on ecological patterns of marine and terrestrial ecosystems. Portis et al. (2001) evaluated the potential seasonality of North Atlantic Oscillation index in a particular simplified framework for optimal applications and studies of monitoring and seasonal variability. Schwing et al. (2002) introduced that an index of climate variability associated with inter-annual variations of El Niño and La Niña event, Northern Oscillation Index, based on the difference in SLP anomalies at the North Pacific High and near Darwin. Cheung et al. (2012) found the connection between Ural-Siberian blocking and the East Asian winter monsoon is regarded as the integrated effects of the Arctic Oscillation and the El Niño-Southern Oscillation.

Among the different aspects of climate changes, precipitation is the principal source of water for agriculture, industry and municipal water use. The consequences of precipitation changes in the timing and amount, such as flooding and droughts, could lead to serious agricultural failure or flood hazards. Advanced diagnosis of precipitation variation becomes more important for water resource management, especially focus on the significant increases in the amount and intensity of precipitation. A number of studies on the annual precipitation of different regions have emphasized the diverse trends depending on the individual region and time periods. Pal et al. (2004) detected the relationship of European summer precipitation trends and extremes with future regional climate projections. Silverman and Dracup (2000) mapped a 1-year periodic time series of the 700-hPa teleconnection indices and ENSO indicators onto the total precipitation of California's seven climatic zones using ANNs. Higgins et al. (2004) produced objective seasonal forecasts of temperature and precipitation for the United States conterminous using tropical Pacific sea surface temperature forecasts for the Niño-3.4 region in conjunction with composites of observed temperature and precipitation keyed to phases of the ENSO cycle. Zhou and Wang (2006) found that the extreme precipitation events dramatically increased every ten years in the Yangtze River basin between 1950 and 1999, and explained that the precipitation variability is following summer East Asian atmospheric circulations as a function of boreal spring Hadley circulation. Jin et al. (2005) detected significant correlations between the Southern Oscillation Index and the transformed precipitation for five stations in South Korea. Wang et al. (2006) identified a significant increasing tendency in the summer precipitation in Seoul between 1778 and 2004, which an increasing trend since the 1950s were likely to be caused by large-scale changes in the circulation of the East Asian summer monsoon system.

The main objective of this study is to investigate the correlation relationships between precipitation variability in Korea (KP) and several climate indices (NAO, SOI, NOI, PDO, WP and NP) using analysis of wavelet image processing approaches. The continuous wavelet transform is used to clearly distinguish the significant features in the wavelet power of the precipitation and these climate indices time series. And the ways of cross wavelet and wavelet coherence are applied to precipitation couple with these climate indices respectively for revealing the associated phase relationship between that couples of time series. Locally potential phase relations between precipitation variability and climate indices in temporal frequency scale will be examined and conducted in these procedure.

2 Methodology and Data

Continuous Wavelet Transform (CWT)

Wavelet analysis can be used for analyzing variations of power within a time series at many different frequencies. It can be characterized by how localized it is in time and frequency. The Morlet wavelet (Torrence and Compo 1998) is defined as:

$$\psi_0(\eta) = \pi^{-\frac{1}{4}} e^{i\omega_0 \eta} e^{-\frac{1}{2}\eta^2} \tag{1}$$

where ω_0 is the dimensionless frequency and η is dimensionless time. The wavelet is extended in time by varying its scale (s), so that $\eta = s \cdot t$, and then normalizing it to have unit energy. A continuous wavelet transforms of the time series $(x_n, n = 1, \ldots, N)$ with uniform time steps δt, is defined as the convolution of x_n with the scaled and normalized wavelet, write as

$$W_n^X(s) = \sqrt{\frac{\delta t}{s}} \sum_{k=1}^{N} x_k \psi_0 \left[(k - n) \frac{\delta t}{s} \right] \tag{2}$$

The $|W_n^X(s)|^2$ is defined as the wavelet power. $W_n^X(s)$ can be explained as the local phase. The continuous wavelet transform has edge stuffs because the wavelet is not completely localized in time. Therefore introduce a cone of influence (COI) in which edge effects aren't neglected. It is regarded as the region in which the wavelet power caused by a discontinuity at the edge has dropped to e^{-2} of the value at the edge (Grinsted et al. 2004).

Cross Wavelet Transform (XWT)

In order to examine whether the time series have a consistent phase relationship and causality, the cross wavelet transform was constructed from two continuous wavelet transforms, which will expose their common power and relative phase in time-frequency space.

The cross wavelet transform of two time series x_n and y_n is defined as

$$W^{XY} = W^X W^{Y*} \tag{3}$$

where $*$ denotes complex conjugation. Define the cross wavelet frequency power as $|W^{XY}|$. The phase angle of W^{xy} can be interpreted as the local relative phase between x_n and y_n in time-frequency space (Torrence and Compo 1998).

Wavelet Coherence

Cross wavelet power reveals a region with high common frequency; to measure the significant coherence of the cross wavelet transforms in time-frequency space, following Torrence and Webster (1998) and Grinsted et al. (2004), define the wavelet coherence of two time series as

$$R_n^2(s) = \frac{|S(s^{-1}W_n^{XY}(s))|^2}{S(s^{-1}|W_n^X(s)|^2) \cdot S(s^{-1}|W_n^Y(s)|^2)} \tag{4}$$

where S is a smoothing operator, it can be written as

$$S(W) = S_{scale}(S_{time}(W_n(s))) \tag{5}$$

where S_{scale} denotes smoothing along the wavelet scale axis and S_{time} smoothing in time.

Wavelet coherence can be thought of as the local correlation between two wavelets in time frequency space.

Data Description

The Korea Meteorological Administration (KMA) provides hourly precipitation measurements recorded covered countrywide. The stations are generally well-distributed across the country and five major river basins. The available station data is processed into monthly data for studying the frequency and intensity of precipitation. Missing stations recorded data are excluded from this analysis. The available station data maintained by the KMA are processed into monthly data and generate an average of stations in 40 years (1973–2012).

Table 1. Description of the climate indices

Climate indices	Description
NAO	North Atlantic Oscillation: is a climatic phenomenon in the North Atlantic Ocean, the fluctuations in the difference of atmospheric pressure at sea level between the Icelandic low and the Azores high
SOI	Southern Oscillation Index: an indication of the development and intensity of El Niño or La Niña events in the Pacific Ocean. The SOI is calculated using the pressure differences between Tahiti and Darwin
NOI	Northern Oscillation Index: an index of climate variability based on the difference in sea level pressure anomalies at the North Pacific High and near Darwin Australia
PDO	Pacific Decadal Oscillation: a pattern of change in the Pacific Ocean's climate. It is the leading primary component of monthly SST anomalies in the North Pacific Ocean
WP	Western Pacific pattern: a primary mode of low-frequency variability over the North Pacific in all months. This pattern reflect pronounced zonal and meridional variations in the location and intensity of the entrance region of the Pacific (or East Asian) jet steam
NP	North Pacific pattern: the area-weighted sea level pressure over the region 30 N-65 N, 160E-140 W

The atmospheric and ocean climate indices monthly time series is attained from the Climate Prediction Center (CPC) of the National Oceanic and Atmospheric Administration (NOAA). In this research, we picked up several indices that have possible connection with precipitation variability in Korea, which are the North Atlantic Oscillation (NAO), Southern Oscillation Index (SOI), Northern Oscillation Index (NOI), Pacific Decadal Oscillation (PDO), Western Pacific pattern (WP) and North Pacific pattern (NP). Table 1 shows some detail description about these indices.

3   Results and Discussion

In order to explore the frequency component in the precipitation variability and these climate indices, Fig. 1 displays the 6 climate indices standardized time series (top) and their wavelet spectrum (bottom), Fig. 2 shows the standardized and deseasonalized time

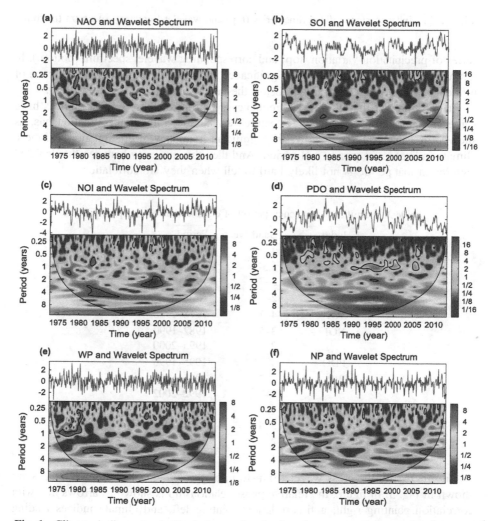

Fig. 1. Climate indices standardized time series (top) and corresponding wavelet frequency spectrum (bottom). (a) NAO, (b) SOI, (c) NOI, (d) PDO, (e) WP, (f) NP.

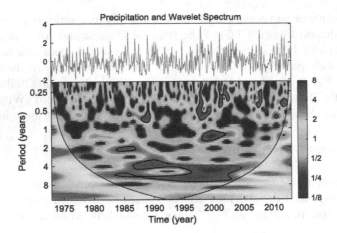

Fig. 2. Precipitation standardized time series (top) and wavelet frequency spectrum (bottom).

series of precipitation variation (top) and corresponding wavelet spectrum (bottom). In figures, the contour assigns the 5% significance level against red noise and the cone of influence where edge effects might distort the picture shown as a lighter pattern.

It is found that the all wavelet frequency spectrum of these climate indices have some similar significant frequency features as the spectrum of precipitation has, as showed in Table 2. Table 2 shows the significant frequency of precipitation and these climate indices in frequency and period. And then the similarity between the depicted patterns in that period is not likely hard to tell when they are correlation.

Table 2. Significant frequency period of climate indices and precipitation

Index	Band (years)	Period
Precipitation	2-	Around 1985
	1.5-	Around 2000
	3–7	1984–2004
NAO	2-	Around 1999
SOI	4–5	1978–1988
NOI	3–5	1982–1990
	2-	1995–2000
PDO	2-	1995–1996
WP	2–3	1981–1989
	3–7	1992–2002
NP	1.5-	1981–1984
	3–4	1978–1984
	5-	1989–1991

In this regard, the cross wavelet transform of precipitation and climate indices are shown in Fig. 3. In figure, the relative phase relationship is shown as arrows that with correlation pointing right, anti-correlation pointing left, and climate indices leading

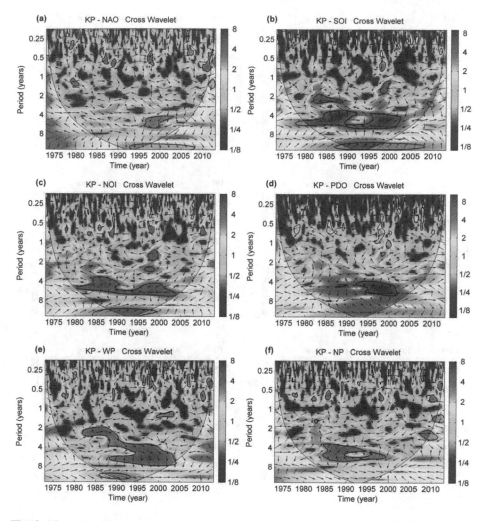

Fig. 3. Cross wavelet transform of precipitation and climate indices. (a) NAO, (b) SOI, (c) NOI, (d) PDO, (e) WP, (f) NP

Table 3. Significant common frequency of precipitation and climate indices in cross wavelet

Index	Band (years)	Period	Index	Band (years)	Period
KP-NAO	4–6	1996–2003	KP-SOI	3–5	1984–1990
				4–6	1995–2002
KP-NOI	4–5	1984–1990	KP-PDO	4–6	1994–2002
	5–6	1995–2001			
KP-WP	2–3	1982–1988	KP-NP	3–5	1985–1994
	4–7	1994–2002		5–6	1985–1999

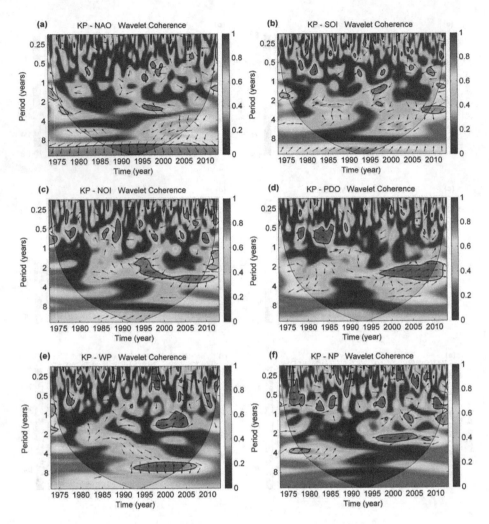

Fig. 4. Wavelet coherence spectrum of precipitation and climate indices (a) NAO, (b) SOI, (c) NOI, (d) PDO, (e) WP, (f) NP.

Table 4. Significant common frequency of precipitation and climate indices in wavelet coherence analysis

Index	Band (years)	Period	Index	Band (years)	Period
KP-NAO	3-	Around 1997	KP-SOI	1.5-	Around 1997
KP-NOI	1.5–3	1994–2000	KP-PDO	0.6-	Around 1984
	2–3	2002–2008		2–3	1998–2008
KP-WP	1–2	1998–2005	KP-NP	2–3	1996–2006
	6–7	1993–2004			

precipitation by 90° pointing straight down. The similar frequency features were noticed from the individual wavelet transforms stand out as being significant at the 5% level. Note that precipitation variability has common significant frequency band in the specified period with these climate indices, as showed in Table 3. Table 3 shows the significant common frequency band of precipitation with each climate indices in cross wavelet. There are significant variations of around 4–6 years of periodicities in all spectrum analysis. It is assumed that there is strong underlying connection between precipitation variability and climate indices that implied by the cross wavelet analysis.

The wavelet coherence of precipitation and climate indices is shown in Fig. 4. Compared with the cross wavelet transforms, the wavelet coherence is slightly different in significant common frequency band regions of precipitation and climate indices. The significant common frequency band of precipitation with each climate indices in wavelet coherence analysis is shown in Table 4. Therefore it further demonstrates that there is a robust relationship between precipitation variability and the climate indices.

4 Conclusion

For investigating of the connection and phase relationships between some climate indices and precipitation in Korea, applying the approach of wavelet spectrum, cross wavelet and wavelet coherence to these climate indices (NAO, SOI, NOI, PDO, WP and NP) with precipitation variability. The wavelet frequency spectrum expands these time series into a time frequency space and presents the significant frequency features of precipitation and climate indices. And the cross wavelet reveals the regions with prominent significant frequency of precipitation variability in common with these climate indices and exhibits the information about the phase relationship. In the wavelet coherence, locally phase features between precipitation variability and the climate indices have been found. It is found that the all wavelet frequency spectrum of these climate indices have some similar significant frequency features as the spectrum of precipitation has. It shows the significant frequency of precipitation and these climate indices in frequency and period. There are significant variations of around 4–6 years of periodicities in all spectrum analysis. It is illustrated that there is strong underlying connection between precipitation variability and climate indices that implied by the cross wavelet and wavelet coherence analysis. Those results simply confirm that the climate indices have the consistent significant correlation relationship with the precipitation in Korea and that might influence on the variation of precipitation.

Acknowledgements. The research funding was provided by the Major Science and Technology Program for Water Pollution Control and Treatment in China (2014ZX07203010), the Chinese Academy of Sciences' key project (KZZD-EW-10-02), and the RCEES "One-Three-Five" project (YSW2013B02-4).

References

Cheung, H.N., Zhou, W., Mok, H.Y., Wu, M.C.: Relationship between Ural-Siberian blocking and the East Asian winter monsoon in relation to the Arctic Oscillation and the El Niño-Southern Oscillation. J. Clim. **25**, 4242–4257 (2012)

Grinsted, A., Moore, J.C., Jevrejeva, S.: Application of the cross wavelet transform and wavelet coherence to geophysical time series. Nonlinear Proc. Geophys. **11**, 561–566 (2004)

Higgins, R.W., Kim, H.K., Unger, D.: Long-lead seasonal temperature and precipitation prediction using tropical Pacific SST consolidation forecasts. J. Clim. **17**, 3398–3414 (2004)

Hurrell, J.W.: Decadal trends in the North Atlantic oscillation regional temperatures and precipitation. Science **269**, 676–679 (1995)

Jin, Y.H., Kawamura, A., Jinno, K., Berndtsson, R.: Detection of ENSO-influence on the monthly precipitation in South Korea. Hydrol. Proc. **19**, 4081–4092 (2005)

Pal, J.S., Giorgi, F., Bi, X.Q.: Consistency of recent European summer precipitation trends and extremes with future regional climate projections. Geophys. Res. Lett. **31**, L13202 (2004)

Portis, D.H., Walsh, J.E., El Hamly, M., Lamb, P.J.: Seasonality of the North Atlantic oscillation. J. Clim. **14**, 2069–2078 (2001)

Schwing, F.B., Murphree, T., Green, P.M.: The Northern Oscillation Index (NOI): a new climate index for the northeast Pacific. Progr. Oceanogr. **53**, 115–139 (2002)

Silverman, D., Dracup, J.A.: Artificial neural networks and long-range precipitation prediction in California. J. Appl. Meteorol. **39**, 57–66 (2000)

Stenseth, N.C., Mysterud, A., Ottersen, G., Hurrell, J.W., Chan, K.S., Lima, M.: Ecological effects of climate fluctuations. Science **297**, 1292–1296 (2002)

Stenseth, N.C., Ottersen, G., Hurrell, J.W., Mysterud, A., Lima, M., Chan, K.S., Yoccoz, N.G., Ådlandsvik, B.: Studying climate effects on ecology through the use of climate indices: the North Atlantic Oscillation, El Niño Southern Oscillation and beyond. Proc. Roy. Soc. Lond. B **270**, 2087–2096 (2003)

Torrence, C., Compo, G.P.: A practical guide to wavelet analysis. Bull. Am. Meteorol. Soc. **79**, 61–78 (1998)

Trenberth, K.E.: The definition of El Ninño. Bull. Am. Meteorol. Soc. **78**, 2771–2777 (1997)

Trenberth, K.E., Stepaniak, D.P.: Indices of El Niño evolution. J. Clim. **14**, 1697–1701 (2001)

Wang, B., Ding, Q., Jhun, J.G.: Trends in Seoul (1778-2004) summer precipitation. Geophys. Res. Lett. **33**, L15803 (2006)

Zhou, B.T., Wang, H.J.: Relationship between the boreal spring Hadley circulation and the summer precipitation in the Yangtze River valley. J. Geophys. Res. **111**, D16109 (2006)

A Study on the Computer Aided English Translation of Local Legal Based on Parallel Corpus

Zhang Zhijie[✉]

Department of Foreign Language Teaching,
Tonghua Normal University, Tonghua, Jilin, China
tt79964094gululi@163.com

Abstract. With the development of modern information technology, people have access to a wider range of information resources and texts, but there are few corpora for local law. Therefore, the computer aided English translation of local legal should meet the basic functional requirements of high efficiency and high quality. The CAT model of parallel corpora WAS constructed to make full use of the storage and computing ability of the software. In order to carry out the practical analysis of the translation method, a concrete evaluation model was constructed. A comparative study of the local laws and regulations in China was conducted in the aspects of the translation time and the quality of the translation by Google, CAT and parallel corpus based computer aided translation. According to the analysis, it can be seen that the computer aided English translation has the characteristics of high quality and high efficiency.

Keywords: Corpus · Computer · Local law · English translation

1 Introduction

After the human society entered in twenty-first century, the way and efficiency of knowledge acquisition have been increasing. The application of corpus and parallel corpus in people's lives is very common. With the deepening of China's economy and global economic exchanges, the translation of local legal provisions in China requires efficient and accurate translation of the original text based on Parallel Corpora (Vîlceanu et al. 2015) [1]. This can strengthen the exchange between China's local economic construction and international competitive development, and help to enrich and improve the computer aided translation based on parallel corpus. China is a country with many local administrative regions. As the national legal system is in the form of Pyramid, local laws based on Constitution are also very complex with relatively large legal provisions (Magoulas 2015) [2]. In order to meet the changing social production environment, the legal provisions of our country are constantly revised and perfected. Especially in recent years, the development of China's social market economy has entered a new stage, and the relevant legal provisions in various fields changed greatly in a short period of time. On this basis, it is very important to carry out the translation of local legal provisions in time and accurately, which is helpful to the development of China's local economy (Hwang 2015) [3]. However, due to the strong professional characteristics of

© Springer Nature Singapore Pte Ltd. 2018
J. C. Hung et al. (Eds.): FC 2017, LNEE 464, pp. 241–256, 2018.
https://doi.org/10.1007/978-981-10-7398-4_26

the legal text, the translation of the terms is also very accurate, otherwise it will produce a large ambiguity, which can impact the interpretation and implementation of the legal provisions (Loponen 2013) [4]. The traditional machine translation MT is lack of professional translation of local law, which seriously affects the accuracy and timeliness of local legal translation in China (Banerjee et al. 2015) [5]. The computer aided translation of local legal English based on parallel corpus can make full use of storage memory function of the parallel corpus, share the terminology, and avoid duplication of translation work, which makes the translation more efficient. After constructing the CAT model based on the parallel corpus, the traditional CAT software is optimized to give full play to its characteristics according to the characteristics of the translation software, and improve its shortcomings. A more appropriate optimization scheme is proposed. Finally, the corresponding evaluation model is established to evaluate the translation quality and efficiency. It is of great significance to the English translation of the local law in China, and it also has some referential significance for the English translation of other fields in our country. The establishment of national and local laws and regulations parallel corpus not only provides a powerful tool for the English translation of laws and regulations, but also can promote the role of China's legal practice.

2 Corpus Linguistics and Translation of Local Laws and Regulations

2.1 Corpus Linguistics

The corpus was new branch of linguistics first developed by Norm Chomsky in the 1960s. After decades of development, the research based on corpus has been becoming more and more extensive. Chomsky's original corpus study emphasizes the ability, not the language. In fact, the method is to suspect a small number of samples, thus representing the limited nature of the real information language data (Yamada 2015) [6]. The study of the initial area of language based on corpus has been focused on five academic fields, such as the compilation of dictionaries, the study of language, the language education the study of grammar and the study of literature. With the development of computer technology, computer corpus came into being. Through computer compilation, storage and processing, a lot of data language has formed the first generation of corpus, in which the most typical one is Brown corpus (Biçici et al. 2015) [7]. Later, new corpora had been developed based on the Brown corpus and Lancaster corpus, which formed the corpus center from the United States to Europe. In the 1980s, great achievements had been made in the compilation of dictionaries, language teaching and natural speech processing. Subsequently, the corpus had entered a new stage of historical development. Although the main ability was limited, the investigation of vocabulary was paid more attention, thus with the emergence of electronic text corpus of spoken and written language (Biçici et al. 2015) [7]. In the 1990s, the focus of the corpus linguistics research was the bilingual corpus, which provided a very important resource for the comparative study of modern translation and language. The last stage of development was the electronic corpus. More and more software annotation and markup language databases were more convenient to process the text data. In China, the earliest corpus was the JDEST corpus compiled by Jiaotong University of Shanghai, which greatly promoted China's foreign language teaching. Now, due to different purposes and applications, the corpus is divided into oral corpus, written corpus, monolingual corpora, parallel corpus, translation corpus and other professional translation (Karim et al. 2015) [8].

In the application of corpus, machine aided translation (MAT or CAT) is considered to be the pioneer of language translation. This machine translation can be traced back to more than 400 years ago. In seventeenth century, Leibniz used the mechanical dictionary to overcome the barriers of different languages, and the true realization of MAT was during World War II (Maclochlainn et al. 2015) [9]. It was not until the 1970s that the corpus based MT system was truly mature. With the application of advanced information technology and artificial intelligence, the new media based on Internet technology was constructed. Taking the distributed system model based on translation assistance theory proposed by Melby as an example, it was shown in Fig. 1. The distributed system was a multi-level translation auxiliary tool. The first layer was on the workstation for the term search and word processing. The second layer was to input data through the computer and add terminology related recommendations. The last layer was the direct selection and translation of sentences (Ebrahim et al. 2015) [10].

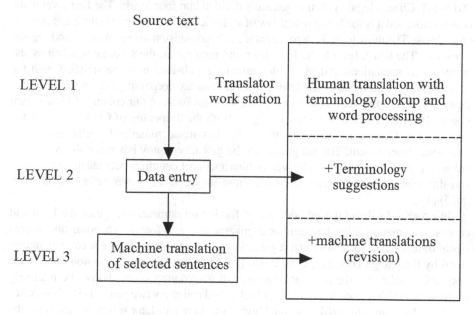

Fig. 1. Machine aided translation model

In the 1990s, CAT became a widely accepted translation workstation, such as SDL Trados translation workbench and IBM translation. After a certain period of development, CAT technology has been rapid development, and relatively mature translation workbench similar to Google translation is widely used (Ding et al. 2015) [11]. The SDL Trados software consists of four components, in which the relatively common components are SDL

Trados Studio and SDL multi word. SDL Trados Studio is the most basic component of the translation suite, which is suitable for the translation of documents, the creation and management of translation memory, and finally the realization of automatic translation. The software integrates project management and CAT tools, which can translate, edit, modify, and support more than 70 different types of documents, such as word, PDF, EXCEL and other formats such as XML, HT and other documents (Chen et al. 2015) [12]. As a professional translation term database, it should be used in different fields of data storage, such as the developing IBM and SDLX translation.

2.2   Current Status of Local Laws and Regulations

Law is the fundamental guarantee of a country's political system and national civil rights. The law has a very important impact on the development of the country, which is related to the stability and peace of the whole country (Magoulas 2015) [2]. According to different legal subjects, the content and scope of the law are also different. Different levels of legal system together constitute a complex legal system. According to the main body of the law in China, it mainly includes nine parts: administrative regulations, ministerial regulations, local laws and regulations, judicial interpretations, international treaties and so on (Loponen 2015) [4]. China's legal system is generally divided into four levels. The first level is the constitution, and it is the fundamental law of all laws. The second level is the basic law and other laws. The third level is departmental rules, administrative regulations and regulations, etc. The fourth level is the local laws and regulations, the separate regulations and autonomous regulations. All of the downstream regulations must be satisfied with the upstream law. In particular, local laws and regulations are deepening and concrete with the characteristics of the various areas based on the legal basis of our country. China's legal system Pyramid structure is shown in Fig. 2. With the deepening of China's reform and opening up, China's accession to the WTO has made remarkable achievements in economic development. The integration of the global economy has made the economic, cultural and military exchanges between China and other countries become more frequent, and the related laws and regulations are more important than ever before (Qian et al. 2015) [13].

In order to facilitate the introduction of foreign investment, strengthen the legal and economic communication between local governments and foreign governments, professional translation of local laws and regulations is required in order to be accurately understood by the foreign (Banerjee et al. 2015) [5]. At present, for the translation of the local laws and regulations, the local governments of our country often choose the relatively professional public relations company, which is updated every two years. This often makes relatively low translation efficiency and high cost. If the translator is non-professional, the accuracy of the translation will be seriously affected (Furukawa 2015) [14]. On the other hand, with the rapid development of computer technology and Internet technology, China's local governments and institutions have built their own legal database. Therefore, it has become an inevitable trend and demand of social development to construct the local legal translation based on corpus (Vann et al. 2015) [15]. For example, in 1999, Chinalawinfo Co. Ltd. established in Zhongguancun was cooperation with Peking University to work out China's effective local regulations and translations. There are foreign companies specifically building an online database for the nine parts of China's laws and regulations, which includes local laws and regulations of the National Bureau of Statistics and the different provinces in different periods, but the some translation of the database information is not complete and accurate (Ramos et al. 2015) [16].

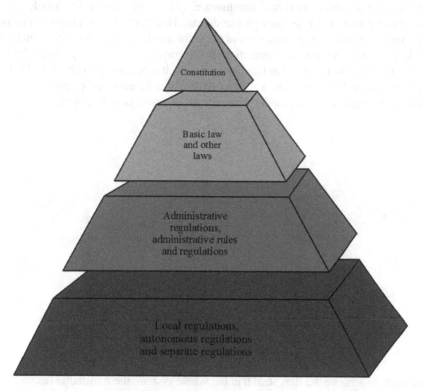

Fig. 2. China law system Pyramid

3   Computer Aided English Translation of Local Legal Based on Parallel Corpus

3.1   Needs Analysis

According to the thought of artificial translation, the translation of language can be divided into three levels: pragmatic level, semantic level and syntactic level. The pragmatic level refers to understand the cultural meaning of the vocabulary based on the meaning of the context and the context of the whole article. The syntactic level refers to simple dictionary translation without the meaning of the context and the context of the whole article. At present, the majority of computer aided translation can only understand and translate in a simple semantic layer, while some computer translation also has a certain level of syntactic translation. However, because of the great differences in the environment, it is difficult to achieve the level of manual translation for computer aided software translation. Therefore, from the perspective of translation accuracy, we need to design the computer aided software based on parallel corpora, and the following problems should be solved: (1) Computer aided translation can better and faster achieve text translation, and effectively improve the quality and efficiency of manual translation; (2) CAT technology is used to ensure translation automation and intelligence; (3) It can achieve the whole process management for the CAT technology translation. Therefore, in the design of computer aided translation software, the functions of the software should be achieved: It can be able to achieve the management and operation of the corpus, automatically search the database, and test the result of the translation. As a whole, it is necessary to give full play to the main function of the computer, so as to realize the function of computer software. Combined with this, the basic model of the corpus designed is shown in Fig. 3.

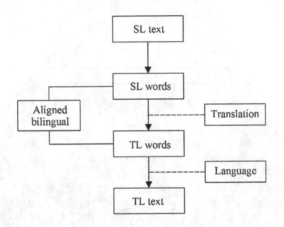

Fig. 3. Corpus based model

In order to achieve the data search in the whole system, the translation results should be stored into the terminology corpus, especially for the local administrative regulations which has a strong feature of the term, in order to help the "memory" of professional translator, and improve the efficiency of translation of translators, which is the so-called TM system. The terms, words, sentences, and phrases translated by the translator are stored, and the translation products and resources can be called out. They can be directly used in the same or similar statements and the environment, so as to avoid the repetition of translation phenomena, save the time of translation dictionary and entry, and guarantee the efficiency of translation. In the 1990s, as a basic machine translation method, the corpus was applied in the MT system. There were two kinds of corpus based methods, and one of the most commonly used was statistical method. This method first appeared in the 1960s, using random statistical techniques of Bias's theorem, but it did not achieve good results. In the 1990s, researchers made a corresponding translation of the source language text in the bilingual corpus based on the statistical method, and achieved matching through a source LAN. Combined with the case study of local laws and regulations, a corpus model based on cases is constructed, which contains the structure of syntactic information and sentence, as shown in Fig. 4.

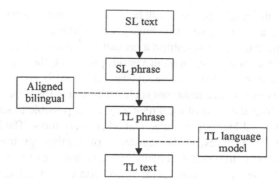

Fig. 4. Case based model

3.2 Construction of CAT Model Based on Parallel Corpus

The SDL Trados is used as the basic tool in CAT software. Based on the present situation of the translation of local laws and regulations and the development of information technology, the translation software based on SDL Trados is studied to adjust the legal texts locally. Due to the existing laws and regulations is more inclined to provide the world's leading legal information, and with the lack of translation of Chinese local laws and regulations, the translation model of CAT local regulations based on parallel corpus is constructed, and the structure model is shown in Fig. 5. In the structural model, the first step of the translation is to write the information obtained from the local laws and regulations of the relevant organizations in China. Then the pretreatment of text is carried out using computer English, and editing and translation is made in accordance with the requirements of SDL Trados. A qualified ED text is generated to ensure the accuracy of the

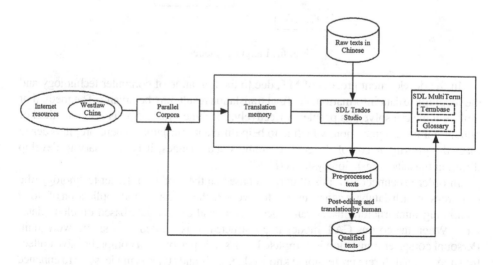

Fig. 5. Corpus based CAT model

results and improve the quality of text translation. Secondly, we annotate the text data and mark the word class, and build a new vocabulary in the corpus.

The most important thing is to establish a parallel corpus for TM design. By collecting the Chinese text of the local regulations in various provinces, the original data which is open to the public are put into the corpus. A static database is established, and we can translate on the basis of network resources such as China and the west. For the establishment of the terminology database of local regulations, it is possible to set up a set of tools by means of professional legal dictionaries and other query tools. The legal language is professional and strong features, so the translation of terminology translation needs of presupposition. In order to improve the quality of a large amount of text data, professional translation is only through manual translation or CAT software. Because for the translation of local laws and regulations, there are many basic terms, words and phrases in specific fields, it is necessary to construct the term database. In order to make the information of corpus into the target language text, we need a rule for the high degree of generalization and conversion, which requires intermediate language as an intermediary between different languages. The language model is shown in Fig. 6.

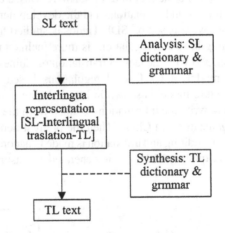

Fig. 6. Language model

In the development process of MT, due to the limitation of computer technology and the incomplete data information, the final translation result may be invalid. The method of multi-engine can solve this problem effectively. Machine Translation is also a method of human-computer interaction, which is to help different translators. Therefore, in order to meet the growing needs of different customers and changes, it is necessary to develop different machine translation systems (Fig. 7).

In order to connect all kinds of corpora based on the modern Internet technology, the resources of each database can be used effectively. In the operation and application of cloud computing infrastructure, users can process, store and extract data based on cloud database. When the corpus CAT model is constructed, it is limited to a single work unit. Personal computer translation is completed by a single person, so a comprehensive collection and use of different professional knowledge, skills and styles is made, so as to enhance

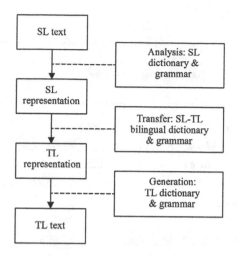

Fig. 7. Transfer model

the quality of translation. Therefore, the parallel corpus, TM and terminology database can be stored on the network client, which can be used to share the resources of the customer and individual groups (Fig. 8).

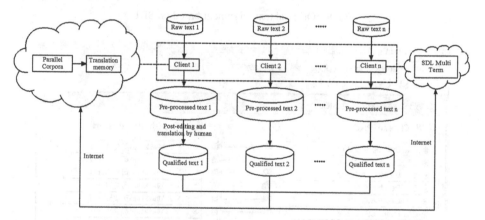

Fig. 8. Extended model based on Cloud Computing

3.3 Construction of Parallel Corpus

Firstly, the text materials are collected to construct the parallel corpus of local laws and regulations. After data collection, text alignment is carried out. The collected text of the local laws and regulations are mainly downloaded from library network system in university, covering all the latest local laws and regulations in 2016, and the detailed information is listed in accordance with the time sequence in the appendix. At the same time, using SDL Trados Studio software to deal with the original text, the specific process is shown in Fig. 9.

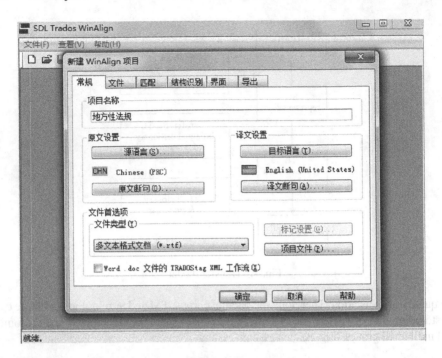

Fig. 9. Original text alignment based on SDL

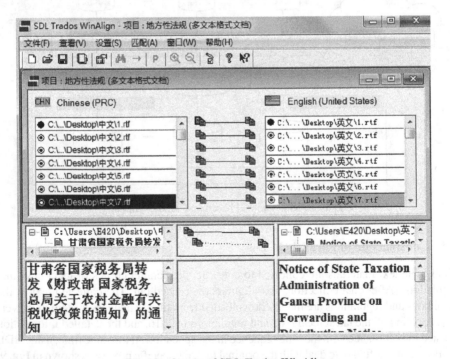

Fig. 10. Output of SDL Trados WinAlign

Select the first item and enter the name, source language and target language. The source language and the target language are matched after inputting English and Chinese, and they are exported in accordance with the TMX format. After the information setting, the output interface is set (Fig. 10).

The next important step is to apply the constructed TM parallel corpus into Trados software (Fig. 11).

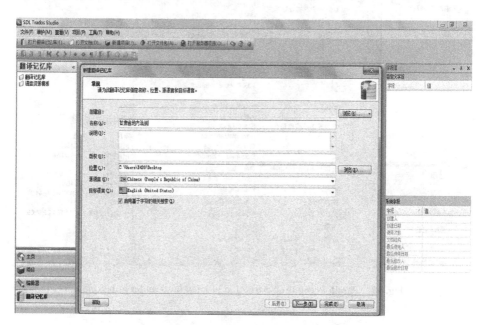

Fig. 11. Construction interface of TM

Finally, TM of local regulations is smoothly constructed. The text is imported according to the menu of SDL Trados. During the import process, the invalid text is automatically filtered so that it cannot be imported. Through the contrastive analysis of Chinese and English, it is easier to translate text materials (Fig. 12).

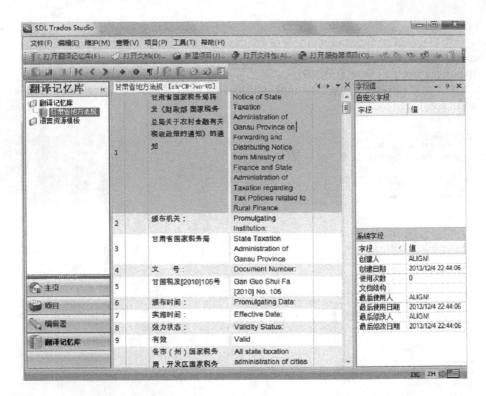

Fig. 12. Parallel corpus construction of local laws and regulations

4   Application Analysis

In order to verify the quality of computer aided local legal translation based on parallel corpora, an evaluation model is constructed to evaluate the quality of translation in accordance with the hierarchical index and the final score. There are three main requirements. The first is to be faithful to the original local laws and regulations. The second is to belong to the translation of the correct unity. And the third is that the translation result of the whole text is smooth and consistent with the background of the text. There are total score of 5 points, and scoring rules and scoring points of each score are shown in Table 1.

As can be seen from Table 1, in the three categories, it is determined by 13 specific indicators. Each specific item is graded according to the scoring criteria. After the comparison between the translated text and the original text, the translation is basically faithful to the original text, and it can get 1.5 points in the first item. If the translation is correct, it can score 1 point. There are very few grammatical errors in the third item, scoring 1 point. So the three points add up to 3.5 points. The method of scoring can be used to evaluate the quality of translation quantitatively and qualitatively.

In order to objectively verify the translation effect of computer aided local laws based on parallel corpus, contrastive analysis is used to analyze the results of Google, CAT and

Table 1. Translation quality assessment form

Objective	Requirements	Standard score	Index evaluation score standard		
			Content evaluation	Evaluation criterion	Score
Translation by corpus-based CAT	First: Faithfulness to the original	2.5	Translation is complete. The original information is accurate. There is no sematic error	There is no complete sentence. The content and structure cannot be conveyed	0
				Only several words match the original	0.5
				Few contents match the original	1
				The translation is basically faithful to the original	1.5
				Most information is conveyed	2
				The original information is accurately and completely expressed	2.5
	Second: Term unification	1	The term is correctly translated. The translation meets the criterion	Absolutely wrong	0
				Inaccuracy	0.5
				Exactly correct	1
	Third: Natural flow of the translated text	1.5	The translation meets standardization of language and writing and expression. The translation is clear and understandable	The translation is hard to understand	0
				There are some seriously grammatical mistakes. The translation text lacks of smoothness	0.5
				There are a few grammatical mistakes	1
				There are no grammatical and usage error. The text reads smoothly and native	1.5

computer aided based on parallel corpora, and experimental indexes include the translation quality and efficiency score. The original text for translation used is relevant local laws and regulations from one province. First of all, three different ways are used to translate it, and then they are scored according to the three indexes. In order to objectively evaluate its score, making the results more objective and fair, experts to score in accordance with the evaluation criteria, the results are as follows:

From Table 2 we can see that in three different ways, CAT quality score based on the corpus is 4.12 points, far higher than the quality of direct use of Google translation, and computer aided translation score based on parallel corpus is higher than that of CAT with a score of 4.5 points. It can be seen that the quality of computer aided translation based on parallel corpora is the best, and it can be used for the English translation of local legal.

Table 2. Comparative experimental translation score scale

Experiment object		1	2	3	4	5	6	7	8	Total	Mean value
Google translation	First	0.5	0.5	1	1	0.5	0.5	1	0.5	9	1.12
	Second	0	0	0.5	0.5	0.5	0	0	0		
	Third	0.5	0.5	0	0.5	0	0.5	0	0		
Corpus-based CAT	First	2.5	2	2	2.5	2	2	2	1.5	33	4.12
	Second	1	1	1	1	1	0.5	1	0.5		
	Third	1	1.5	1	1	1.5	1	1.5	1		
Existing material	First	2	2.5	2.5	2	2	2	2	2.5	36	4.5
	Second	1	0.5	1	1	1	1	1	1		
	Third	1	1.5	1.5	1.5	1.5	1..5	1.5	1		

The second is the statistical analysis of the time spent on translation. The time spent on Google translation, the intervention of CAT and the required computer aided translation based on parallel corpora is collected randomly selecting 6 translators, and the results are shown in Table 3.

Table 3. Three different translation methods usually time-consuming contrast

Experiment object	Completion time (day)						
	1	2	3	4	5	6	Time per request
Google translation	4	5	5	4	6	5	4.83
Corpus-based CAT	4	7	6	4	5	4	5.00
Existing material	3	5	4	5	4	4	4.17

From Table 3 we can see that The time required for the adoption of three different translation methods of 6 translators from long to short is: CAT > Google translation > computer aided translation based on parallel corpus. Therefore, computer aided translation based on parallel corpus can effectively improve the efficiency of translation of translators. According to the quality score of translation results in three different ways, it can be seen that computer aided translation based on parallel corpus not only can effectively improve the efficiency of the translation of local laws, but also can ensure the quality of translation. What's more, computer aided translation based on parallel corpus can be applied to the translation of the local law, and it can also be used as a reference for the translation of similar texts.

5　Conclusions

Research on computer aided English translation of local legal based on parallel corpus is helpful for the construction of parallel corpus in China, and it also can improve the accuracy and timeliness of Chinese English translation of local law. Through the analysis of its functional requirements, the CAT model based on parallel corpus is constructed, and the SDL Trados is used as the basic software of CAT tool to optimize the design. In order to verify the practicability of the software translation, an evaluation model of translation quality score is constructed, which is used to make quantitative analysis of the quality of translation by score. In order to objectively verify the translation effect of computer aided local laws based on parallel corpus, contrastive analysis is used to analyze the results of Google, CAT and computer aided based on parallel corpora. It can be seen from the experimental results that the translation quality score of computer aided local laws based on parallel corpus is 4.5 points, with the best quality, relatively short time consuming, and he average time of about 4.17 days. As a result, it can be seen that the computer aided translation based on parallel corpus has the characteristics of high efficiency and high quality. However, because there are only a small number of samples in a single area of our country in this experiment, the conclusion has some limitations. Therefore, the follow-up study can increase the number of subjects, in order to obtain more accurate conclusions.

Acknowledgements. The author acknowledges the Project of Advisory Committee of Foreign Language Teaching in Vocational Education, Ministry of Education, P. R. China "Research on Vocational College English Teachers' Professional Development based on the Course Construction" (GZWYJXGG-018); The Jilin Social Science Project "Semantic Prosody Problems in English-Chinese Translation" (No. 387, 2014); "Research on the Cultivation and Improvement of Teachers' Comprehensive Quality in Normal Colleges and Universities in Jilin Province" (2016B343); "College English Teachers' Professional Development Based on the Requirement of Educational Informatization" (GH16345).

References

1. Vîlceanu, T.: Developing evaluation skills with legal translation trainees. Acta Universitatis Sapientiae, Philologica **7**(3), 5–13 (2015)
2. Magoulas, C.: Kawadias' and Mutis' freighters: intertextuality and intersemiosis in different literary traditions but common forms of life: chinese semiotic studies. Chin. Semiot. Stud. **2**(1), 141–147 (2015)
3. Hwang, M.: Filled translation for bootstrapping language understanding of low-resourced languages (2015)
4. Loponen, M.: Translating irrealia – creating a semiotic framework for the translation of fictional cultures: chinese semiotic studies. Chin. Semiot. Stud. **2**(1), 165–175 (2013)
5. Banerjee, P., Rubino, R., Roturier, J., et al.: Quality estimation-guided supplementary data selection for domain adaptation of statistical machine translation. Mach. Transl. **29**(2), 77–100 (2015)
6. Yamada, M.: Can college students be post-editors? an investigation into employing language learners in machine translation plus post-editing settings. Mach. Transl. **29**(1), 49–67 (2015)
7. Biçici, E., Way, A.: Referential translation machines for predicting semantic similarity. Lang. Resour. Eval. 1–27 (2015)
8. Karim, N., Latif, K., Anwar, Z., et al.: Storage schema and ontology-independent SPARQL to HiveQL translation. J. Supercomputing **71**(7), 1–26 (2015)

9. Maclochlainn, S.: Divinely generic: bible translation and the semiotics of circulation. Signs Soc. **3**(2) (2015)
10. Ebrahim, S., Hegazy, D., Mostafa, M.G.M., et al.: English-arabic statistical machine translation: state of the art, vol. 9041, pp. 520–533 (2015)
11. Ding, S.H.H., Fung, B.C.M., Debbabi, M.: A visualizable evidence-driven approach for authorship attribution. ACM Trans. Inf. Syst. Secur. **17**(3), 1–30 (2015)
12. Chen, X., Liu, Z., Sun, M.: Estimating translation probabilities for social tag suggestion. Expert Syst. Appl. **42**(4), 1950–1959 (2015)
13. Qian, Q., Jin, R., Yi, J., et al.: Efficient distance metric learning by adaptive sampling and mini-batch stochastic gradient descent (SGD). Mach. Learn. **99**(3), 1–20 (2015)
14. Furukawa, H.: Intracultural translation into an ideological language: the case of the Japanese translations of Anne of Green Gables. Neohelicon **42**(1), 297–312 (2015)
15. Vann, R.E.: Language exposure in catalonia: an example of indoctrinating linguistic ideology. Word J. Int. Linguist. Associations **50**(2), 191–209 (2015)
16. Ramos, F.P.: Quality assurance in legal translation: evaluating process, competence and product in the pursuit of adequacy. Int. J. Semiot. Law - Revue internationale de Sémiotique juridique **28**(1), 11–30 (2015)

Improvement Method of Full-Scale Euler Angles Attitude Algorithm for Tail-Sitting Aircraft

Yang Liu[1(✉)], Hua Wang[1], Feng Cheng[1], Menglong Wang[1], and Xiaoyu Ni[2]

[1] Beihang University, Astronautics, Beijing, China
m5415823chuizho@163.com
[2] Hebei University of Architecture, Mechanical Engineering, Zhangjiakou, China

Abstract. In this paper, an engineering algorithm that can overcome singularity of Euler Equation is adopted to adapt to particularity of tail-sitting aircraft. According to the practical significance and reference to the other algorithms, we expand its definitions and verify numerical calculation. The results demonstrate that the method is simple and quite effective. With the application of this method in attitude computation system for tail-sitting aircraft, satisfactory results are obtained. As a result, there is no requirement for rotatable parts and corresponding control units and the whole aircraft is in simple structure and small mass. Besides, the course of flight is simplified into fixed-wing aircraft maneuvering, thus it is easy to operate and especially suitable for the unmanned vehicles, for which there is no need to consider physical limitations of flight attendants. After redefining the value range of Euler angles, this method can be perfectly applied in attitude computation for tail-sitting aircraft which is also proved to be feasible through using experimental verification with universal application and reference value in engineering practice.

Keywords: Tail-sitting · singularity · Full-scale Euler angle
Attitude algorithm

1 Introduction

Aircraft can be divided into fixed-wing aircraft and rotor aircraft by structure, among which, the former is characterized by high speed, high efficiency and long endurance but also requires necessary conditions such as runway and airspace while landing. Thus, it can only be applied in limited areas; however, rotor aircraft is with vertical take-off and landing (VTOL), spot hover and good low-speed motor function, so it can be successfully used in small space but only under a low speed, flight efficiency and short voyage. In order to meet the increasingly flight missions, numerous new VTOL aircraft have been produced as the times require. Those aircraft can not only cruise horizontally at a high speed like fixed-wing aircraft but also realize VTOL, spot hover and fly steadily at a low speed [1–3]. Tail-sitting aircraft falls into such aircraft mentioned above, among which, the direction of thrust is fixed at vertical axis of the

© Springer Nature Singapore Pte Ltd. 2018
J. C. Hung et al. (Eds.): FC 2017, LNEE 464, pp. 257–269, 2018.
https://doi.org/10.1007/978-981-10-7398-4_27

fuselage, indicating that the thrust and fuselage rotate simultaneously while changing of direction of thrust.

Due to the flight characteristics of tail-sitting aircraft, its pitch angle remains 90° or so for quite a long period during take-off, hovering and landing. Under some complicated circumstances, the angle of pitch may be greater than 90°, bringing about two problems in the confirmation of aircraft attitude information, namely, problems in aircraft attitude solution arising from definition and singularity of Euler angles and problems in the accuracy of normal data fusion algorithms given rise to by complicate flight modes and flight envelope.

With the application and improvement of an engineering method that can solve singularity of Euler Equations, this paper makes it possible to keep the continuity during the whole process of numerical simulation of Euler angles and also realizes numerical solution of full-scale attitude angles. Besides, the method can meet the actual flight attitude of the aircraft. In addition, this paper also introduces a kind of self-adapting complementary filters based on a uniform definition of attitude angles obtained from gyroscope, accelerometer and geomagnetic sensor. Thus, it is possible to realize the solution of high-precision attitude information under various working conditions based on relatively simpler calculation. Accordingly, the two problems are perfectly solved in practical application.

2 Singularity and Full-Scale Attitude Angle Computation

2.1 Singularity of Euler Equation

When it comes to flight mechanics, attitude motion of aircraft in space is defined by Euler's concept as: movement towards a new position through changing the angle for three times in the sequence of yawing, tilting and rolling when aircraft body axis system coincides with the earth-fixed axis system. Accordingly, the corresponding three angles are named as Euler angles, recorded respectively as ψ, θ and φ. Besides, the relation among the angle velocity and aircraft autorotation angle velocity w_x, w_y and w_z is shown in Formula (1), namely, Euler Equation [13].

$$\begin{bmatrix} \dot{\varphi} \\ \dot{\psi} \\ \dot{\theta} \end{bmatrix} = \begin{bmatrix} w_x - \dot{\psi} \sin \theta \\ (w_y \cos \varphi - w_z \sin \varphi)/ \cos \theta \\ w_y \sin \varphi - w_z \cos \varphi \end{bmatrix} \tag{1}$$

In this case, the transition matrix from the earth-fixed axis system to the aircraft body axis system is [14]:

$$A_d^t = \begin{bmatrix} \cos \theta \cos \psi & \sin \theta & -\cos \theta \sin \psi \\ \sin \varphi \sin \psi - \cos \varphi \sin \theta \cos \psi & \cos \varphi \cos \theta & \sin \varphi \cos \psi + \cos \varphi \sin \theta \sin \\ \cos \varphi \sin \psi + \sin \varphi \sin \theta \cos \psi & -\sin \varphi \cos \theta & \cos \varphi \cos \psi - \sin \varphi \sin \theta \sin \psi \end{bmatrix}$$

$$\tag{2}$$

It can be seen from Formula (1) that, if $\theta = 90°$, the equation cannot be solved where singularity of Euler Euqation arises.

2.2 Solution to Singularity

At present, the most popular methods to overcome singularity of Euler angles refer to quaternion method and dual Euler method and so on [8, 9]. The former is usually adopted in strap down inertial navigation system to calculate attitude angle. However, in practical application, due to round-off error (to represent approximately decimals with many digits or unlimited decimals by limited decimals), it is impossible to unitize the quaternion in all cases, consequently leading to error in such algorithm and the error is in an increasing trend. Therefore, the calculation accuracy of this algorithm is lower than that of Euler Equation [10, 11]. The dual Euler method can provide higher calculation accuracy but massive complicate trigonometric function calculations are involved in solving positive and reflex Euler angle equations. Thus, it causes great load for the micro aircraft in which embed system is usually involved. In addition, the domain of definition for pitch by traditional dual Euler method is also $[-\pi/2, \pi/2]$ and the adjustment [12] of domain of definition is extremely complicated and indirectly perceived. Therefore, such method is not suitable for working conditions of tail-sitting aircraft. In the following, an engineering algorithm is used to solve singularity of Euler equation.

Formula (3) can be obtained through taking the derivatives of elements in the transition matrix A_d^t in Formula (2):

$$\dot{a}_{11} = -\sin\theta\cos\psi \cdot \dot{\theta} - \sin\psi\cos\theta \cdot \dot{\psi}$$

$$\dot{a}_{12} = \cos\theta \cdot \dot{\theta}$$

$$\dot{a}_{13} = \sin\theta\sin\psi \cdot \dot{\theta} - \cos\psi\cos\theta \cdot \dot{\psi}$$

$$\dot{a}_{21} = \cos\varphi\sin\psi \cdot \dot{\varphi} + \cos\psi\sin\theta \cdot \dot{\psi} + \sin\varphi\sin\theta\cos\psi \cdot \dot{\varphi}$$
$$- \cos\theta\cos\varphi\cos\psi \cdot \dot{\theta} + \sin\psi\cos\varphi\sin\theta \cdot \dot{\psi}$$

$$\dot{a}_{22} = -\sin\varphi\cos\theta \cdot \dot{\varphi} - \cos\varphi\sin\theta \cdot \dot{\theta}$$

$$\dot{a}_{23} = \cos\varphi\cos\psi \cdot \dot{\varphi} - \sin\varphi\sin\psi \cdot \dot{\psi} - \sin\varphi\sin\theta\sin\psi \cdot \dot{\varphi} \quad\quad (3)$$
$$+ \cos\varphi\cos\theta\sin\psi \cdot \dot{\theta} + \cos\varphi\sin\theta\cos\psi \cdot \dot{\psi}$$

$$\dot{a}_{31} = -\sin\varphi\sin\psi \cdot \dot{\varphi} + \cos\varphi\cos\psi \cdot \dot{\psi} + \cos\varphi\sin\theta\cos\psi \cdot \dot{\varphi}$$
$$+ \sin\varphi\cos\theta\cos\psi \cdot \dot{\theta} - \sin\varphi\sin\theta\sin\psi \cdot \dot{\psi}$$

$$\dot{a}_{32} = -\cos\varphi\cos\theta \cdot \dot{\varphi} + \sin\varphi\sin\theta \cdot \dot{\theta}$$

$$\dot{a}_{33} = -\sin\varphi\cos\psi \cdot \dot{\varphi} - \cos\varphi\cos\psi \cdot \dot{\psi} - \cos\varphi\sin\theta\sin\psi \cdot \dot{\varphi}$$
$$- \sin\varphi\cos\theta\sin\psi \cdot \dot{\theta} - \sin\varphi\sin\theta\cos\psi \cdot \dot{\psi}$$

Formula (4) can be obtained on the basis of Formula (2) by substituting formula (1) into formula (3):

$$
\begin{aligned}
\dot{a}_{11} &= a_{21}\omega_z - a_{31}\omega_y \\
\dot{a}_{12} &= a_{22}\omega_z - a_{32}\omega_y \\
\dot{a}_{13} &= a_{23}\omega_z - a_{33}\omega_y \\
\dot{a}_{21} &= a_{31}\omega_x - a_{11}\omega_z \\
\dot{a}_{22} &= a_{32}\omega_x - a_{12}\omega_z \\
\dot{a}_{23} &= a_{33}\omega_x - a_{13}\omega_z \\
\dot{a}_{31} &= a_{11}\omega_y - a_{21}\omega_x \\
\dot{a}_{32} &= a_{12}\omega_y - a_{22}\omega_x \\
\dot{a}_{33} &= a_{13}\omega_y - a_{31}\omega_x
\end{aligned}
\tag{4}
$$

Time-stepped fourth-order Runge-Kutta method is used to solve formula (4), so the transition matrix can be figured out along with formula (5):

$$
\begin{aligned}
a_{11} &= \cos\theta\cos\psi \\
a_{13} &= -\cos\theta\sin\psi \\
a_{22} &= \cos\theta\cos\varphi \\
a_{32} &= -\cos\theta\sin\varphi \\
a_{12} &= \sin\theta
\end{aligned}
\tag{5}
$$

Then, the three Euler angles can be determined:

$$
\begin{aligned}
\theta &= \arcsin(a_{12}) \\
\varphi &= arctg(-a_{32}/a_{22}) \\
\psi &= arctg(-a_{13}/a_{11})
\end{aligned}
\tag{6}
$$

In any circumstance, formula (4) and (6) make sense and can be used to solve singularity of Euler equation.

2.3   Definition of Full-Scale Attitude Angle in Euler Equation

In formula (6), the domain of the definition of the three Euler angles is the same, namely, $[-\pi/2, \pi/2]$, which obviously does not conform to the actual situation. Both the traditional dual Euler method and quaternion method clearly define value of the pitch angle as the intuitive value, namely, $[-\pi/2, \pi/2]$, and the roll angle and yaw angle are in the scope of $[-\pi, \pi]$ so as to obtain reliable Euler angle values corresponding to certain attitudes of the aircraft. However, in case the actual pitch angle of some kind of aircraft exceeds 90°, such as, the tail-sitting aircraft and the computed value of roll angle would be reduced to a 180-degree change. According to Fig. 1, that change is equivalent to 180° rolling of the aircraft [12, 16], which does not occur in actual flight.

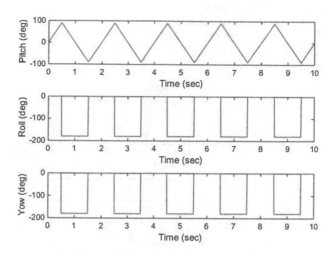

Fig. 1. Conventional attitude algorithm diagram

If the limits on pitch angle are canceled, there will be two sets of Euler angles corresponding to a certain attitude of the aircraft (including the intuitive value and miracle value) so that it is hard to adopt or discard one of the two angles [13]. Therefore, in view of the actual working conditions and environment of tail-sitting aircraft through specifying the roll angle being $[-\pi/2,\ \pi/2]$, we expand the definition range of pitch angle and yaw angle into full scale, namely, $[-\pi,\ \pi]$ to meet the practical requirement and ensure that there is only one set of Euler angle corresponding to one attitude. In this case, the true values of pitch and yaw angle can be determined as per formula (5) and (6) based on the method shown in Figs. 2 and 3.

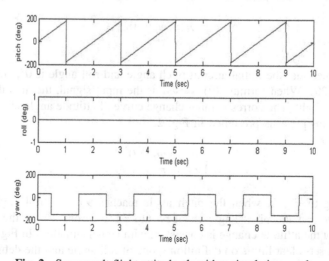

Fig. 2. Somersault flight attitude algorithm simulation results

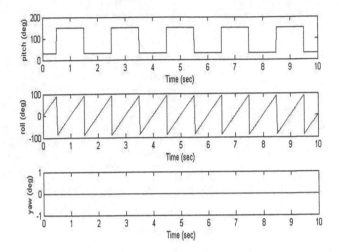

Fig. 3. Rolling flight attitude algorithm simulation results

In program language, the pitch angle in formula (6) can be modified into the following expression so as to improve calculation accuracy of pitch angle.

$$\theta = \arctan(a_{12} \cdot \cos \varphi / a_{22}) \tag{7}$$

Under some special circumstances,simulation is carried out to verify correctness the algorithm. It is assumed that the initial aircraft pitch angle and roll angle is 0°, and the initial yaw angle is 30°. When formula (8) is used as the input signal, the aircraft is in pure somersaulting flight. The corresponding change curve of attitude angle obtained by the algorithm in this paper is shown in Fig. 3.

$$\begin{cases} \omega_x = \omega_y = 0 \\ \omega_z = \pi \end{cases} \tag{8}$$

It is assumed that the initial aircraft pitch angle and roll angle is 0°, and the initial yaw angle is 30°. When formula (9) is used as the input signal, the aircraft is in pure somersaulting flight. The corresponding change curve of attitude angle obtained by the algorithm in this paper is presented in Fig. 2.

$$\begin{cases} \omega_y = \omega_z = 0 \\ \omega_x = \pi \end{cases} \tag{9}$$

According to Fig. 2, when the pitch angle reaches 90°, the roll angle is not in abrupt change and the pitch angle changes into −150° from 30° at the same time, indicating that the attitude change is consistent with actual situation. In Fig. 5, because the algorithm sets clear limits on definition scope of roll angle and the definition range of pitch angle is expanded into full scope, it demonstrates that when the roll angle exceeds 90°, the pitch angle turns into 150° from 30° with no change in the yaw angle. This is because such algorithm puts limits on definition scope of the roll angle, and the definition range of pitch angle is expanded into full scope. Accordingly, the simulation conforms to the actual flight attitude.

2.4 Full-Scope Attitude Computation Based on Accelerometer and Geomagnetic Sensor

The attitude algorithm mentioned above and the definition of full-scope is based on gyroscope data. However, in actual engineering application, accelerometer and geomagnetic sensor data is needed for correction of the results in order to eliminate the accumulative error of the gyroscope. If the accelerometer output is $[a_x \quad a_y \quad a_z]$, following formulas can be obtained:

$$\begin{bmatrix} a_x \\ a_y \\ a_z \end{bmatrix} = \begin{bmatrix} g \cdot \sin\varphi\cos\theta \\ -g \cdot \sin\theta \\ -g \cdot \cos\varphi\cos\theta \end{bmatrix} = \begin{bmatrix} -g \cdot a_{32} \\ -g \cdot a_{12} \\ -g \cdot a_{22} \end{bmatrix} \tag{10}$$

$$\begin{cases} \theta = \arcsin(-a_y/g) = \arcsin a_{12} \\ \varphi = \arctan(-a_x/a_z) = \arctan(-a_{32}/a_{22}) \end{cases} \tag{11}$$

If the geomagnetic sensor output is $\begin{bmatrix} B_x^t & B_y^t & B_z^t \end{bmatrix}$, following formulas can be obtained:

$$\begin{cases} B_x^d = B_y^t \cdot \cos\varphi - B_z^t \cdot \sin\varphi \\ B_y^d = B_x^t \cdot \sin\theta\cos\varphi + B_y^t \cdot \cos\theta - B_z^t \cdot \sin\theta\cos\varphi \end{cases} \tag{12}$$

$$\begin{cases} \psi_s = \arctan(B_y^d/B_x^d) \\ \psi = \psi_s + \tau \end{cases} \tag{13}$$

In formula (13), τ represents the magnetic declination between the magnetic north and the true north.

Determination of true value of θ, ψ in formula (11) and (13) is similar to the third part. Please refer to Figs. 2 and 3 for detailed information. The definition range is adjusted into $[-\pi, \pi]$ to realize full-scale attitude computation on the basis of accelerometer and geomagnetic sensor.

In program language, the pitch angle in formula (11) can be modified into the following expression so as to improve calculation accuracy of pitch angle.

$$\theta = \arctan(-a_y \cdot \sin\varphi/a_x) \tag{14}$$

3 Data Fusion Algorithm

After the determination of full-scope attitude algorithm for the gyroscope, accelerometer and magnetic sensor, fusion algorithm for attitude computation results of all sensors is to be carried out for final attitude information. Currently, the typical data fusion algorithm include two categories: The first one is to distinguish and eliminate noise by frequency domain regarding statistical characteristic of noise, such as, various improved complementary filter algorithms based on complementary filter, including classic complementary filter (CF), explicit complementary filter (ECF), and the complementary filter based on gradient descent method (GDCF) [17–21]; The second one is to design filter in time domain with the use of status space, like various extended and derivative algorithms based on Kalman filtering algorithm, such as extended Kalman filter (EKF), unscented Kalman filter (UKF) and federated Kalman filter etc. [22–26]. Among them, Kalman filter can provide higher calculation accuracy but requires complicated principle and massive calculation; compared with Kalman filtering algorithm, the complementary filtering algorithm does not require due consideration given to the statistical characteristics of signals, or precise modeling of errors but distinguishes and eliminates disturbance [27] by frequency domain. As a result, it is simple and easy to realize, and more suitable for low-precision MEMS inertial integrated navigation.

3.1 Basic Principles of Complementary Filter

Attitude angle can be obtained through integral operation of the gyroscope output, thus, accumulative error would occur. In low frequency range, the measurement data is in large error but in high frequency range, which can clearly reflect the dynamic changes of attitude with satisfactory measurement effects; However, measurement error of accelerometer does not accumulate along with time but it can be significant influenced by mechanical vibration, acceleration and noise. Therefore, it can realize dynamic response in low frequency range but produce large error in high frequency range. From the perspective of frequency domain, through complementary filter high and low pass filter are introduced to gyroscope and accelerometer respectively to eliminate the low-frequency noise in the gyroscope and high-frequency noise in the accelerometer. Then, after data fusion, the dynamic response advantages of both the gyroscope and accelerometer in their own frequency range are completely reflected to reduce measurement and estimation error. Its basic principles are as shown in Fig. 4, in which, $F_1(s) = \frac{K}{s+K}$, $F_2(s) = \frac{s}{s+K}$, and $F_1(s) + F_2(s) = 1$, all the requirements are met to form complementary filter.

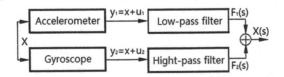

Fig. 4. Block diagram of complementary filter basic principles

3.2 Determination of Cut-off Frequency

The effects of complementary filter lie in the determination of the cut-off frequency K of high and low pass filter, In which, the geomagnetic sensor error in estimated attitude angle mainly comes from measurement noise. Therefore, it is better to select the smallest cut-off frequency under the premise of accurate estimation of the attitude angle based on the meeting gyroscope so as to filter out noise in the measurement among a larger range of frequency [28]. The range of cut-off frequency of the geomagnetic sensor can be figured out through using static tests. In this paper, $K_\psi = 0.1$.

Measurement error of accelerometer mainly originates from maneuvering acceleration of the carrier. Therefore, fixed parameters cannot meet all the situations especially large flight envelope and complex motion. As such, this paper provides a self-adaptive parameter adjustment scheme which can select proper cut-off frequency according to the actual situation on the basis of fuzzy module.

Define $e = \left| \sqrt{a_x^2 + a_y^2 + a_z^2} - g \right|$, it refers to the difference between acceleration vector modulus length and gravitational acceleration, which also reflects motion of the carrier. When the carrier is in state of rest, or of uniform motion, e approaches 0; if the carrier is in state of maneuvering acceleration, e becomes bigger. When e is regarded as the parameter to adjust the input of fuzzy module, the triangular membership function is obtained as shown in Fig. 5 (1); when cut-off frequency K is used as the parameter to adjust output of the fuzzy module, the membership function is obtained as presented in Fig. 7 (2).

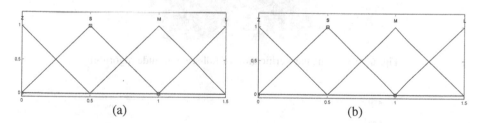

(a)   (b)

Fig. 5. Input E membership function and Output K membership function

The fuzzy reasoning rules are:

if e Z, then K L;
if e S, then K M;
if e M, then K S;
if e L, then K Z;

Normalized gain g_e, g_K is selected according to the actual operation status of the object, here $g_e = 0.58$, $g_{K_\theta} = 1.7$, $g_{K_\varphi} = 1.6$.

4   Experimental Verification

In this part, the actual effects of such algorithm are verified. With the platform of flight control board made by myself, the full-scale attitude algorithm is used. Besides, by attitude computation through collecting gyroscope data and information fusion with other sensor signals, three Euler angles are achieved. The test is carried out in the way of manual operation, during which, the aircraft pitch angle is gradually increased along with swinging back and forth in the vertical direction. changes in the three Euler angles are recorded and compared to the results obtained by a mature flight control system (MWC) based on conventional attitude algorithm. As indicated by Fig. 6, the left part represents MWC flight control, and the right part is the self-made flight control board. The results are provided in Figs. 7 and 8 which are basically consistent with the simulation results.

Fig. 6. Experimental verification of full-scale attitude algorithm

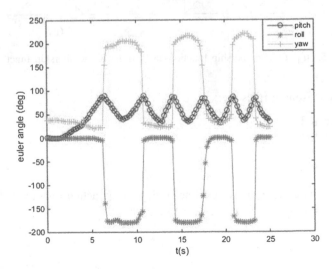

Fig. 7. Euler angles outputted by the attitude system through conventional attitude algorithm

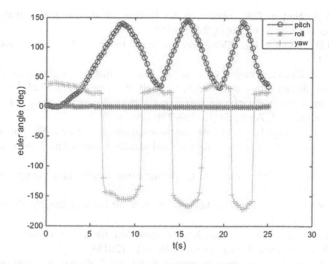

Fig. 8. Euler angles outputted by the attitude system through full-scale attitude algorithm

5 Conclusion

To conclude, the attitude algorithm provided in this paper combines the advantages of Euler Equation and quaternion method respectively, namely, visual representation and simple calculation, and successfully solves the problem of singularity of Euler Equation without any flaw in principles or error in method, implying that the calculation accuracy is as high as that of Euler Equation. After redefining the value range of Euler angles, this method can be perfectly applied in attitude computation for tail-sitting aircraft which is also proved to be feasible through using experimental verification with universal application and reference value in engineering practice. In this part, the actual effects of such algorithm are verified. With the platform of flight control board made by myself, the full-scale attitude algorithm is used. Besides, by attitude computation through collecting gyroscope data and information fusion with other sensor signals, three Euler angles are achieved. The test is carried out in the way of manual operation, during which, the aircraft pitch angle is gradually increased along with swinging back and forth in the vertical direction. changes in the three Euler angles are recorded and compared to the results obtained by a mature flight control system (MWC) based on conventional attitude algorithm.

References

1. Janota, A., Nemec, D.: Improving the precision and speed of euler angles computation from low-cost rotation sensor data. Sensors **15**(3), 7016 (2015)
2. Garcia, V., Roberta, K., et al.: Unscented Kalman Filter applied to the spacecraft attitude estimation with Euler Angles. Math. Probl. Eng. **2012**(1), 161–174 (2011)

3. Phillips, W.F., Hailey, C.E., Gebert, G.A.: Review of attitude representations used for aircraft kinematics. J. Aircr. **40**(1), 223 (2015)
4. Shuster, M.D., Markley, F.L.: Generalization of the Euler Angles. J. Astronaut. Sci. **51**(2), 123–132 (2003)
5. Garcia, O., Castillo, P., Wong, K.C., et al.: Attitude stabilization with real-time experiments of a tail-sitter aircraft in horizontal flight. J. Intell. Rob. Syst. **65**(1), 123–136 (2012)
6. AIAA. Experimental validation of the hybrid airfoil design procedure for full-scale ice accretion simulation. J. Aircr., **36**(5), 769–776 (1998)
7. Liu, F., Li, J., Wang, H., et al.: An improved quaternion Gauss-Newton algorithm for attitude determination using magnetometer and accelerometer. Chin. J. Aeronaut. **27**, 986–993 (2014)
8. Levy, R.H., Petschek, H.E., Siscoe, G.L.: Aerodynamic aspects of the magnetospheric flow. Aiaa J. **2**(12), 1502 (1963)
9. Shearer, C.M., Cesnik, C.E.S.: Nonlinear flight dynamics of very flexible aircraft. J. Aircraft **44**(5), 1528–1545 (2007)
10. Nemec, M., Zingg, D.W.: Newton-Krylov algorithm for aerodynamic design using the Navier-Stokes Equations. Aiaa J. **40**(6), 1146–1154 (2015)
11. Janota, A., Nemec, D.: Improving the precision and speed of euler angles computation from low-cost rotation sensor data. Sensors **15**(3), 7016 (2015)
12. Phillips, W.F., Hailey, C.E., Gebert, G.A.: Review of attitude representations used for aircraft kinematics. J. Aircr. **40**(1), 223 (2015)
13. Cong, L., Li, E., Qin, H., et al.: A performance improvement method for low-cost land vehicle GPS/MEMS-INS attitude determination. Sensors **15**(3), 5722–5746 (2015)
14. Shandor, M., Stone, A.R., Walker, R.E.: Secondary gas injection in a conical rocket nozzle. Aiaa J. **1**(2), 334–338 (2015)
15. Mourcou, Q., Fleury, A., Franco, C., et al.: Performance evaluation of smartphone inertial sensors measurement for range of motion. Sensors **15**(9), 23168–23187 (2015)
16. Valenti, R.G., Dryanovski, I., Xiao, J.: Keeping a good attitude: a quaternion-based orientation filter for IMUs and MARGs. Sensors **15**(8), 19302–19330 (2015)
17. Kempe, T., Lennartz, M., Schwarz, S., et al.: Imposing the free-slip condition with a continuous forcing immersed boundary method. J. Comput. Phys. **282**, 183–209 (2015)
18. Drela, M., Giles, M.B.: Viscous-Inviscid analysis of transonic and low Reynolds number airfoils. Aiaa J. **25**(10), 1347–1355 (2015)
19. Levin, D., Katz, J.: Vortex-Lattice method for the calculation of the nonsteady separated flow over delta wings. J. Aircr. **18**(12), 1032–1037 (2015)
20. Chen, P.C., Zhang, Z., Sengupta, A., et al.: Overset Euler/Boundary-layer solver with panel-based aerodynamic modeling for aeroelastic applications. J. Aircr. **46**(6), 2054–2068 (2015)
21. Renno, J.M., Inman, D.J., Chevva, K.R.: Nonlinear control of a membrane mirror strip actuated axially and in bending. Aiaa J. **47**(3), 484–493 (2015)
22. Zhang, J., Wu, M., Li, T., et al.: Integer aperture ambiguity resolution based on difference test. J. Geodesy **89**(7), 667–683 (2015)
23. Zhao, Y., Ji, J., Huang, J., et al.: Orientation and rotational parameters of asteroid 4179 Toutatis: new insights from Chang'e-2's close flyby. Mon. Not. R. Astron. Soc. **450**(4), 3620 (2015)
24. Evangelista, D., Cam, S., Huynh, T., et al.: Shifts in stability and control effectiveness during evolution of Paraves support aerial maneuvering hypotheses for flight origins. Peerj **2**(2), e632 (2015)

25. Baù, G., Bombardelli, C., Peláez, J., et al.: Non-singular orbital elements for special perturbations in the two-body problem. Mon. Not. R. Astron. Soc. **454**(3), 2890–2908 (2015)
26. Stanway, M.J., Kinsey, J.C.: Rotation identification in geometric algebra: theory and application to the navigation of underwater robots in the field. J. Field Rob. **32**(5), 632–654 (2015)
27. Guha, A.K.: Attitude stability of dissipative dual-spin spacecraft. Aiaa J. **10**(7), 851–852 (2015)
28. Elmandouh, A.A.: New integrable problems in rigid body dynamics with quartic integrals. Acta Mech. **226**(8), 2461–2472 (2015)

A Research on the Developing Platform of Interactive Location Based Virtual Learning Application

Edgar Chia-Han Lin[✉]

Department of Information Communication, Asia University, Taichung, Taiwan
edgarlin@asia.edu.tw

Abstract. Due to the development of mobile devices and wireless networks, mobile applications are more and more popular in our daily lives. Since most mobile devices have the GPS positioning function, it makes the Location Based Services become an important research issue. The early applications of location based services focused on GIS-related services, such as mobile trajectory records, maps, navigation and other services. Pokemon GO is a well-known LBS related game. In addition, in the past few years, the usage of IT technology in teaching and learning environment has become more popular. Due to the popularity of mobile devices, the design of teaching applications can been implemented through various mobile devices. It provides students with ubiquitous learning area.

Therefore, how to combine the LBS application and mobile learning system such that the student can learn anywhere with various learning activities become an important research issue. Furthermore, it will form a whole new learning method for students to learn.

The main purples of this paper is to develop an interactive learning platform for LBS learning applications. Based on this platform, the teacher can design different kinds of content, such as teaching activities, AR applications, interactive games, multi-level tournament and so on, for different locations. The corresponding designed content will be triggered when the user or student pass some particular location. Moreover, the nearby users can help the other users to complete the learning activities which is truly a kind of collaboration learning.

Keywords: Location based service · Mobile learning · Database management
Spatial databases

1 Introduction

With the popularity of mobile devices and wireless networks, mobile applications are increasingly popular in people's daily lives. Because GPS function is popular in modern mobile devices, the location based services (LBS) applications become more popular recently. At first, the LBS applications focus on GIS based service such as moving trajectory records, maps, navigation and other services. Pokemon Go shows a totally different way of LBS applications. We finally realize that the LBS services in closely

© Springer Nature Singapore Pte Ltd. 2018
J. C. Hung et al. (Eds.): FC 2017, LNEE 464, pp. 270–277, 2018.
https://doi.org/10.1007/978-981-10-7398-4_28

related to our daily life. Moreover, in the past few years, the IT technology is highly integrated to the teaching and learning applications. Due to the popularity of mobile devices, the design of teaching applications can be implemented through various mobile devices. It provides students with ubiquitous learning area.

Therefore, how to combine the LBS application and mobile learning system such that the student can learn anywhere with various learning activities become an important research issue. Furthermore, it will form a whole new learning method for students to learn.

The main purples of this paper is to develop an interactive learning platform for LBS learning applications. Based on this platform, the teacher can design different kind of content, such as teaching activities, AR applications, interactive games, multi-level tournament and so on, for different locations. The corresponding designed content will be triggered when the user or student pass some particular location. Moreover, the nearby users can help the other users to complete the learning activities which is truly a kinds of collaboration learning.

In this paper, a content management module is designed to process the content, which including the development and establishment of the learning content, content database construction and management, the integration of positioning information, and the establishment and management of spatial database. Moreover, the user management module is also designed to record the learning activities of each students. At last, a learning application management system is constructed to combine the user information and the LBS learning activities.

2   Related Work

With the popularity of mobile devices and Internet access, people can access information via mobile devices anytime, anywhere, and the GPS positioning system on mobile devices also makes the application LBS [1] more popular. The LBS application is used to provide particular service for users based on the geographic information. Therefore, most of the early applications are focused on track records, map navigation applications, and commercial navigation applications which are geographically relevant. In addition, some studies focus on the integration of geographical location to carry out the information recommended [2–4].

At the same time, learning through mobile devices has become very important for both teaching and learning [5, 6]. Because of the mobility of mobile device, the learning activities can be perform outside the classroom. It may be good to enhance the learning performance of the students.

In our previous works [7–12], we focus on the various types of multimedia database index structure, query processing mechanism and management direction, In this paper, the construction and management of different types of databases also play a very important role. How to effectively find the relevant learning content to users based on the location information is an important research topics.

3 Proposed Approach

An Interactive Location-Based Learning Application Development Platform is proposed in this paper. Users can get the learning material based on their location. The learning material including the teaching content, AR application, interactive games and so on. The user account can be created via the app and the interested content will be provided based on the user profile. The learning material can be designed and uploaded through the platform along with the GPS coordinates. The system architecture is shown in Fig. 1.

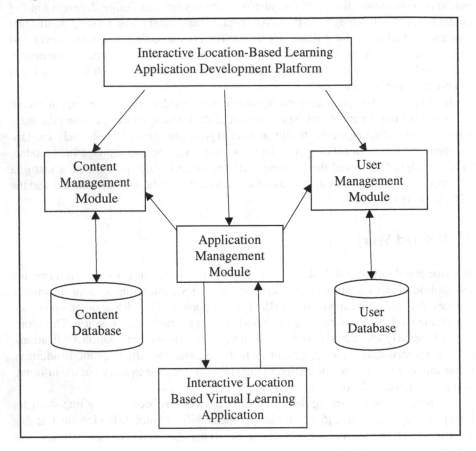

Fig. 1. System architecture

3.1 Content Management Module

In this module, the interactive learning activities can be created along with the location properties. The learning activities can be designed through text, multimedia content or even AR application. Furthermore, some key-test can be designed to verify the learning performance which can be recorded in the user profile. Moreover, since the learning activities are related to the location properties, the construction of spatial database should be also considered and integrated to the content database. The proposed content management module is shown in Fig. 2, and described in detail as follows.

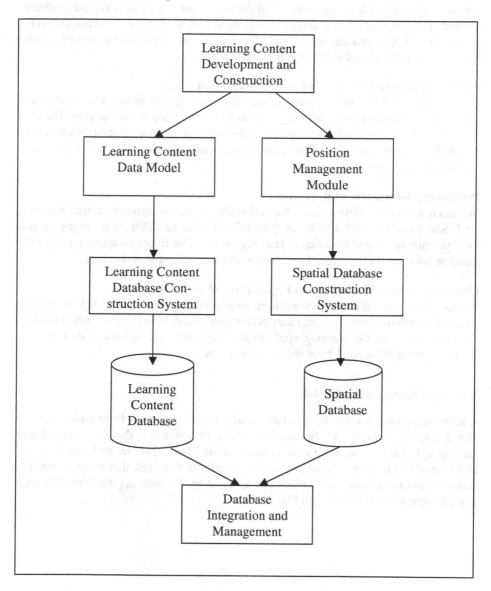

Fig. 2. Content management module building and system processes

The Development and Establishment of Learning Content
In this subsection, different types of learning content and the corresponding presentation method are defined. Since we are constructing a development platform in this paper, all the accepted types of learning content and the corresponding formats should be clearly defined. The learning content designed only need to design the multimedia materials, exams, or AR applications, in a standardized way. And the learning content can be integrated into the platform for further management.

Content Database Construction and Management
After the design of the learning content, since the learning content may exist in different forms, the data model of the learning content will be defined in this section. The data model includes the semantic description of the relevant learning content. Each item in the database (each learning content material) is recorded based on the data model which is described as a metadata.

Positioning Information Integration
In this subsection, the positioning information along with the learning content is considered. Since one learning activity may be related to some different locations, or one location may be related to different learning contents, an integration approach for the location information is designed and used to construct the spatial database.

Construction and Management of Spatial Database
In this subsection, all the location information is integrated and recorded in a spatial database. The index structures and the development of database integration applications are constructed. In the learning applications, the learning content can be found by searching the user location from the spatial database.

3.2   User Management Module

Another important research topic is the management of user data. To provide a personalized learning experience, the user's learning profile must also be recorded and analyzed. Therefore, we will focus on the analysis of user profiles and their learning history and build a user database to record the information such that it can be quickly accessed and integrated with the content database by learning applications. The user management module is shown in Fig. 3.

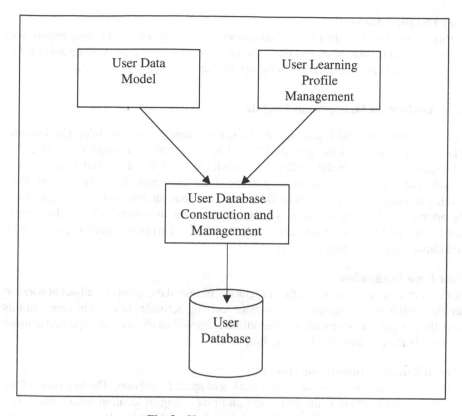

Fig. 3. User management module

User Profile

The key issue of user management module is how to establish and manage the user data. Therefore, the user data model will be created to describe the basic information of the user and can be used as a basis for collaborative learning. In the learning application, all the learning activities, such studying, solving problems, collaborating with other users, will be recorded in the user profile.

Learning Interest and Learning Profile

In this subsection, the user's learning process will be recorded and appropriately managed it. The user can select the subject of interest at the beginning of the registration, and in the course of the application, the learning activity will be triggered according to the location of the user, and the learning process will be recorded. In the design of learning content, teachers can also create cross-regional learning activities, such as the compilation of information and activities of some special elementary schools around the Taiwan. When the user participate the learning activities of certain number of schools, he can get certain learning resources, or unlock other learning activities.

User Database Design
In this subsection, user data model introduced previously and the learning history will be recorded in the database, and the corresponding index structure will be designed to provide fast and efficient query processing and data management.

3.3 Application Management Module

In this section, we will focus on the application management module. The learning Application will access the user profile and learning content through the application management module. In the application module, the location of the mobile device will be detected and the user information and learning history recorded in the user database and the learning activities stored in the learning content database will be integrated and the appropriate learning activities will be applied in the user mobile device. In order to achieve the effect of collaborative learning, we will use the management module to track and cluster users' location regularly.

User Data Integration
In this subsection, we focus on the integration of user data, grouping adjacent users to provide collaborative learning to complete learning activities or examinations. In this step, the geographic information of the online users will be checked and updated in order to provide the collaborative learning function.

Spatial Database Integration Query
In this subsection, we focus on the integration of spatial databases. The key issue of the location based learning is the extraction and processing of location information. The nearby learning content can be efficiently found through the index structure of the spatial database and can be provided to interested users. Moreover, the online users' location management is also considered to efficiently cluster the users for collaborative learning.

4 Discussion and Conclusions

In this paper, a content management module is designed to process the content, which including the development and establishment of the learning content, content database construction and management, the integration of positioning information, and the establishment and management of spatial database. Moreover, the user management module is also designed to record the learning activities of each students. At last, a learning application management system is constructed to combine the user information and the LBS learning activities.

The experimental results shows that the platform works well, the application can access the user profile and the learning activities through the platform and presented in the mobile device.

We are currently apply the platform to local learning environment. The geometric range and user amount is limited. We are currently working on extend the application range and encourage more teachers to create their own learning content for students to learn anytime anywhere.

References

1. Schiller, J., Voisard, A. (eds.): Location-Based Services. Elsevier, San Francisco (2004)
2. Husain, W., Dih, L.Y.: A framework of a personalized location-based traveler recommendation system in mobile application. Int. J. Multimedia Ubiquit. Eng. 7(3), 11–18 (2012)
3. Savage, N.S., et al.: I'm feeling loco: a location based context compliance recommended system. In: Advances in Location-Based Services, pp. 37–54. Springer, Heidelberg (2012)
4. Kuo, M.-H., Chen, L.-C., Liang, C.-W.: Building and visiting a location-based service recommendation system with a preference adjustment mechanism. Expert Syst. Appl. 36(2), 3543–3554 (2009)
5. Chen, Y.-S., Kao, T.-C., Sheu, J.-P.: A mobile learning system for scaffolding bird watching learning. J. Comput. Assist. Learn. 19(3), 347–359 (2003)
6. Chen, Y.-S., et al.: A mobile butterfly-watching learning system for supporting independent learning. In: 2004 Proceedings of the 2nd IEEE International Workshop on Wireless and Mobile Technologies in Education. IEEE (2004)
7. Lin, C.-H., Chen, A.L.P.: Indexing and matching multiple-attribute strings for efficient multimedia query processing. IEEE Trans. Multimedia 8(2), 408–411 (2006)
8. Lin, C.-H., Chen, A.L.P.: Approximate video search based on spatio-temporal information of video objects. In: The First IEEE International Workshop on Multimedia Databases and Data Management (2006)
9. Lin, C.-H., Chen, A.L.: Approximate video search based on spatio-temporal information of video objects. Asian J. Health Inf. Sci. 3(1–4), 52–68 (2008)
10. Lin, E.C.-H.: Research on sequence query processing techniques over data streams. Appl. Mech. Mater. 284–287, 3507–3511 (2013)
11. Lin, E.C.-H.: Research on multi-attribute sequence query processing techniques over data streams. Appl. Mech. Mater. 513–517, 575–578 (2014)
12. Lin, E.C.-H.: A research on 3D motion database management and query system based on Kinect. In: Lecture Notes in Electrical Engineering, vol. 329 (2015)

The Next Generation of Internet of Things: Internet of Vehicles

Wei-Chen Wu[1(✉)] and Homg-Twu Liaw[2]

[1] Hsin Sheng College of Medical Care and Management,
Taoyuan, Taiwan
`wwu@hsc.edu.tw`
[2] Department of Information Management, Shih Hsin University,
Taipei, Taiwan
`htliaw@cc.shu.edu.tw`

1 Introduction

The IoV integrates many important features of the IoT in order to provide numerous new services with advantages for humans and society, in which many sensor nodes are developed for the IoT, such as home appliances, including televisions and refrigerators, and health-related equipment, including heart rate sensors and foot pod meters. However, the IoV products are less developed at present. Hence, we should survey related sensor devices to implement on the vehicles in the future works and then integrate the current research on SDMS and MFSS to achieve vehicle-to-sensor interactions. In order to achieve the desired goal of zero traffic accidents, all information in the vehicle is collected through sensor devices, included in safety messages and forwarded to other vehicles or drivers. The future scheme should be extented to gather, share, process, compute, and release secure information to other humans, vehicles, devices and roadside services. The development and deployment of fully connected vehicles requires a combination of various IoT technologies. From the perspective of commercial value we can see that the future of economies of scale is to be expected, and from the perspective of extending technological development we can see that there are further contributions to be made.

2 Commercial Value

The research will expect to create commercial value in the future. We will create contribution from each of the following perspectives: government, service provider, vehicle industry, software developer and end user.

- Government

From the social costs of government, traffic accidents cause losses of about NTD 475 Billion in Taiwan according to a statistical report [1]. The report represents that for each person who dies in a car accident, the cost to society is about NTD 16 Million loss, and each person who is injured in a car accident, causes a loss of about NTD 1 Million.

© Springer Nature Singapore Pte Ltd. 2018
J. C. Hung et al. (Eds.): FC 2017, LNEE 464, pp. 278–282, 2018.
https://doi.org/10.1007/978-981-10-7398-4_29

According to the statistics of the Ministry of Transportation and Communications [2], 1,928 people died in a car accident and 373,568 people were injured in 2013. This resulted in losses of about NTD 30.3 Billion from car accident deaths and NTD 444.5 Billion in car accident injuries. The total loss was NTD 474.8 Billion, about 3.3% of the Gross Domestic Product [1]. Figure 1 shows the number of deaths and injuries in traffic accidents between 2007 and 2015. If the scheme can provide an efficient way to deliver and forward safety messages to many vehicles to decrease traffic accidents, that would decrease the social cost of government.

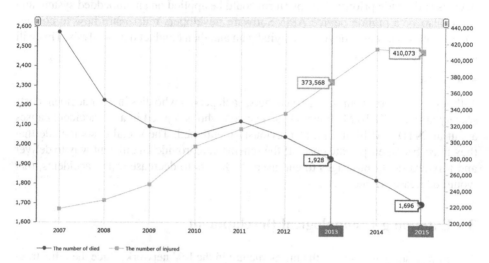

Fig. 1. The number of died and injured in traffic accident

- Service provider

For the sake of service providers, various kinds of vehicle status data will be collected through sensor devices and given to service providers to produce valuable information. When the data are collected in the IoT network, big data will be generated by service providers. This valuable information may be messages of safety or lack of safety. Some valuable information will be provided to vehicles or drivers for free and other valuable information will be sold.

A service provider can build strategic alliances with another service provider. For example, Far Eastern Electronic Toll Collection Co. (FETC) can collaborate with Chunghwa Telecom (IPTV [3] service provider). When a vehicle is equipped with eTag [4] and is traveling on the highway, it can play multimedia file, and multimedia can be provided in the vehicle on demand. This will attract more people to become equipped with eTag and create more economic benefits.

- Vehicle industry

For the sake of the vehicle industry, many related sensor devices that must have Ethernet modules in vehicles will be developed. Various kinds of vehicle's status data will be collected through sensor devices and interact with others sensor devices. For example:

A vehicle speed sensor can be [5] used to measure the speed of the vehicle, a wheel speed sensor can be [6] used for reading the speed of a vehicle's wheel rotation, a tire-pressure monitoring sensor can be [7] used to monitor the air pressure inside the tires, and an engine coolant and oil temperature sensor can be [8] used to measure engine and oil temperature.

- Software developer

For software developers, developing a platform to integrate all the data from sensor devices is their top priority. The platform could be applied on an embedded system and could also be a mobile device APP. Software developers must learn how to convert analog data from sensor devices to digital data and then conduct data analysis. This will be a substantial market.

- End user

Although end users is our major consumers, each person who dies in a car accident costs society about NTD 16 Million and each person who is injured in a car accident causes about an NTD 1 Million loss, as previously mentioned. The social costs include that they must bear their private costs. If the scheme can provide an efficient way to deliver and forward safety messages to end users or drivers to decrease traffic accidents, that would decrease private costs.

3   Extending Technological Development

The current study focuses on the infrastructure in the IoV network. Once the infrastructure is complete, the future research will extend to cloud computing and big data. Each vehicle will have to obtain a valuable information in the IoV by analyzing information from big data and cloud computing. This valuable information will achieve the goal of self-driving cars by providing a wide range of traffic information. Figure 2 shows the future research.

- Big data & Cloud computing

It will require a massive database to collect all the data from the sensor devices in many vehicles, RSUs, drivers and other traffic-related entities. The management of all data and operations will be centralized. Hence, big data & cloud computing in the IoV will be a good topic for ITS or IoT.

- Self-driving cars

In recent years, the issue of self-driving cars has been presented [9–12]. In particular, there have been heated discussions on Google's self-driving cars [11, 12]. However, these techniques rely on related sensor devices to determine the direction of the road and sensor devices simply to do sensing work, which do not have a data collection function. If the sensors that have the function of data collection were produced by the industry and auto suppliers, it would contribute to building up big data in the IoV.

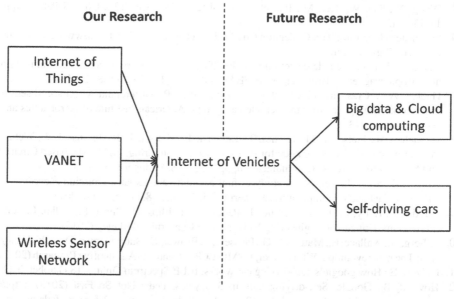

Fig. 2. Future research

Although self-driving cars still seem like science fiction, the vehicle industry projects that such cars will be available in the future.

4 Conclusion

The development and deployment of fully connected vehicles requires a combination of various IoT technologies. From the perspective of commercial value we can see that the future of economies of scale is to be expected, and from the perspective of extending technological development we can see that there are further contributions to be made. The research will expect to create commercial value in the future. We will present create contribution from each of the following perspectives: government, service provider, vehicle industry, software developer and end user. The current study focuses on the infrastructure in the IoV network. Once the infrastructure is complete, the future research will extend to cloud computing and big data. Each vehicle will have to obtain a valuable information in the IoV by analyzing information from big data and cloud computing. This valuable information will achieve the goal of self-driving cars by providing a wide range of traffic information.

References

1. MOTC Institute of Transportation: Traffic accident loss about 475 billion annually (2014). https://www.motc.gov.tw/uploaddowndoc?file=news/201408210828550.docx
2. Ministry of Transportation and Communications: Statistics of ministry of transportation and communications (2016). http://stat.motc.gov.tw/

3. Wang, Y., Shu, W., Her, M.: Internet protocol television, May 12 2006. US Patent App. 11/433,142
4. Far Eastern Electronic Toll Collection Co. (FETC). etag card (2016). http://www.fetc.net.tw/product/eTag-card.html
5. Yang, B., Lei, Y.: Vehicle detection and classification for low-speed congested traffic with anisotropic magnetoresistive sensor. IEEE Sensors J. **15**(2), 1132–1138 (2015)
6. Hazlett, A.C., Crassidis, J.L., Fuglewicz, D.P., Miller, P.: Differential wheel speed sensor integration with gps/ins for land vehicle navigation. American Institute of Aeronautics and Astronautics (2013)
7. Dielacher, M., Flatscher, M., Herndl, T., Lentsch, T., Matischek, R., Prainsack, J., Weber, W.: A robust wireless sensor node for in-tire-pressure monitoring. In: MEMS-based Circuits and Systems for Wireless Communication, pp. 313–328. Springer (2013)
8. Kim, H.-I., Shon, J., Lee, K.: A study of fuel economy and exhaust emission according to engine coolant and oil temperature. J. Therm. Sci. Technol. **8**(1), 255–268 (2013)
9. Hudda, R., Kelly, C., Long, G., Luo, J., Pandit, A., Phillips, D., Sheet, L., Sidhu, I.: Self-driving cars. College of Engineering University of California, Berkeley (2013)
10. Silberg, G., Wallace, R., Matuszak, G., Plessers, J., Brower, C., Subramanian, D.: Self-driving cars: The next revolution. White paper, KPMG LLP & Center of Automotive Research (2012)
11. Guizzo, E.: How google's self-driving car works. IEEE Spectrum Online, 18 October 2011
12. Howard, B.: Google: Self-driving cars in 3–5 years. Feds: Not So Fast (2013). http://www.extremetech.com/extreme/147940-google-self-driving-cars-in-3-5-years-feds-not-so-fast

Using Data Mining to Study of Relationship Between Antibiotic and Influenza

Hsin-Hua Kung[1], Jui-Hung Kao[2,3(✉)], Chien-Yeh Hsu[1], Homg-Twu Liaw[4], and Chiao-Yu Yang[4]

[1] Department of Information Management, National Taipei University of Nursing and Health Sciences, Taipei, Taiwan
[2] Graduate Institute of Biomedical Electronics and Bioinformatics, National Taiwan University, Taipei, Taiwan
kao.jui.hung@gmail.com
[3] Computer Center, National Taipei University of Nursing and Health Sciences, Taipei, Taiwan
[4] Department of Information Management, Shih Hsin University, Taipei, Taiwan

1 Introduction

Since the National Health Insurance was applied in 1995, the main goal was to offer better medical care to the general public, and to reduce the medical fare as well. Even with the general influenza, the general public would rather choose large medical center instead of local clinic since large hospital is recognized as better and more accurate. As the result, people would directly visit large operation for illness.

According to Antibiotic Usage in Hospitalized Patients in Taiwan: 2000 to 2004 (Zhang 2006). The main content indicates that the use of antibiotic in medical center has occupied the highest ratio in all the medical operations. In addition, it introduces the happening of antibiotic use for hospitalized patients in Taiwan from 2000 to 2004 based on the analysis on open data provided by Bureau of National Health Insurance. From the report, it is obvious that the antibiotic prescription charge has been increased from 6.7 billion in 2000 to 9.3 billion in 2004 which is 39% more in five years.

From many previous researches, they indicate that it is not proper to apply antibiotic to those uncomplicated influenza.

1. Use RFM analysis model to figure out the high usage in Influenza group.
2. Apply Two Step Cluster model to find the characteristics of antibiotic prevention.

Thus, the study will mainly focus on the usage of influenza cases, analyzing the needs of applying antibiotic since most of the cases are caused by virus.

When the topic is confirmed, we had started to collect the data, information and reference, finding the corresponding code for influenza. Finally, we would convert and analyze the characteristics based on the database. The research process is shown as Fig. 1:

© Springer Nature Singapore Pte Ltd. 2018
J. C. Hung et al. (Eds.): FC 2017, LNEE 464, pp. 283–297, 2018.
https://doi.org/10.1007/978-981-10-7398-4_30

Fig. 1. Research process

2   Previous Studies

This research targets patients with Influenza. During the outpatient visits, antibiotics are prescribed by physicians. We look into the likelihood of patients during the outpatient care receiving hospitalization. The primary method applied in this research is Data Mining, which aims to find the correlations between influenza and the use of antibiotics.

International Statistical Classification of Diseases and Related Health Problems (ICD) is a statistical classification of disease diagnoses developed by World Health Organization (WHO). It provides numbers for diseases, symptoms, syndromes, abnormalities, discomfort and injuries, and calculate international fatalities. ICD code is essential, for it provides a common language and report to measure diseases and keep them under surveillance. ICD benefits the world with a fast, consistent and shared statistics.

Flu virus is disseminated as droplets through air, human-to-human contact or virus-contaminated item contact. Flu is similar to the general upper respiratory infection in symptoms but treatments are not the same. Antibiotics are used to treat flu rather than the general upper respiratory infection. Therefore, two genres have to be segmented during the treatment.

Antibiotics are defined as metabolites or other derivatives from microorganism with the function of inhibiting other microorganisms. Since many scholars researched and developed chemically-synthesized antibiotic drugs, antibiotics have been further defined as drugs with the function of inhibiting or killing bacteria, no matter whether the drug substances are productions from microorganism or chemical syntheses, can be referred to as antibiotics.

Dong's research demonstrates that the behavior of strolling around the hospital is to blame for most of the wastes of the medical resource. Children are among the main culprits that stroll around the hospital. The affecting factors of children's strolling the hospital are patients themselves, physicians and other factors, in the order of influence level (Dong 2008).

Regarding the issue of antibiotics usage, it is inappropriate to treat non-complicated UTRI since antibiotics are of no remedial utility and worse still, they can cause bacterial resistance. Positive incidence between antibiotics usage and antibiotic resistance has been verified by many studies and journals.

In order to reduce antibiotics use, National Health Insurance Administration Ministry of Health and Welfare added a new drug benefit regulation on February 1st in 2001 regulating that UTRI patients of normal colds or viral infections should not be provided with antibiotics.

It is consultation physicians that have direct contact with patients who have expectations. In other words, patient expectations have an impact on physicians' prescription.

It is also likely that when patients start to be anxious about their health problems, physicians' judgment of medical prescription and referral is affected.

Research has found that children with respiratory diseases use drugs repeatedly and take Antihistamine underage in the outpatient division of Western medicine. The inappropriate prescriptions are related to physicians' age, specifications and urbanization degree. In the future, safety literacy about children's medication can be imparted to physicians of different specifications. It is also suggested that hospitals take on more assessments of medication safety indicators to reduce the children medication-derived risks (Wu 2011).

Key data are utilized in the so-called database, which enables us to retrieve needed data by performing search, addition and modification functions of the key data. There are many data storage varieties among which the most well-known one is tabular data storage. Data items are listed in columns and rows. Data items can accumulate to an uncountable extent. Huge database is formed by the accumulation of data items.

As personal computers and databases are more advanced, RFM model is again reapplied in Data Mining as one of the good approaches to mining and analyzing data. Although RFM model is just among the numerous Data Mining tools, it has the simplest and fast feature of classification, which is why RFM model is commonly used. RFM has three indicators: Recency, Frequency and Monetary Value, the combination of which is used to decide the value of every customer and plan useful customer relations strategies so that entrepreneurs can effectively manage valuable customers. The main purpose of RFM is to provide a simple framework to quantify customer expenditure behavior. Once customers are distributed to the RFM fraction, they can be efficiently segmented into different profit-making groups. The analysis of profit forms the foundation of contact frequency with customers (Miglautsch 2000).

The calculation method raised by Miglautsch is comparatively simple. Better still, the division of RFM into five rating scores can effectively prevent extreme values from affecting the results and causing margin of errors. Therefore, many scholars resort to Quintiles to calculate RFM.

The main purpose of Cluster Analysis is to cluster data based on distance. The closer the relative distance of two items, the more similar the two items. Then the two items can be clustered together. Cluster Analysis includes Agglomerative and Divisive.

(1) Agglomerative: Cluster the closest two clusters together so that the clusters keep decreasing. Ultimately, all clusters are subsumed into one cluster.
(2) Divisive: Divide a cluster into several items.

TwoStep Cluster analysis is primarily designed to demonstrate data cluster as a clustering pursuit tool. It has been already used on continuous variables and nominal variables. The primary feature of TwoStep Cluster analysis is the capabilities of coping with nominal variables and continuous variables and of automatically selecting and classifying clusters. The comparison of different cluster models and selection criteria values can automatically determine the optimum cluster number. TwoStep Cluster Analysis has high expansibility which can be used to analyze large data.

In other researches, computer-assisted Electronic Medical Records (EMRs) is expected to replace traditional paper work. Medic-related industries make use of informational systems in the hope of improving the quality of medical institutions, ensuring medical safety of patients, and reducing hospital costs (Zeng 2012).

It is a shortcut to comprehending national health insurance database to first know the data structures and relations in the database. Below are variables of national health insurance as shown in Fig. 2:

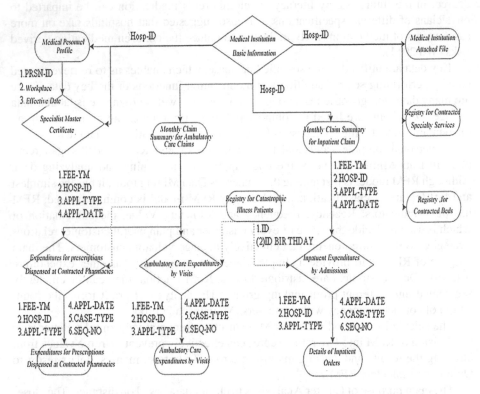

Fig. 2. Systematic sampling on NHIRD website

3   Research Methods

We have clearly introduced the influenza and antibiotic in previous chapter. The next would be seeking for proper research methods. Firstly, we are going to use the SQL Server to gather the data from the National Health Insurance (NHI) database, linking them in series. One thing to be noticed, ambulatory care expenditures by visits and details of inpatient orders must set in series. Finally, we would execute the longitudinal cohort with series file, analyzing the longitudinal cohort data with SPSS Modeler 14.1.

The first layer of this research is to grab the influenza data from the NHI database. The second step is to apply the SPSS Modeler structure RFM and Tow Step Method to analyze the data. The main structure is shown in Fig. 3:

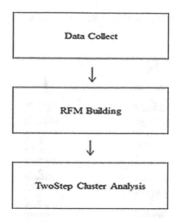

Fig. 3. Research structure

Since this research is in series with antibiotic, we need to add the antibiotic filter while inputting OO file into CD file and DD file into DO file within the NHI database. Since the series drugs are all antibiotic, the drug name would not be displayed in the NHI database. Thus, the code to represent this drug is applied.

The following NHI files are used in this research and divided into 5 types.

(1) CD: Ambulatory care expenditures by visits
(2) DD: Inpatient expenditures by admissions
(3) OO: Details of ambulatory care orders
(4) DO: Details of inpatient orders
(5) Longitudinal Cohort: Using ID and birthday to cumulate the count.

This research mainly apply Hughes (1994) RFM model to analyze. Since RFM is the tool used to evaluate customer value, it can be applied to any customer with purchase history. RFM is generally formed by purchase time, frequency and amount. Thus, we are able to filter the large patient list through the RFM technology.

(1) Latest Purchase Date: It means the closet date of visiting hospital for patients. Thus, if the score is higher, the closer visit date it is. In addition, the score will use quintiles to divide into R1, R2, R3, R4, and R5. The R5 represents the closest purchase date of the patient.
(2) Purchase Frequency: It means total visits among certain period of time for a patient. On the other hand, higher frequency represents the higher loyalty and customer value. Its quintiles are F1, F2, F3, F4, and F5. The F5 is the highest part.
(3) Purchase Amount: It means a total expense in a certain time period for a particular patient. The higher amount represents the better customer value. Its quintiles are M1, M2, M3, M4, and M5. The M5 indicates the highest purchase amount.

This study tends to use RFM from IBM SPSS Modeler as analysis tool. It is functioning in 5 parts. When SPSS Modeler is executing the RFM model, it would automatically generate 3 numbers from 555, 554, 553, 552, 551, 545, and 544. After encoding, we can calculate the contribution for each group of client and the figure is shown as Fig. 4:

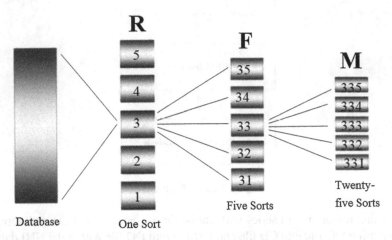

Fig. 4. RFM code construction

Cluster analysis is a method to organize the observed data into groups. It is also another way to group the unidentified data which is similar. Cluster analysis has two main types: Hierarchical method and non-hierarchical method. As the result the TwoStep clustering is combined with these two methods.

Hierarchical method mainly works with the small-scale data, applied to analyze the single, complete and average linkage. The main method is Ward's Method. On the other hand, the nonhierarchical is well-known as K-Means method. It is often used to process the large data. As the result, when the number of group is classified, the TwoStep Cluster would automatically generate the best group set. Thus, the TwoStep Cluster method is applied in this paper.

4   Data Analysis and Result

According to the research method in Sect. 3, we have applied TwoStep Cluster model and used SPSS Modeler to run RFM when the influenza data of the NHI database is transferred out by the longitudinal cohort at the beginning.

This research firstly needs to filter out the influenza patients from Million Longitudinal Cohort from 1997 to 2010 and select those patients that have used the antibiotic.

This study matches the corresponding drug code to the name of antibiotic. Thus, we export the influenza data from the NHI database. The following is the procedure of gathering data:

Firstly, we open the IE browser.

(1) Enter the main page.
(2) We select the import option to filter out those influenza patients that have taken antibiotic.

(3) We can select the years and choose the international classification disease codes. Then we input ICD-9 and A-code code here.

(4) Then select the research year we would like to include. We will include date from 1997 to 2010. After the data importing is finished, it will display the required time and status.

(5) After importing the visits data, we can move on to import the inpatient data and simply switch the option from visits to inpatients, and using the same ICD number to collect the influenza data.

(6) If both visits and inpatients data importing are finished, the longitudinal cohort can be generated by clicking the enter button.

We use the IBM SPSS Modeler 14.1 manual to configure the variables. Some variable settings would not be included in this illustration:

(7) SPSS Modeler cannot read the date column in the NHI database. Thus, we need to convert the file into .tex and reformat into DD-MM-YYYY for SPSS Modeler to read the data.

(8) This paper sets the Excel node as the source for building the RFM mode, configuring the SPSS Modeler to read all the data within this node since it is needed to process the column data into configurable data.

(9) We need to change the name of data reading format for ID and ID_SEX and set them as input. For other measurements, we need to modify them to be serial.

(10) In the section of node type, we need reconfirm the setting of node if it is same as previous setting. If yes, then we can jump to the next step.

(11) RFM node is mainly used to set the ID, data, and value. ID would be set in ID cell, date would be choosing sick date, and value would be used as selecting the total amount.

(12) In the analysis cells in RFM, we would set them as default and set the value to good. Finally, we would configure the Bin Threshold value to read from the Bin value tag.

(13) There are three types in Bin value: recent, times, and cost. We usually set them as default and not going to change the setting.

(14) The calculated cost setting result from RFM model would be exported to the cell. We firstly select the output them choose the save to file. After that, we would input the file destination as CSV file.

(15) The view of completing the final exporting is shown in Fig. 5:

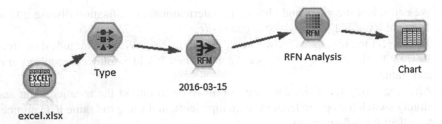

Fig. 5. Final flow chart

(16) We open the exported Excel from the pointed saved destination and introduce the RFM value. A1 cell is the ID; B2 cell is the R value which is also the gap between two visits; C1 cell is F value which is also the frequency distribution and total amount of consumption; D1 is the M value and means the total expense for the patient, shown in Fig. 8:

(17) E1 is the R value with score range 1–5, F1 is the F value with score range 1–5, G1 is the M value with score range 1–5, H1 is the RFM value which is the sum of R, F, and M value, shown in Fig. 6:

	A	B	C	D	E	F	G	H
1	ID	Recent	Times	Cost	Recent Score	Times Score	Cost Scpre	RFM Score
2	fb0f175a2e3d806a0c3782!	5778	1	2400	2	1	5	215
3	fb0ef3dfa38d03a816ef60c.	5239	1	46386	4	1	5	415
4	fb0ef27221f6616df944a39	6179	1	1057	1	1	4	114
5	fb0e8180a0b7e1f98ed7eba	4186	1	386	4	1	3	413
6	fb0e6f5c4ec30c88f53f57b!	6244	1	4255	1	1	5	115
7	fb0e6809292f59f4b17fd77	6086	1	1193	1	1	4	114
8	fb0e52a40dccc0c463f3ab4	5714	1	6583	2	1	5	215
9	fb0e40996f4cb1477e0209(	6385	1	2385	1	1	5	115
10	fb0df7929b0660c776164a(	4268	1	396	4	1	3	413
11	fb0de6056afaee72f329e06	5136	1	301	4	1	3	413
12	fb0de2b5b75d2cd6da7b37	5093	1	331	4	1	3	413
13	fb0d8020903ca65f7f9e31c	4897	1	243	4	1	3	413
14	fb0d7513a52be6a6d71e28	5658	1	6082	3	1	5	315
15	fb0d544e6ed9d04c00b094	6385	1	19535	1	1	5	115
16	fb0d468ce80c37e9b78d3a!	6245	1	2239	1	1	5	115
17	fb0d2e1fb96b37d9ecaf4d5	5565	1	1686	3	1	5	315
18	fb0ce9c47414426af5ed70]	6010	1	4058	2	1	5	215
19	fb0cde0eddd9f3112a0a7d?	4737	1	311	4	1	3	413

Fig. 6. Patient data value

(18) In the RFM data screen, the cell with $null$ means the empty value. Thus, while exporting the data, this paper would not accept the any data with one or more empty value, shown in Fig. 7:

◢	A	B	C	D	E	F	G	H
1	ID	Recent	Times	Cost	Recent Score	Times Score	Cost Score	RFM Score
2	9642	SnullS	1	1	SnullS	1	1	SnullS
3	32209	SnullS	1	1	SnullS	1	1	SnullS
4	36170	SnullS	1	2	SnullS	1	1	SnullS
5	5651	SnullS	1	1	SnullS	1	1	SnullS
6	20685	SnullS	1	1	SnullS	1	1	SnullS
7	7332	SnullS	1	1	SnullS	1	1	SnullS
8	13410	SnullS	1	1	SnullS	1	1	SnullS
9	3433	SnullS	1	2	SnullS	1	1	SnullS
10	12303	SnullS	1	1	SnullS	1	1	SnullS
11	19108	SnullS	1	1	SnullS	1	1	SnullS
12	11169	SnullS	1	1	SnullS	1	1	SnullS
13	16344	SnullS	1	1	SnullS	1	1	SnullS
14	15740	SnullS	1	1	SnullS	1	1	SnullS
15	22266	SnullS	1	1	SnullS	1	1	SnullS
16	463731	SnullS	1	4	SnullS	1	1	SnullS
17	3989	SnullS	1	1	SnullS	1	1	SnullS
18	16494	SnullS	1	1	SnullS	1	1	SnullS
19	12329	SnullS	1	1	SnullS	1	1	SnullS
20	19244	SnullS	1	1	SnullS	1	1	SnullS
21	4545	SnullS	1	1	SnullS	1	1	SnullS
22	5174	SnullS	1	1	SnullS	1	1	SnullS

Fig. 7. Patient data value

(19) While IBM SPSS Modeler 14.1 is exporting the data in RFM, the result may be in fraction. Thus, this study tends to divide the RFM score indicator into 5–4 as high and 1–3 as low, shown in Table 1:

Table 1. Classification for each RFM score

1	1HHH	555, 554, 545, 544, 445, 444, 455, 454
2	HHL	553, 552, 551, 543, 542, 541, 453, 452, 451, 443, 442, 441
3	HLH	515, 525, 535, 514, 524, 534, 415, 425, 435, 414, 424, 434
4	HLL	511, 512, 513, 521, 522, 523, 531, 532, 533, 411, 412, 413, 421, 422, 423, 431, 432, 433
5	LHH	155, 154, 145 , 144, 255, 254, 245, 244, 355, 354, 345, 344
6	LHL	151, 152, 153, 141, 142, 143, 251, 252, 253, 241, 242, 243, 351, 252, 353, 341, 342, 343
7	LLH	114, 115, 124, 125, 134, 135, 214, 215, 224, 225, 234, 235, 314, 315, 324, 325, 334, 335
8	LLL	111, 112, 113, 121, 122, 123, 131, 132, 133, 211, 212, 213, 221, 222, 223, 231, 232, 233, 311, 312, 313, 321, 322, 323, 331, 332, 333

(20) We divide the patient contribution into eight groups. Since each group has RFM indicator, each indicator will be divided into high cost and low cost. This, we can realize the value of each group based on the high and low cost, shown in Table 2:

Table 2. Each RFM group indicator

Group	Item				
	Patients with type	R Indicator	F Indicator	M Indicator	Group value
First	Recently continued high cost patient	High cost	High cost	High cost	High
Second	Recently high cost patient	High cost	Low cost	High cost	High
Third	Recently continued visit patient	High cost	High cost	Low cost	Low
Fourth	Continued high cost patient	Low cost	High cost	High cost	High
Fifth	High cost patient	Low cost	Low cost	High cost	High
Sixth	Continued visit patient	Low cost	High cost	Low cost	Low
Seventh	Recently visit patient	High cost	Low cost	Low cost	Low
Eighth	General visit patient	Low cost	Low cost	Low cost	Low

After exporting RFM data with IBM SPSS Modeler 14.1, we import the Excel file into RFM data. By using the TwoStep Cluster model, we can find the importance of predictable variable.

1. Exporting Excel file from RFM, we can build the TwoStep Cluster model, shown in Fig. 8:

rfm.xlsx   Two Steps

Fig. 8. Screen of building TwoStep cluster

2. The next would be introducing the TwoStep Cluster setting. The main setting is to exclude the outlier's value and set others as default.

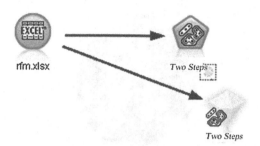

rfm.xlsx

Two Steps

Two Steps

Fig. 9. Screen of TwoStep cluster setting

3. The last step would be clicking the execute and wait until finish, shown in Fig. 9:
   (1) There are total 7 inputs in the gold brick mode: recent, times, cost, recent score, times score, cost score, and RFM score. TwoStep Cluster model would further automatically divide into two group, shown in Fig. 10:

Model Summary

Algorithm	TwoSteps
Inputs	7
Clusters	2

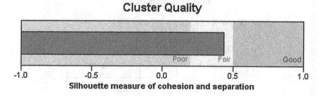

Cluster Quality

Poor   Fair   Good

-1.0   -0.5   0.0   0.5   1.0

Silhouette measure of cohesion and separation

Fig. 10. Screen of building TwoStep cluster

   (2) In the partial detail information of generated group, the smallest frequency is defined as −1 which is 40.8% of total; the highest frequency is defined as −2 which is 59.2% of total. On the other hand, the difference between these two is 1.45%, shown in Fig. 11:

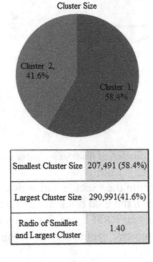

Cluster Size

Smallest Cluster Size	207,491 (58.4%)
Largest Cluster Size	290,991(41.6%)
Radio of Smallest and Largest Cluster	1.40

Fig. 11. Distribution for biggest and smallest group

(3) Comparing these two groups with seven input data, input with 1.0 importance are recent, recent score, cost score, and RFM score. On the other hand, cost has 0.1 importance. The rest does not meet the correlation, shown in Fig. 12:

Cluster

Feature Importance

■ 1.0 ■ 0.8 ▦ 0.6 ▦ 0.4 ▦ 0.2 ▦ 0

Cluster	1	2
Label		
Dsecription		
Size	58.4%	41.6%
Features	Recent	Recent
	Times	Times
	Cost	Cost
	Recent Score	Recent Score
	Times Score	Times Score
	Cost Score	Cost Score
	RFM Score	RFM Score

Fig. 12. Importance of group predictable input variable

(4) As we view from the importance of predictable variables, the ones with 1.0 score are RFM score, cost score, and recent, score. In addition, the RFM score plays the important role, and the least important role is cost, shown in Fig. 13:

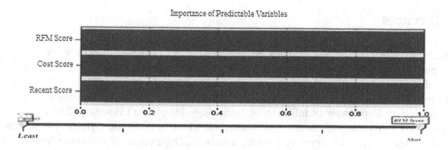

Fig. 13. Importance of predictable variables

After applying the TwoStep Cluster model, we predict three important variables: RFM score, cost score and recent score. With these three variables, we can construct 3D cubic print. The x-coordinate is set as RFM score, y-coordinate is set as cost score and z-coordinate is set as recent score. The result of 3D cubic print is shown in Fig. 14:

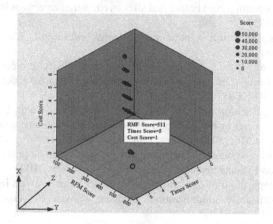

Fig. 14. RFM cubic print 511

5   Conclusion

RFM is the foundation of this research. RFM, the foundation of this research. It was applied to find the highest value group which have most expense on antibiotic use. And to find the connection of patients who 1. keep using Antibiotics or 2. use the high cost of Antibiotics, in order to reduce Antibiotic use. Consumption behavior of influenza patients is quantified and normalized to find consumption orientation. RFM data indicators worked out by SPSS Modeler distribute patients into 8 facets which reveal high consumption and low consumption of patients.

References

Angela, R., Branche, E.E.W., Formica, M.A., Falsey, A.R.: Detection of respiratory viruses in sputum from adults using automated multiplex Polymerase Chain Reaction (PCR). In: Risk Taking and Information Handling in Consumer Behavior, pp. 23–33 (1967)

Chang, T.P.: Analyzing consumer's behavior of medical cosmetology by applying data mining technology. Department of Information Technology, Ling Tung University (2012)

Chou, C.-I.: Applying clustering and classification in discovering symptom's attributes classification rules for upper respiratory infections. Department of Computer Science and Information Management, Providence University (2008)

Cheng, C.-Y.: An adaptive product recommendation system based on RFM method. Department of Information Management, Chaoyang University of Technology (2010)

Chu, C.-H.: Comparisons of inpatient mortality of pneumonia patients between military and non-military hospitals. School of Health Care Administration, Taipei Medical University (2010)

Cockburn, J., Pit, S.: Prescribing behavior in clinical practice: patients expectations and doctors perceptions of patients expectations-a questionnaire study. Br. Med. J. **315**, 520–523 (1997)

Curry, M., Sung, L., Arroll, B., Goodyear-Smith, F., Kerse, N., et al.: Public views and use of antibiotics for the common cold before and after an education campaign in New Zealand. NZ Med. J. (2006). U1957

Low, D.: Reducing antibiotic use in influenza: challenges and rewards. Clin. Microbiol. Infect. **14**, 298–306 (2008)

Goosens, H., Ferech, M., Vander Stichele, R., Elseviers, M., ESAC Project Group: Outpatient antibiotic use in Europe and association with resistance: a cross-national database study. Lancet **365**, 579–587 (2005)

Lin, H.C., Hsu, Y.T., Kachingwe, B.H., Hsu, C.Y., Uang, Y.S., Wang, L.H.: Dose effect of thiazolidinedione on cancer risk in type 2 diabetes mellitus patients: a six-year population-based cohort study. J. Clin. Pharm. Ther. **39**(4), 354–360 (2014)

Wu, H.-Y.: The association between continuity of care and doctor-shopping behavior. Institute of Health Policy and Management, National Taiwan University (2011)

Huang, H.-H.: The application of data mining to customer value analysis in the food industry. Department of Business Administration, Southern Taiwan University of Science and Technology (2011)

Hung, L.P.: A personalized recommendation system based on product taxonomy for one-to-one marketing online. Expert Syst. Appl. **29**(2), 383–392 (2005)

Huang, J.-D.: Impacts of characteristics of medical care organizations and physicians on medical resources utilization of children with asthma. School of Health Care Administration, Taipei Medical University (2012)

Jorgensen, L.C., Friis Christensen, S., Cordoba Currea, G., Llor, C., Bjerrum, L.: Antibiotic prescribing in patients with acute rhinosinusitis is not in agreement with European recommendations. Scand. J. Prim. Health Care **31**, 101–105 (2013). https://doi.org/10.3109/02813432.2013.788270

Kamber, M., Han, J.: Data Mining: Concepts and Techniques. Morgan Kaufmann Publishers, Waltham (2006)

Kung, K., Wong, C.K., Wong, S.Y., Lam, A., Chan, C.K., Griffiths S., et al.: Patient presentation and physician management of upper respiratory tract infections: a retrospective review of over 5 million primary clinic consultations in Hong Kong. BMC Fam. Pract. **15**(95) (2014). https://doi.org/10.1186/1471-2296-15-95

Kuo, C.-J.: Antibiotics prescription patterns and medical utilization analysis for outpatients with upper respiratory diseases by central medical Centers. Department of Public Health, Kaohsiung Medical University (2002)

Kyoung-Jae Kim, H.A.: A recommender system using GA K-means clustering in an online shopping market. Expert Syst. Appl. **34**(2), 1200–1209 (2008)

Liu, C.-N.: A multilevel model analysis of antibiotic prescribing behavior in unspecific upper respiratory infections and acute bronchitis among ambulatory care physicians. Institute of Health Policy and Management, National Taiwan University (2003)

Lloyd, M., Webb, S.: Prescribing and referral in general practice: a study of patients' expectation and doctors' actions. Br. J. Gen. Pract. **44**, 165–169 (1994)

Marroquí, L., Alonso-Magdalena, P., Merino, B., Fuentes, E., Nadal, A., Quesada, I.: Nutrient regulation of glucagon secretion: involvement in metabolism and diabetes. Nutr. Res. Rev. **27**, 48–62 (2014)

Anvari, M.S., Naderan, M., Boroumand, M.A., Shoar, S., Bakhshi, R., Naderan, M.: Microbiologic spectrum and antibiotic susceptibility pattern among patients with urinary and respiratory tract infection. Int. J. Microbiol. (2014)

McNulty, C.A., Nichols, T., French, D.P., Joshi, P., Butler, C.C.: Expectations for consultations and antibiotics for respiratory tract infection in primary care: the RTI clinical iceberg. Br. J. Gen. Pract. **63**, e429–e436 (2013). https://doi.org/10.3399/bjgp13X669149

Miglautsch, J.R.: Thoughts on RFM scoring. J. Database Mark. **8**(1), 67–72 (2000)

Monnet, D., Mölstad, S., Cars, O.: Defined daily doses of antimicrobials reflect antimicrobial prescriptions in ambulatory care. J. Antimicrob. Chemother. **53**, 1109–1111 (2004)

Owrang O, M.M., Grupe, F.H.: Using domain knowledge to guide database knowledge discovery. Expert Syst. Appl. **10**(2), 173–180 (1996)

Hsieh, P.-H.: An RFM model analysis of consumer value and supplier dependence - an example of o company. Department of Information and Communication, Shih Hsin University (2013)

Chen, S.-H.: Using hierarchical clustering and different classification methods to analyze human normal tissue specific genes. Institute of Systems Biology and Bioinformatics, National Central University (2011)

Shih, Y.Y., Liu, D.-R.: Integrating AHP and data mining for product recommendation based on customer lifetime value. Inf. Manag. **42**(3), 387–400 (2005)

Tseng, S.-H.: Hospital management and patient safety–the study of antimicrobial stewardship in Taiwan. National University of Kaohsiung-EMLBA Exclusive Master of Law and Business Administration (2012)

Smith, S.M., Fahey, T., Smucny, J., Becker, L.A.: Antibiotics for acute bronchitis. Cochrane Database Syst. Rev. **3**, CD000245 (2014). https://doi.org/10.1002/14651858.CD000245.pub3

Data Mining Technology Combined with Out-of-Hospital Cardiac Arrest, Symptom Association and Prediction Model Probing

Chih-Chun Chang[1], Jui-Hung Kao[2,3(✉)], Chien-Yeh Hsu[1],
Homg-Twu Liaw[4], and Tse-Chun Wang[4]

[1] Department of Information Management, National Taipei University
of Nursing and Health Sciences, Taipei, Taiwan
[2] Graduate Institute of Biomedical Electronics and Bioinformatics,
National Taiwan University, Taipei, Taiwan
kao.jui.hung@gmail.com
[3] Computer Center, National Taipei University of Nursing and Health Sciences,
Taipei, Taiwan
[4] Department of Information Management, Shih Hsin University, Taipei, Taiwan

1 Background and Motivation

Since the first ambulance has been introduced to Taiwan, the members of fire departments have been dedicating themselves to emergency service training. Though all the emergency rescue measures have been improved, the causes and problems of Out-of Hospital-cardiac Arrest (OHCA) caused are still actively being explored by the experts from emergency medicine. The main causes of Out-of-Hospital Cardiac Arrest are classified into two categories: medical causes and surgical causes, the later ones mainly caused by car accidents or falls. There are some chronic disease such as heart disease (myoeardial infaracriton), chronic kidney disease, and diabetes are highly associated with Out-of-Hospital-Cardiac Arrest medical causes (Patel et al. 2014). However, in Taiwan, heart disease, hypertension, and diabetes mellitus are recognized as the three main medical causes for Out-of-Hospital-Cardiac Arrest.

The Ministry of Interior composes eight strategies which involves many different types of people and agencies including emergency medical services, Out-of Hospital-cardiac Arrest patients are categorized under emergency medical services (EMS), and ambulances are the responsibilities of the fire departments (Ministry of Interior 2013).

Emergency medical services (EMS) are usually activated by immediate response to bystanders' emergency calls. Followed by dispatching relevant medical personnel: Emergency Medical Technicians (EMT), who deliver the initial treatments such as Advance Life Support (ALS) and Basic Life Support (BLS) based on the seriousness of individual case. Emergency Medical Technicians are also divided into three different degrees to suit for different emergency situation, from typical levels of emergency to critical ones: Emergency Medical Technician-1, Emergency Medical Technician-2, and Emergency Medical Technician-Paramedic. Each level performs for specifically level of intensity to serve of different degrees of emergency medical care required in the emergency circumstances. According to rule 18, Regulations of Emergency Medical

© Springer Nature Singapore Pte Ltd. 2018
J. C. Hung et al. (Eds.): FC 2017, LNEE 464, pp. 298–309, 2018.
https://doi.org/10.1007/978-981-10-7398-4_31

Services (Ministry of Health and Welfare 2013), there should be at least two EMTs in the ambulance while perform the emergency medical services or sending patients to hospitals. Intensive ambulance service requires two EMTs, one of whom need to be a medically trained and qualified doctor, and another one need to be a medical trained staff to perform the duty as EMT-2.

According to Cai's study (2009), the levels and numbers of Emergency Medical Technician would not affect the outcomes of ROSC (Return Of Spontaneous Circulation), since most ROSC care were integrated shorter on site treatment time and early cardiopulmonary resuscitation (CPR).

Sending EMT-P should have been regarding on the life threatening situations. After Intelligent Computer Aided Dispatch System was activated, ALS dispatch rate, ALS excessive usage rate, BLS dispatch rate were raised under a significant flow. Nevertheless, ALS idling rate at the same time also raised sharply. The results within correct case studies suggested further adjustments are indeed required for a better outcome. Emergency Medical Service (EMS) is categorized into two categories: Pre-Hospital Emergency Care and in Hospital Emergency Medical Services on the other hand belongs to all hospitals' emergency departments.

However, the emergency factors which have been discovered are quite different and there is not enough data from the emergency ambulance table containing the Out-of Hospital-cardiac Arrest patients with symptoms which has been analyzed by mining techniques. Most survive to discharge patients had complications such as hypoxic brain lesions and in very few cases demonstrated the ability of fully restored to their original conditions. This study tried to find out whether this presents emergency care has missed its prime time or the patients' pre-conditions are highly related to the issue.

According to the researches from Taiwan, the average Out-of Hospital-cardiac Arrest survival rate is 1.0–13.8% in Taiwan (Sutton et al. 2015). There is an urgent need for an integrated emergency care to responds to a more complex medical needs such as cardiac arrest. The objective of this research is to understand how the variables affect OHCA survival rate through OHCA data analysis in order to set up a complete and objective evaluation to assign or improve all the resources in EMS system across the nation.

Purpose of the Study

- To understand OHCA patients' medical characteristics and danger factors could assist the fire-departments, Emergency Medical Technicians, and hospitals to make appropriate decisions in order to increase patients' survival rate.
- By using Biometrical method to get frequency distribution in order to understand data characteristic value.
- By using Apriori analysis to seek the association rules underneath the emergency ambulance record forms.
- By using Classification and Regression Tree analyzes data to seek a risk factor model.

Research Critireas

This study accumulated data from the moment when Emergency unit receives the emergency phone calls until the time when emergency medical technicians returning to the fire department which has been classified into three stages: dispatching, Out-of-hospital emergency medical treatments, and In-hospital follow up. Conditions for dispatching included: OHCA occurring seasons, OHCA Occurring Hours, OHCA Types, OHCA occurrence, sending EMT/EMTP, and administrative districts (areas). Out-of-hospital emergency medical treatments would need to consider each individual factors of each individual emergency call which includes gender, age, location of arrest, medical history, medical history combination, allergy history, injured area, respiratory disposal, general treatment or intensive treatment, ACLS, dispatching reaction time, EMT scene treatment time, transporting (hospital) time, transfer hospital, first aid capabilities hospital, ventilation ratios for CPR, bystander CPR, sending EMT-P, medication, Applying Laryngeal Mask Airway or Endotracheal Tube Intubation, ROSC on site, Utilizing automated external defibrillator, automated external defibrillator, and automated external defibrillator result. Hospital follow up would focused on the two hour survival rate.

2   Literature Review

2.1   Common Causes of OHCA

According to Wu and Yu's study (2011), nationwide average annual incidence rate (cases per 100,000 population) of Out-of Hospital-cardiac Arrest was 73 and male incidence rate was 2.5 times higher than female's.

Boller (2013) stated 80% of adult Out-of Hospital-cardiac Arrest patients and 55% of In Hospital Cardiac Arrests (IHCA) adult patients failed to return to spontaneous circulation (ROSC) after receiving CPR. This result is highly possible associated with the fact that a lot of patients who experienced Out-of Hospital-cardiac Arrest had a pre-medical condition or pre-medical history of heart disease, hypertension, and diabetes. (Li 2010; Yan 2010; Jian 2011).

1. Heart diseases

Heart related diseases such as Coronary Artery Disease (CAD), septal defect, heart failure, abnormal position of cardiac valve, and Cardio vascular diseases are all recognized as heart diseases. Other researchers (Huang et al. 2009) explained that most Out-of Hospital-cardiac Arrest patients had Coronary artery disease, (CAD); acute myocardial infraction was the main cause of cardiogenic sudden death.

2. Hypertension

According to the data observed from Ministry of Health and Welfare on June 6th, 2013, hypertension is the top eight cause of death in Taiwan. Hypertension, is a long term medical condition with very miner symptoms, which is also commonly known as high blood pressure. Hypertension is classified as primary (essential) high blood pressure or secondary high blood pressure. The causes of primary high blood pressure are still remain unidentifiable. It is highly possible associated with lifestyle, environment or genetic factors. Social and environment shocks or genetic problems might trigger the essential high blood pressure. Hypertension is only be able to controlled and improved by the changes in daily routines and lifestyle activities, there is no such cure could be offered solely by modern medicine. Patients can only rely on life changes and some minor medications to assist and to maintain their own wellbeing.

3. Diabetes

According to Ministry of Health and Welfare (2011), diabetes is the top five in the causes of death in Taiwan. Diabetes mellitus (DM), commonly referred to as diabetes, is a chronic disease of metabolic disorder in which high blood sugarlevels or high sugar levels in urine occurred, which in relations to the failure as in sufficient insulin produced by pancreas or body cells' resistance to insulin produced. Diabetes also causes protein and fat metabolic disorder. The possible as common comprehensible causes are lifestyle, environment or genetic factors.

2.2 Success Rate of OHCA Patients

Since American Heart Association promoted Chain of Survival (2016), (Szpliman et al. 2014), emergency medical system has been promoting this as a useful concept of Emergency Cardiac Care (ECC) system.

There are 5 links in the adult out-of-hospital Chain of Survival: recognition of cardiac arrest and activation of the emergency response system; early cardiopulmonary resuscitation (CPR) with an emphasis on chest compressions; rapid defibrillation belong lay resources; basic and advanced emergency medical services; and advanced life support and post-cardiac arrest care. It is suggested that a well performed Chain of Survival can increase the chances of survival and recovery for cardiac arrest victims.

Abrams et al. (2011) revealed the importance of reacting time, public place, bystander, and age towards common Out-of Hospital-cardiac Arrest survival predictive value up to 11%. Thus common hospital discharge rates were among $1 \sim 10\%$.

Lee (2011) stated that OHCA patients with VF or VT had higher survival chances than patients with PEA.

Aso (2011) used Genetic Algorithm combined age statistics to learn the best locations of setting AEDs were at 7–11 s.

On the basis of the experiment results, Li (2010) concluded all the infield emergency medical treatment factors in Table 1. In the out of hospital emergency medical services factors, Gender in this case only affected by Out-of Hospital-cardiac Arrest occurring locations. However, other factors significantly affected the result of out of hospital emergency medical services, such as Out-of Hospital-cardiac Arrest occurring locations and bystander CPR treatment in result. Among all the hospital emergency medical service factors, age also deeply influenced the outcomes towards Cardiac Arrest, Initial Rhythm, and Inotropic Agent Dose. Age factor was related to whether patients would admit to ICU under the prognosis results.

2.3 Prognosis Factors of OHCA Patient

The deification of "prognosis" refers to the expected medical results and the percentage of these results occurring (Wu and Yu 2011). In other words, prognosis includes the sequelas, recovery and the possibility of death within the danger factors. Some researchers (Liu et al. 2009) discovered among the patients who suffered from Out-of Hospital-cardiac Arrest, a large group of patients had cardiac etiology and aged from 66 to 80. Cardiac etiology in Out-of Hospital-cardiac Arrest patients had higher shock able rhythms and better prognosis.

Table 1. Connections of out of hospital emergency treatment impact factors, hospital emergency treatment impact factors, and prognosis performance (Chen 2013)

Dimension		ROSC Attempted	ROSC Time	Hospital Admission	Dates /Hospital Admission	Discharged Alive
OHC A EMT Factors	Location	★		★		
	Arrest witnessed	★		★		
	Transformation					
	Ambulance Arrival time					
	Treatment Time	★				
	Transformation Time					
	Total React Time	★				
	AED			★		★
	Bystander CPR	★		★		
	EMT CPR					
Hospital EMT Factors	Reason of Cardiac Arrest			★		
	Initial Rhythm	★		★		
	Body Temperature Level	★		★		
	Pupil Size	★				★
	Inotropic Agent Dose		★	★		
	AED	★				
	ER EMT Time	★	★	★		
	Treatment Cost	★		★	★	★
★= direct correlation						

3   Methodology

3.1   Utstein Style

The research related to cardiac arrest and CPR, especially the epidemiological studies in particular, each difference of the relevant variables definition has created different results in report, which contained important prognostic indicators, such as the possibility of recovery from circulation and survival rates. Furthermore, Utstein unified reporting system has also been cited to the usage of hospital emergencies: the authorities suggested using the common format (Utstein style) for CPR out of the hospital to uniform the styles and reduce the differences in academic papers.

3.2   OHCA Patients of the Risk Factors

Wu and Yu (2011) noted in their research the so-called post-diseased patients' prognosis referred to the expected results and the variables that might occur. In other words, the prognosis had some risk factors: patients might have sequelas, the possible progress or the possibilities of death. However, whether the poor prognoses were created by the differences in the intensive care unit (ICU), or the policy differences in care treatment are still remain unclear. (Liu et al. 2009) Patients who had cardiac arrest caused by psychogenic illness in the pre-hospital history or after hospital have higher possibility of shock able rates compared to the non-psychogenic patients, and the former had a better prognoses as well. Some researchers discovered (Laish-Farkash et al. 2007) that there were four main factors closely influencing the results of therapeutic hypothermia, such as age, comorbidities, low temperature, and the recovery time of returning to spontaneous circulation. It was stated in a study (Bray et al. 2012) after the differences of pre-hospital factors were adjusted, female patients had higher survival rates than male the patients. In a group studies (Soholm et al. 2012) the researchers published their research on Out-of Hospital-cardiac Arrest patients' that improving survival situations in higher comparative of Copenhagen area hospitals and non-tertiary hospitals. The researchers indicated that pre-hospital factors such as age, initial rhythm, and period from the alarm to the emergency operations units' arrival, early defibrillation, witness cardiac arrest, and bystander cardiopulmonary resuscitation (CPR) were known to be the important predictors of outcome factor.

3.3   Patients and Study Area

All the Out-of Hospital-cardiac Arrest records in this study, were obtained from New Taipei City emergency unit, which filtered through the cases occurred to Out-of Hospital-cardiac Arrest through the emergency cases: emergency medical personnel rescued cardiac arrest patients. Cardiac arrest represents and define as hearts suddenly stop beating without warning and it terminates the functions of blood flowing into brains and other organs; within merely few seconds, the patient heat stopped breathing and had no pulse as the result of cardiac arrest. These conditions were called "clinically dead", and the patients might be actively rescued after apply adequate emergency measures and medical treatments. This study analyzed survival prognoses of Out-of Hospital-cardiac Arrest patients in comprehension to their survival rates, conditions and in extension to understand the survival factors that impact towards patients.

3.4   Data Processing

The Out-of Hospital-cardiac Arrest data within this paper used was based on the international standard UteStein formula, which enabled the researcher to cross evaluate and compare national, regional and hospital-based emergency medical services.

4   Result and Discussions

4.1   Statistical Analysis

The researchers applied decision tree classification algorithm to analyze Out-of Hospital-cardiac Arrest multivariate, which explains the survival rate with risk factors set forth under the joint effects of these variables in the model.

4.2   Linear Regression Analysis

The Linear Regression Analysis represents as a feature selector. Schuller, Friedmann, and Eyben stated that the Linear Regression was an effective selector with higher overall accuracy rates.

4.3 Logistic Regression Analysis

To Apply Linear Regression Analysis aimed to find out the OHCA potential factors, and the purpose of apply the Logistic Regression Analysis was to identify risk factors accuracy.

4.4 Apriori Analysis

Apriori algorithm was mainly applied for researchers to seek the association rules in the backup project through frequent patterns as the calculated indicator for the probability of each backup item appearance.

4.5 Classification and Regression Tree

While using CART data to analyze data, the data attributions would be classified into two different groups: continuous and discontinuous. Continuous data could be apply to predict and estimate weight values into variables, and discontinuous data could be applied to the categories of identification.

4.6 Results

As the research paper shows, Logistic regression for survival Out-of Hospital-cardiac Arrest occurrence of risk factors and a two-hour analysis were found that the risk of death of the OHCA patients age 65 increased to 84.4% comparing to the OHCA patients less then age 65. Sending the patients to a third stage emergency hospital, the survival rate was 15.75 times then the ones sent to non-emergency responsibility hospital; 2.4 times of sending the patients to a first stage emergency responsibility hospital emergency and 1.3 times to send the patients to a second emergency responsibilities hospital. When Rhythm encoded without using AED defibrillation shock (VT/VF) increased 26.1% the risk of mortality. Surgical OHAC patients in comparison with medical OHA patients, the formers' death rate increased to 67.7%. The survival rate on the other hand of receiving EMT-P emergence care is 1.25 times higher than receiving the emergency care of general EMTs.

5   Summary of the Study

5.1   Discussion

This research paper has discovered many intriguing but insightful circumstances that interrelated with emergency care procedures, regulations, scientific and clinical measures that applied into Taiwan's daily rescue missions. The medical community has acknowledged that the acute traumatic death could be separated into three peak catergorise. They are immediate death, early deaths, and late deaths. Most of the patients died immediately after the fatal injuries. Patients died in a few seconds/a few minutes after the injury, such as traumatic aortic rupture and Medulla oblongata lacerations. The rate of patients discharging from hospital of those Out-of Hospital-cardiac

Arrest patients with serious injury were significantly less. Henceforth, to dispatch those patient to the suitable emergency care hospital has become essentially critical after the serious injuries has occurred. To prevent possible delay in this case has influence the chances of patient's survival. Nevertheless Apriori could only perform as the indicator to prove the possibility of mutable chronic diseases. Apriori on the other hand was not capable to identify the accurate causes among those diseases listed.

As far as for the matters that concerns related to the ambulance units, in this thesis also found that EMT-P have provided the most appropriate method to collaborate with emergency personnel that provides the most economically efficient and effective first aid service which at the same improved the opportunities for patients' survival before reaching to the hospital for those pre-hospital cardiopulmonary arrest cases. The results under the study also indicated that after EMT-P arrived at the emergency scene, they could provide immediate and necessary first aid treatments at first, then send the patients to hospital in the second move. However in conjunctions to the survival rate of Out-of-Hospital-cardiac Arrest patients, the result could not be increased when patients helped by the EMT-P's rescue treatments.

This study did found some positive outcomes occurred when advanced first aid technicians of the rescue team have been send to the rescue especially to the patients who were escorted in particular. Pre-hospital first aid and care services has been extended from the primary rescue projects in the past, to intubation, intravenous administration, electric shock, and other senior first aid service under the current setting. Yet the advanced emergency services were rarely applied in terms of improving the qualities of the overall rescue that consolidated the effectiveness which demonstrates in first-aid effectiveness evaluation indicators; most significantly could be view as the saving rate and the survival are of the Out-of Hospital-cardiac Arrest patients.

Many variables in the cases achieved had been performed further assessment which proved that ROSC and the rest ones are unable to provide the definite results in perdition. Stepwise regression analysis on the other hand was applied to select an independent factors. Kaplan-Meier method was used to analyze the patients who received the hospital Advanced Life Support (ALS) to seek for the possibility of survival. ALS have given the benefit with greater chance to survive which also share more significant results for cardiopulmonary resuscitation patients prior reach to the hospital.

Based on the capability of the hospital to accommodate for the above mentioned emergency cases, three levels of hospital were set in response to these emergency cases. The purpose to classifying the responsibility hospitals is to dispatch the patients that requires specific disease treatment.

These hospitals has the competency to accommodate for the matters such as stroke, myocardial infarction. It was a critical grading standard to rate for the responsibility hospitals to perform for the specific needs. Gonçales et al. (2015) under their studies have divided patients who discharged from hospital into different color groups (red, yellow, orange, blue and green), a referral method to some external health units, comparing the progress of the patients till death. In order to define the danger of each group; a large difference was found among the groups.

Their analysis result illustrated that the priority patients required more attention, planning, and caring measures in their medical services. When the higher death risk increased, the higher priority medical services would require to apply and the longer time period of caring to the unit.

Thus, when hospital medical staff applies Manchester Triage System, it created a better emergency medical services for the patients in need. Based on these results, the researchers could developed an appropriate care strategy to benefit the patients who needed to receive the suitable medical services under the hospital care measure.

5.2   Conclusions

Prognosis refers to the expected results and the possibilities may happen after the patients suffering from the diseases.

In another word, prognosis means certain risk factors, the complication, recovery rates and mortality after the patients suffered from the diseases. Risk factors indicate the variables for better or worse prognoses.

Through Out-of Hospital-cardiac Arrest risk factors, Classification and Regression Tree, the study indicates of Out-of Hospital-cardiac Arrest patients' survival rates are affected by age, surgical/medical causes, and the levels of the emergency hospitals transferred. Patients with surgical causes are exposed in higher risks than the patients suffered from medical causes.

5.3   Suggestions for Future Studies

This research discovered the connections and survival path among symptom factors and Apriori criteria. Researchers also found the characteristics of Advance Life Support (ALS) and Basic Life Support (BLS) within the cases observed. As the common practices, emergency unit dispatching was based on the reporter's descriptions from emergency call center, yet there were professional discrepancies found in some cases which leads to exaggerate descriptions that interference the correct judgment for emergency unit dispatching duties.

As in the usage of Classification and Regression Tree, four main risk factors that affected to an OHCA patients' survival rate were discovered. First, Patients' age was the indicator factor. Second, patients with surgical causes were exposed in higher risks than the patients purely suffered from medical causes. Third, survival rates could be potentially higher if the Out-of Hospital-cardiac Arrest patient had been delivered to mid or high stage emergency hospitals. Fourth, accurate EMT-P dispatching would result a more rapid evacuation, more supportive to the respiratory tract, ventilation maintaining, administered intravenous fluids, medications, and more effective CPR and other first aid medical care measures professionally, which it means the survival rates of Out-of Hospital-cardiac Arrest patients would increase (25.9% vs. 21.3%.).

The critical indicators of a better survival rate from EMT-P are to provide patient's Logistic regression extraction risk factors that used in the decision making within the thesis are: Age, Out-of Hospital-cardiac Arrest Type, First Aid Capabilities Hospital, Sending EMT-P, AED Initial Heartbeat Coding, and Ventilating Ratios for CPR. Out-of Hospital-cardiac Arrest patients can be segmentized into three groups, which help first aid medical dispatchers (Emergency medical dispatcher, EMD) to allocate Out-of Hospital-cardiac Arrest patients in the pre-hospital medical resources.

References

Patel, A.A., Arabi, A.R., Alzaeem, H., Al Suwaidi, J.A., Singh, R., Al Binali, H.A.: Cinical Profile, Management, and Outome in patients with out of hospital cardic arrest: insights from a 20-year regesry. Int. J. General Med. **7**, 373–381 (2014)

Davies, H., Loosely, A., Dolling, S., Eve, R.: Predicting survival in patients admitted to intensive care following out-of-hospital cardiac arrest using the Prognosis after Resuscitation score. Crit. Care **18**(Suppl. 1), P491 (2014)

van Genderen, M.E., Lima, A., Akkerhuis, M., Bakker, J., van Bommel, J.: Persistent peripheral and microcirculatory perfusion alterations after out-of-hospital cardiac arrest are associated with poor survival*. Crit. Care Med. **40**(8), 2287–2294 (2012)

Maupain, C., Bougouin, W., Lamhaut, L., Deye, N., Diehl, J.-L., Geri, G., et al.: The CAHP (Cardiac Arrest Hospital Prognosis) score: a tool for risk stratification after out-of-hospital cardiac arrest. Eur. Heart J., ehv556 (2015)

Sutton, R.M., Case, E., Brown, S.P., Atkins, D.L., Nedkarni, V.M., et al.: A quantitive analysis of out-of-hospital pediatric and adolescent resuscitation quality - a report from the ROC epistry-cardiac arrest. Resuscitation **93**, 150–157 (2015)

Lai, S.C., Li, C.J., Hsiao, M.H., Chou, C.C., Chang, C.F., Liu, T.A., et al.: Demographic and outcome analysis in adult patients with non-traumatic out-of-hospital cardiac arrest in central Taiwan. J. Emerg. Med. **11**(1), 1–9 (2009)

Wu, Z.P., Yu, C.Y.: A study on resuscitative outcome of out-of-hospital cardiac arrest patients-the implementation experience of the fire bureau of Taipei County. J. Crisis. Management. **8**(1), 9–18 (2011)

Cummins, R., Chamberlain, B., Hazinski, M., Nadkami, V., Kloeck, W., Kramer, E., et al.: Recommended guidelines for reviewing, reporting, and conducting research on in-hospital resuscitation: the in hospital 'Ustein Style'. Ann. Emergency Med. **20**(5), 650–679 (1997)

Danciu, S., Klein, L., Hosseini, M., Ibrahim, L., Coyle, B., Kehoe, R., et al.: A predictive model for survival after in-hospital cardiopulmonary arrest. Resucatation **62**(1), 35–42 (2004)

Boller, M.: Will models of naturally occurring disease in animals reduce the bench-to-bedside gap in biomedical research? Chin. Crit. Care Med. **1**, 5–7 (2013)

Aso, S.-I., Imamura, H., Sekiguchi, Y., Iwashita, T., Hirano, R., Ikeda, U., et al.: Incidence and mortality of acute myocardial infarction a population-based study including patients with out-of-hospital cardiac arrest. Int. Heart J. **52**(4), 197–202 (2011)

Ambulatory Blood Pressure Monitoring, Lifestyle Modifications: The Seventh Report of the Joint National Committee on Prevention, Detection, Evaluation, and Treatment of High Blood Pressure (2013)

Szpliam, D., Webber, J., Quan, L., Bierens, J., Morizor-Leite, L., Lamgerdprfer, S.T., Beerman, S., Løfgren, B.: Creating a drowning chain of survival. Resuscitation **85**(9), 1149–1152 (2014). https://doi.org/10.1016/j.resuscitation.2014.05.034

Abrams, H.C., Moyer, P.H., Dyer, K.: A model of survival from out-of-hospital cardiac arrest using the boston EMS arrest registry. Resuscitation **82**(8), 999–1003 (2011)

Council, A.R., Chamberlain, D.A.: Recommended guidelines for uniform reporting of data from out-of-hospital cardiac arrest: the Utstein style. Ann. Emerg. Med. **20**(8), 861–874 (1991)

Markusohn, E., Sebbag, A., Aronson, D., Dragu, R., Amikam, S., Boulus, M., et al.: Primary percutaneous coronary intervention after out-of-hospital cardiac arrest: patients and outcomes. Hypertension **12**, 48 (2007)

Laish-Farkash, A., Matetzky, S., Kassem, S., Haj-Iahia, H., Hod, H.: Therapeutic hypothermia for comatose survivors after cardiac arrest. Israel Med. Assoc. J. IMAJ **9**(4), 252–256 (2007)

Bray, J.E., Stub, D., Bernard, S., Smith, K.: Exploring gender differences and the "oestrogen effect" in an Australian out-of-hospital cardiac arrest population. Resuscitation (2012)

Søholm, H., Wachtell, K., Nielsen, S.L., Bro-Jeppesen, J., Pedersen, F., Wanscher, M., et al.: Tertiary centres have improved survival compared to other hospitals in the Copenhagen area after out-of-hospital cardiac arrest. Resuscitation (2012)

Zhou, J., Zhou, Y., Cao, S., Li, S., Wang, H., Niu, Z., et al.: Multivariate logistic regression analysis of postoperative complications and risk model establishment of gastrectomy for gastric cancer: A single-center cohort report. Scand. J. Gastroenterol. 51(1), 8–15 (2015)

Han, J., Kamber, M., Pei, J.: Data Mining: Concepts and Techniques. Elsevier (2011)

Mukhopadhyay, A., Maulik, U., Bandyopadhyay, S.: A novel biclustering approach to association rule mining for predicting HIV-1–human protein interactions. PLoS ONE 7(4), e32289 (2012)

Win, S., Hussain, I., Hebl, V., Redfield, M.M.: Mortality and readmissions after heart failure hospitalization in a community based cohort: estimating risk using the acute decompensated heart failure national registry (ADHERE) classification and regression tree (CART) algorithm. Circulation 132(Suppl. 3), A12302-A (2015)

Dumas, F., Rea, T.D.: Long-term prognosis following resuscitation from out-of-hospital cardiac arrest: role of aetiology and presenting arrest rhythm. Resuscitation 83(8), 1001–1005 (2012)

Husemoen, L., Jorgensen, T., Borch-Johnsen, K., Hansen, T., Pedersen, O., Linneberg, A.: The association of alcohol and alcohol metabolizing gene variants with diabetes and coronary heart disease risk factors in a white population. PLoS ONE 5(8), e11735 (2010)

Li, C.-J., Syue, Y.-J., Lee, C.-H., Kung, C.-T., Chou, C.-C., Chang, C.-F., et al.: Predictors of sustained return of spontaneous circulation in patients with blunt traumatic out-of-hospital cardiac arrest in Taiwan. Life Sci. J. 10(1) (2013)

Ettehad, D., Emdin, C.A., Kiran, A., Anderson, S.G., Callender, T., Emberson, J., et al.: Blood pressure lowering for prevention of cardiovascular disease and death: a systematic review and meta-analysis. Lancet (2015)

Qiu, M., Shen, W., Song, X., Ju, L., Tong, W., Wang, H., et al.: Effects of prediabetes mellitus alone or plus hypertension on subsequent occurrence of cardiovascular disease and diabetes mellitus longitudinal study. Hypertension 65(3), 525–530 (2015)

Muntner, P., Whittle, J., Lynch, A.I., Colantonio, L.D., Simpson, L.M., Einhorn, P.T., et al.: Visit-to-visit variability of blood pressure and coronary heart disease, stroke, heart failure, and mortality: a Cohort Study. Ann. Intern. Med. 163(5), 329–338 (2015)

Gonçales, P.C., Júnior, D.P., de Oliveira Salgado, P., Chianca, T.C.M.: Relationship between risk stratification in emergency medical services, mortality and hospital length of stay. Investigación y Educación en Enfermería. 33(3) (2015)

Chen, H.-I.: Using data mining technologies to extract the relations of out-of-hospital cardiac arrest with symptoms: Shih Hsin University-Department of Information Management (2012)

Ministry of the Interior Department of Statistics (2013). http://statis.moi.gov.tw/micst/stmain.jsp?sys=100

Ministry of Health and Welfare (2013). http://www.mohw.gov.tw/cht/DOS/Statistic_P.aspx?f_list_no=312&fod_list_no=2425&doc_no=13717

Law and Regulations Database of The Republic of China (2013). http://law.moj.gov.tw/LawClass/LawAll.aspx?PCode=L0020045

Tsay, H.-J.: A Study on Resuscitative Outcome of Out-of-Hospital Cardiac Arrest Patients (2009)

Chien, D.-K.: Analysis of predictors of hospital outcome in Non-Traumatic Death-On-Arrival adult patients (2007)

Kung, C.-T.: Pre-arrest factors influencing survival after in-hospital cardiopulmonary resuscitation on the general wards (2011)

Cheng, M.-T.: An Evaluation of Emergency Medical Dispatch System in Taipei (2011)

Yen, Z.-S.: Association between Particulate Matter and the Incidence of Out-of-Hospital Cardiac Arrest (2012)

Lee, C.-K.: The Analysis of Ilan's Out-of-Hospital Cardiac Arrest (OHCA) Patients (2010)

Yen, T.H.: The outcomes of CPR and related factors among non-traumatic patients with in-hospital cardial arrest in a crowded emergency department (2010)

Chien, T.-Y.: The analysis of delayed-action response interval in the fire branch of Taoyuan County - by using RCA (2011)

Huang, C.-H., Chen, W.-J.: Focused on PCI and hypothermia treatment (2009)

Li, C.-J.: Predictors of sustained return of spontaneous circulation in patients who suffer traumatic out-of-hospital cardiac arrest in Taiwan (2011)

Critical Perspectives on Learning Interface Design of Music Sight-Singing: Audition vs. Vision

Yu Ting Huang[1(✉)] and Chi Nung Chu[2]

[1] Department of Music, Shih Chien University, No. 70 Ta-Chih Street,
Chung-Shan District, Taipei, Taiwan, R.O.C.
yutingll@mail.usc.edu.tw
[2] Department of Management of Information System,
China University of Technology, No. 56, Sec. 3, Shinglung Rd.,
Wenshan Chiu, Taipei 116, Taiwan, R.O.C.
nung@cute.edu.tw

Abstract. The intention of this paper is to construct the constitutive senses of audition and vision in the design of learning interface for music sight-singing. Studies in the learning interface design for music staff notation, pitch recognition and sight-singing in music education are discussed. The aim is to establish current human-computer interface study results with the senses of audition and vision in music sight-singing acquisition and what is needed for further progress in this field of research. It is argued that the coming digital learning environment provides an effective and efficient field in music sight-singing research. It is also pointed out that the researches on the learning interface design with the combination of audition and vision benefit both children and adults. The paper highlights what learning interface design with the senses of hearing and vision in music sight-singing instruction can learn from research and where future research may provide further advancements.

Keywords: Learning interface · Music sight-singing · Audition
Vision

1 Introduction

Having a well cognition of music sight-singing is essential to being a capable music learner and acculturated musician [2, 31, 39]. Sight-singing is an important indicator of an overall level of musicianship. The elements for music sight-sing consist of musical notes reading, ear-training, and sight-singing [10, 25]. It requires higher cognitive processing demand to read and perform with precision from musical rhythm notation [37, 38].

There is a dual coding theory referring both visual and verbal information as to be processed differently and along distinct channels with the human mind [27]. The use of computer technology to support multisensory stimuli has been an emerging development in the recent music pedagogy [22, 23]. It is a theory of cognition. Both visual and verbal codes for representing information are used to organize incoming information into knowledge that can be acted upon, stored, and retrieved for subsequent use.

© Springer Nature Singapore Pte Ltd. 2018
J. C. Hung et al. (Eds.): FC 2017, LNEE 464, pp. 310–314, 2018.
https://doi.org/10.1007/978-981-10-7398-4_32

The interaction of different sensory approaches is particularly effective for learning music [Sharman]. Therefore it is the multisensory approach possible in learning music through the eye, ear, and body coordination. Multimedia learning system provides learners with separated information processing systems for visual and verbal representations. The interaction mechanism can engage learners in active learning.

2 Difficulties in Music Sight-Singing

The music perception and cognition, error detection, and individual instruction in the music sight-singing are the three main difficulties covered by the previous researches. Music sight-singing is a process of converting special visual symbols of music notation into sounds. When scanning the music staff, the accurate identification and interpretation of clef, notes, pitch, and rhythm in mind are a challenging task [19, 20, 28, 30]. Errors occur inevitably in the learning process, and how to identify the learners' individual errors is one of the important components to refine students' music sight-singing skills. The difficulty with identifying the individual error types may be compounded by instructors' lack of a specified way of teaching it [12]. Error detection is an efficient pedagogy to guide error correction that involves knowing what, when, and how to bring about positive changes for accurate sight-singing, especially in the early stages of training [15, 16, 33]. It will focus learners on developing their sense of misunderstanding. However the error detections in aural training are regarded as a tedious procedure in themselves but as a means to attain individual learning success [7, 35].

Generally sight-singing is taught in group arrangement. There are always questions in the instructor's mind that how many students are really following the music reading [6]. There are individual instructions for music sight-singing had experienced outstanding group sight-singing [4, 11]. Although the individual instruction could compensate the limitations as the instructor lacks time and resources for regular, ongoing assessment in group teaching [26]. There exists difficulty in providing reliable and efficient individual evaluation procedures for sight-singing. Therefore most music sight-singing instructions proceed with doing group evaluation or informal observation only [5].

3 Beneficial Perspectives in Music Sight-Singing Learning Interface Design

Singing music at sight is a complex skill, requiring the singer to identify pitch and rhythm simultaneously. Researches indicate that training with suitable multisensory stimuli on a simultaneity perceiving task is capable of eliciting meaningful, lasting changes in the learning process [1, 17, 29, 34]. The ability of human sensory faculties to integrate information across the different modalities provides a wide range of behavioral and perceptual benefits. This integration process is dependent upon the relationship of the different sensory signals, with stimuli occurring close together in the learning process typically resulting in the largest behavior changes.

In the learning process of identifying pitch, intervals and rhythms, learner need more immediate error feedback to review, reinforce, and develop their own music perceptions. To bridge the gap between the cognition load of novice and graph of music notation, the Spoken Music Browser adopts scaffold learning strategy with aural technologies. With the help of requested automatic transformation of music notation in Chevé system, vocal descriptions, solmization and melody, the novice can build his own music perceptions himself [3]. The transformation of spacial music notation through Chevé system is effective to start novice on notation learning by giving them opportunity to aurally identify what they see on a Chevé system. That is able to provide a clearly mental map to be made between sounds and a mark on Chevé system which can then be extended to the more complicated music notation related to specific tones. Beyond the graphic representations of the individual cognitive patterns, the alternative graphic fashion that allows for direct listening pattern comparisons with events occurring in the music stimuli.

The inherent abstract music sound of extraction and identification in musical ear-training is hard for learners to comprehend and distinguish. Many learners in traditional learning environment have limited immediate learning feedback which allows them to review, reinforce, and develop such aural skills. The Adaptive Error Feedback Music Ear-Training System is designed to provide immediate diagnostic feedback on the melodic line assessment with pitch recognition, interval recognition and rhythm recognition [13]. As the difficulty of the music texture increases, so does their frustration. Therefore providing mistake analyses in music ear-training learning is needed to consolidate their own error.

The process of sight-singing involves the conversion of musical information from sight to sound [9, 18]. It is hard for the learners to distinguish their sound of soundness by themselves. Multiple cognitive procedures are involved concurrently when learners read music by sight [8, 25, 36].

There is a design with self-generated visualization on pitch recognition for music sight-singing [14]. Each pitch of music note sang by a user could be explicitly recognized and self-generated visualization on pitch recognition and responded visually with the compared results of accuracy in three ways: visualizing sight-singing music notes, tuning errors in sight-singing visually with quantified scale, and Transforming hearing into vision results in individualized sight-singing learning. Therefore erroneously sung music notes during sight-singing can be identified and depicted with self-generated visualization on pitch recognition that is visually presented with quantified scale in waveforms with each corresponding music notes. The learner could specifically adjust his/her errors during sight-singing.

4   Conclusion

Music sight-singing is a complex skill mastered by many musicians. The abstract concept hinders those who commence music study from moving forward smoothly at a satisfactory level. Achieving accurate music sight-sing skills limits many music learners' motivation to music study. Researches show that accurate proficiency in music sight-sing is not simply a matter of repeated practice and instruction but rather a process requiring the specified individual difference of learning needs in musical understanding.

This study shows that learning to acquire music sight-singing skills should involve the types of visual, aural, and intermixed forms of alternative learning pathways, and that these multisensory forms of learning interface designs are a convenient means by which learners facilitate coordinating ear, eye, and hand on what they see in notation and hear or imagine in their mind. At a fundamental level, this study expects the long-term benefits that would take shape if all the visual and aural learning interfaces were introduced appropriately during the earlier stages.

References

1. Altieri, N.: Multisensory integration, learning, and the predictive coding hypothesis. Front. Psychol. **5**, 257 (2014)
2. Boyle, D., Lucas, K.: The effect of context on sightsinging. Bull. Counc. Res. Music Educ. **106**, 1–9 (1990)
3. Chu, C.N., Huang, Y.T., Li, T.Y.: Web-based aural skills learning system with spoken music browser for the elementary school students. In: World Conference on e-Learning in Corporate, Government, Healthcare, & Higher Education, 24–28 October 2005, Vancouver BC, Canada
4. Demorest, S.M., May, W.V.: Sight-singing instruction in the choral ensemble: factors related to individual performance. J. Res. Music Educ. **43**(2), 156–167 (1995)
5. Dwiggins, R.: Teaching sight-reading in the high school chorus. Appl. Res. Music Educ. **2**(2), 8–11 (1984)
6. Farenga, J.: Arizona high school choral educators' attitudes toward the teaching of group sight singing and preferences for instructional practices. Doctoral dissertation, Arizona State University (2013)
7. Foulkes-Levy, L.: Tonal markers, melodic patterns, and musicianship training: part i: rhythm reduction. J. Music Theor. Pedagogy **11**, 1–24 (1997)
8. Grutzmacher, P.A.: The effect of tonal pattern training on the aural perception, reading recognition, and melodic sight-reading achievement of first-year instrumental music students. J. Res. Music Educ. **35**(4), 171–181 (1987)
9. Hagen, S.L., Cremaschi, A., Himonides, C.S.: Effects of extended practice with computerized eye guides for sight-reading in collegiate-level class piano. J. Music Technol. Educ. **5**(3), 229–239 (2013)
10. Harrison, C.S., Asmus, E.P., Serpe, R.T.: Effects of musical aptitude, academic ability, music experience, and motivation on aural skills. J. Res. Music Educ. **42**(2), 131–144 (1994)
11. Henry, M., Demorest, S.M.: Individual sight-singing achievement in successful choral ensembles: a preliminary study. Appl. Res. Music Educ. **13**(1), 4–8 (1994)
12. Hoffman, R., Pelto, W., White, J.W.: Takadimi: a beat-oriented system of rhythm pedagogy. J. Music Theor. Pedagogy **10**, 7–30 (1996)
13. Huang, Y.T., Chu, C.N.: Making music meaningful with adaptive immediate feedback drill for teaching children with cognitive impairment: a dual coding strategy to aural skills. Lect. Notes Comput. Sci. **8547**, 459–462 (2014)
14. Huang, Y.T., Chu, C.N.: Visualized comparison as a correctness indicator for music sight-singing learning interface evaluation—a pitch recognition technology study. Lect. Notes Electr. Eng. **375**, 959–964 (2016)

15. Killian, J.N.: The relationship between sightsinging accuracy and error detection in junior high singers. J. Res. Music Educ. **39**(3), 216–224 (1991)
16. Kostka, M.J.: The effects of error-detection practice on keyboard sight-reading achievement of undergraduate music majors. J. Res. Music Educ. **48**(2), 114–122 (2000)
17. Lee, H., Noppeney, U.: Long-term music training tunes how the brain temporally binds signals from multiple senses. Proc. Nat. Acad. Sci. **108**(51), E1441–E1450 (2011)
18. Lehmann, A., McArthur, V.: Sight-Reading: The Science and Psychology of Music Performance: Creative Strategies for Teaching and Learning. Oxford University Press, New York (2002)
19. Lucas, K.V.: Contextual condition and sightsinging achievement of middle school choral students. J. Res. Music Educ. **42**(3), 203–216 (1994)
20. Margulis, E.H.: A model of melodic expectation. Music Percept. **22**, 663–714 (2005)
21. Middleton, J.: Develop choral reading skills. Music Educ. J. **68**(7), 29–32 (1984)
22. Mayer, R.E., Moreno, R.: Nine ways to reduce cognitive load in multimedia learning. Educ. Psychol. **38**, 43–52 (2003)
23. Mayer, R.E., Moreno, R., Boire, M., Vagge, S.: Maximizing constructivist learning from multimedia communications by minimizing cognitive load. J. Educ. Psychol. **91**, 638–643 (1999)
24. Mishra, J.: Factors related to sight-reading accuracy: a meta-analysis. J. Res. Music Educ. **61**(4), 452–465 (2014)
25. Mishra, J.: Improving sightreading accuracy: a meta-analysis. Psychology of Music (2013). https://doi.org/10.1177/0305735612463770
26. Norris, C.E.: A nationwide overview of sight-singing requirements of large-group choral festivals. J. Res. Music Educ. **52**(1), 16–28 (2004)
27. Paivio, A.: Mental Representations: A Dual Coding Approach. Oxford University Press, Oxford (1986)
28. Paney, A.S., Buonviri, N.O.: Teaching melodic dictation in advanced placement music theory. J. Res. Music Educ. **61**(4), 396–414 (2014)
29. Powers, A.R., Hillock, A.R., Wallace, M.T.: Perceptual training narrows the temporal window of multisensory binding. J. Neurosci. **29**(39), 12265–12274 (2009)
30. Rinck, M., Denis, M.: The metrics of spatial distance traversed during mental imagery. J. Exp. Psychol. Learn. Mem. Cogn. **30**, 1211–1218 (2004)
31. Rogers, M.: Aural dictation affects high achievement in sight singing, performance and composition skills. Aust. J. Music Educ. **1**, 34 (2013)
32. Sharman, E.: The impact of music on the learning of young children. Bull. Counc. Res. Music Educ. **66**(1), 80–85 (1981)
33. Sheldon, D.A.: Effects of contextual sight-singing and aural skills training on error-detection abilities. J. Res. Music Educ. **46**(3), 384–395 (1998)
34. Stevenson, R.A., Wallace, M.T.: Multisensory temporal integration: task and stimulus dependencies. Exp. Brain Res. **227**(2), 249–261 (2013)
35. Waggoner, D.T.: Effects of listening conditions, error types, and ensemble textures on error detection skills. J. Res. Music Educ. **59**(1), 56–71 (2011)
36. Wolf, T.E.: A cognitive model of musical sight-reading. J. Psycholinguist. Res. **5**(2), 143–171 (1976)
37. Wurtz, P., Mueri, R.M., Wiesendanger, M.: Sight-reading of violinists: eye movements anticipate the musical flow. Exp. Brain Res. **194**, 445–450 (2009)
38. Zatorre, R.J., Chen, J.L., Penhune, V.B.: When the brain plays music: auditory–motor interactions in music perception and production. Nat. Rev. Neurosci. **8**(7), 547–558 (2007)
39. Zhukov, K.: Evaluating new approaches to teaching of sight-reading skills to advanced pianists. Music Educ. Res. **16**(1), 70–87 (2014)

The Research of the Seven Steps of Normalized Object Oriented Design Class Diagram

Yih-Chearng Shiue, Sheng-Hung Lo[(⊠)], and Kuan-Fu Liu

Department of Information Management, National Central University,
Taoyuan, Taiwan
ycs@mgt.ncu.edu.tw, shenghung.lo@gmail.com,
tofull3@yahoo.com.tw

Abstract. The first of these seven steps is to eliminate multivalued attributes, composite attributes, and composite operations. The second is to eliminate the partial dependency and transitive dependency among the attributes, as well as shared operations. The third is to eliminate homogeneous operations among the classes to meet the requirements of inheritance and polymorphism. The fourth and fifth steps involve establishing classes for encapsulation, and the sixth and seventh steps eliminate multivalued dependency and operations with multivalued dependency. These steps create normalised concrete classes and control classes in the object-oriented class diagram, and they also maintain favourable consistency, completeness, and accuracy of the data. As such, they can provide effective guidelines and practical reference for system analysis and development.

Keywords: Object-oriented · Normalisation · Class diagram · System analysis
Unified Modelling Language

1 Introduction

Since the emergence of object-oriented programming, the information technology and related industries have employed it in the development and design of system and process concepts to such an extent that it is worth reorganising and rethinking its logic and rules to resolve the problems in system development posed by normalisation.

In object-oriented system design, the consistency, completeness, and accuracy of class diagrams determines the success, or lack thereof, of a system. Clear and specific principles and procedures have been set up for conventional structured system design through the use of entity relationship diagrams and database normalisation, yet when it comes to object-oriented system design, although which has proposed shared operation as the object-oriented principle for the second normal form, there are still no specific rules for the first, third, and fourth normal forms, nor are there considerations for object-orientedness.

Therefore, the present study investigates the theoretical basis of database normalisation from object-oriented programming's characteristics of encapsulation, inheritance, and polymorphism to propose principles for the first to fourth normal forms, and to translate such principles into specific modelling procedures in class diagrams, for the purpose of generating a normalised database and operable object classes.

© Springer Nature Singapore Pte Ltd. 2018
J. C. Hung et al. (Eds.): FC 2017, LNEE 464, pp. 315–325, 2018.
https://doi.org/10.1007/978-981-10-7398-4_33

2   Literature Review

2.1   Class Diagrams

The class diagram is a type of static structure diagram in software engineering. It uses the Unified Modelling Language (UML) to describe a system's classes and their attributes, the structure of operations, and the interclass relationships [1, 2]. Classes constitute the core of an object-oriented system, for they help translate the model into scripts [3–5].

Various types of relationships can exist between the classes, either hierarchical or parallel, each of which can be denoted by a distinct symbol [6, 7]. The multiplicity in the relationships must also be considered: depending on the number of objects involved, multiplicity values must be assigned to signify the number of instances of the associated classes, which can be one-to-one, one-to-many, or many-to-many.

2.2   Object-Oriented Programming

Object-oriented system development takes into account both data and operations, with the primary task of identifying the relationships between the participating objects to solve problems through the collaboration of said objects. The concept of object-oriented programming is to enhance the repeatability of program components and view the objects as basic elements of a program that encapsulate sets of attributes and operations to enhance flexibility and expandability. Object-oriented programming has three primary features: encapsulation, inheritance, and polymorphism.

3   Object-Oriented Features and Object Normalisation

3.1   Research Framework

The present study primarily relies on analysing the logic behind database normalisation and the theoretical basis of object normalisation, as well as object-oriented design techniques and characteristics, to propose logic rules for the normalisation of object-oriented design. The core of the research framework has been to propose a whole new set of methods and procedures for the normalisation of class diagrams, as shown in Fig. 1.

3.2   Research Process

The primary purpose of the present study has been to propose procedures for the normalisation of object-oriented design by integrating object-oriented characteristics into the normalisation of databases. The literature on object-oriented programming, database normalisation, and object normalisation was reviewed to clarify and establish the logic behind and between object-oriented programming and object normalisation, then translate this logic into procedures for drawing class diagrams. Next, a course selection system derived from Wu [8] was adopted for modelling to examine the

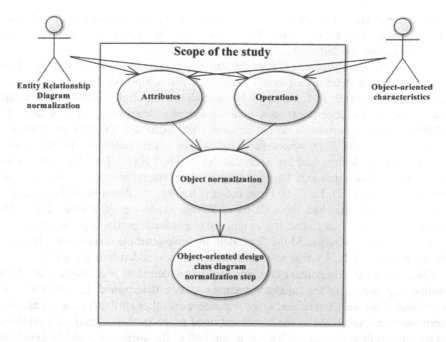

Fig. 1. Research framework

normalisation logic for specific objects. Finally, the results are discussed and their limitations laid out for reference in future studies.

3.3 Dependency Between the Determinants, Attributes, and Operations

The database normalisation method by Wu [8] was adapted for class normalisation. Because the structure of the system as a whole is determined by the dependency between system attributes and operations, the establishment of dependency was recognised as the core procedure for object normalisation.

3.3.1 Attribute-Dependent Determinants

The master/slave relationship between system attributes can be learned from descriptions of the attributes. If A represents the set of system attributes, and D represents the set of system determinants, $D \subseteq A$; and if N represents the nondeterminants dependent on D, $N \subseteq A$. Furthermore, when $Ax \subseteq A$, $Ay \subseteq A - Ax$; the change in an attribute Ax determines the change in another attribute Ay and identifies the change in Ay'. This attribute can be named the determinant Dx and the dependent Ay' named Nx, where $Dx \subseteq D$ and $Nx \subseteq N$. When a Dx can be used to determine Nx, it can be referred to as the attribute dependent on the determinant $Dx \rightarrow Nx$. Therefore, system attributes could be divided from a master/slave viewpoint into the determinants (Ds) and nondeterminants (Ns). Attributes that were found to be neither D nor N were coded L, and $L = A - D - N$, meaning that the Ls are not dependent on either of the attributes, nor can they determine other attributes [8].

Per the preceding description, the first step of normalisation should be the division of the attributes into three groups: the determinants, the attributes dependent on the determinants, and the leftovers.

So far only the situation of a single attribute becoming a determinant has been considered, but a determinant could also be a set comprising multiple attributes. Moreover, even though an L might not be dependent on a single-attribute determinant, it could in fact be dependent on a multiple-attribute determinant. The nature of a multiple-attribute determinant would involve partial dependency between attributes, that is, $Nx \subseteq N$, $Dx \subseteq D$, where the attribute Nx is dependent on the subset of a multiple-attribute determinant Dx' such that $Nx \rightarrow Dx \wedge Nx \rightarrow Dx'$. Therefore, of the total number of determinants Dn, the known determinants would be combined and permutated as $Dy \subseteq D$, $Ly \subseteq L$. From the smallest set, the determinant Dy would be brought out to be matched from $CDn1$ to $CDnDn-1$, to verify whether $Ly \rightarrow Dy$. Whenever $Lx \rightarrow D'$ was found, it was ignored to eliminate partial dependency; that is, if and only if $Lx \rightarrow Dy$ would the dependent multiple-attribute determinants be added into the set D, and the Ly that satisfies the dependency added into the set N.

After all of the determinant combinations were identified and assigned to D, the absolute dependency of the nondeterminants could be determined. L, which includes the attributes that were not related to, or dependent on, other attributes, was set as a new determinant and subjected to the aforementioned procedures to identify the attributes and operations that were dependent on it, until all of the attributes could be classified. Through this method, all of the determinant sets in the system could be identified, and all of the nondeterminant sets were dependent on the determinant sets. The dependencies between the nondeterminants were eliminated after this reorganisation of the nondeterminants, thus eliminating the transitive dependency between the attributes [9].

3.3.2 Operation-Dependent Determinants

The dependency on a determinant is established through the identification of dependent attributes, whereas the dependency of operations is established through access to the determinant. Unlike the attributes of an operation, which are readily known, the determinant can only be defined through the dependency between attributes. The fact that an operation cannot define a determinant means that the process of system modelling has to define the determinants based on interattribute dependencies first, and then determine the operations' dependencies on the determinants. If O represents the set of system operations, and $Ox \subseteq O$ if Ox can only access the determinant Dx and the nondeterminant Nx that is dependent on Dx (i.e., Ox cannot access any other attributes that are not dependent on Dx), the determinant is referred to as operation-dependent determinant $Dx \rightarrow Ox$. Through the classification process addressed in Sect. 3.3.1, every time a new determinant is found, the operations dependent on it can also be found from the corresponding dependent attributes.

Through this classification process, each determinant corresponds to multiple operations, whereas each operation only corresponds to one determinant. When examining the set of operations dependent on the same determinant, if functional overlapping is found among the operations, the overlapping part would be set aside as a new operation so that the original operation no longer includes the function of the new operation. However, not all of the operations are dependent on one determinant, and if the range of an operation's access is found to exceed Dx + Nx, this means that the range exceeds the access to a class, and further action is required.

3.4 Class Generalisation and Specialisation

After establishing the dependency between attributes and operations, the system's preliminary data and operational structures are ready for further refinement based on the system's needs. Generalisation involves extracting the common parts of disparate classes as a basic class that is a level higher than other classes; it corresponds with the concept of inheritance in object-oriented programming, whose purpose is to simplify system structure while still maintaining the design intent of the original structure. Specialisation, however, forcefully breaks up the existing logic and rules of the class structure through processes or outputs that are unique in use case diagrams. It fulfils the special needs of the exceptional cases in object-oriented programming with minimal overriding of existing scripts.

3.5 Class Encapsulation

After the adjustments in Sect. 3.2, and after the inheritance of different classes has been determined though the homogeneous operation, the class attributes and operations can be encapsulated into official classes. The encapsulation of a class can be realised through its visibility; a concrete class must come with a name, including the attributes and operations of the association (which must fit to a predetermined visibility) and actual data type [10].

3.6 Control Classes

Given the types of dependencies discussed in Sect. 3.1, examining whether the range of an attribute accessed by an operation is wide enough to encompass a determinant set and its dependent nondeterminants, and then adding the operation dependent on the determinant set into a concrete class in the fourth step, can leave some leftover operations (Ls) untreated since the second step. These Ls could be dependent on any determinant (D), and the range that these operations access does not follow the dependent attributes, placing it beyond the attribute set dependent on one single determinant. Moreover, although the dependency on a single determinant cannot be determined for the class operation, disqualifying it for a concrete class, this also means that the access authority of this type of operation is higher than ordinary operations of concrete classes. An operation of this type is able to control attributes in multiple classes, which conforms with the shared operation of 2ONF, and should be classified under the control class.

3.7 Multiplicity in Interclass Relationships

After establishing the concrete and control classes, the sixth step is to define the relationship between the classes. Notation could be used to indicate the hierarchy and active/passive statuses in such relationships. A class could maintain a relationship with multiple other classes, each of which exhibits multiplicity in the relationship. The relationship corresponding to a single-determinant class can vary considerably from one to a multiple-determinant class; the former can only be inferred from use case diagrams and logic, whereas the latter can be inferred from the set of class determinants. A multiple-determinant class is a concrete class made up of the relationship between two classes; each sub-determinant in a multiple-determinant class corresponds to the relationship with one other single-determinant class [6–8].

Therefore, the relationships with single-determinant classes should be established first, and then a multiple-determinant class can be inserted between two single-determinant classes. Because the control classes do not have a determinant, they can access the concrete class attributes of the aforementioned single-determinant classes and multiple-determinant classes. The relationship where control classes use concrete classes and the used concrete classes influence the control classes is a temporary weak relationship; each time an attribute in a concrete class is accessed, a "use" relationship would be established with that class.

When the class determinant set is composed of more than three determinants, the class is in multiple relationships with more than two other classes. Because such a class does not conform to 4NF multivalued dependency, the class determinant has to be dismantled into a set of two or fewer determinants, which should still maintain the multiplicity of the corresponding relationships. Multiplicity in relationships can lead to dependency problems, however. In the following sections, the study discusses the multiplicity problems of one-to-one and many-to-many relationships according to the records of various classes that were paired by determinants. When there is a multivalued dependent determinant in a class, there are also operations dependent on it; therefore, when eliminating multivalued dependencies, the operations accessing the multivalued dependent determinant must also be examined for adjustments. When a new attribute is assigned to a class, the class operations must be adjusted as a result, and if the adjusted determinant is to form a multiple-attribute determinant with the original class determinant, all the operations accessing the determinant must be adjusted accordingly. Conversely, if the adjusted determinant is to serve only as a class nondeterminant instead of forming a multiple-attribute determinant with the original class determinant, and the original class operations do not access it either, then only the operations accessing the determinant have to be altered in response to the change.

3.8   Seven Steps for Object Normalisation

Before object normalisation, information on the object's class must be obtained to grasp the logical relationship between the attributes. This information includes names and descriptions of attributes, data types, and examples (Table 1). The operational information required includes the names, descriptions, and parameters of the

Table 1. List of class diagram attributes

Description	Name	Data Type	Examples

operations, as well as accessed attributes (Table 2). The information addressed here must be ready before object normalisation can proceed.

Table 2. List of class diagram operations

Description	Name	Parameters	Attributes

As discussed in preceding sections, the object normalisation procedure has seven steps.

Step 1: Classification of Attributes and Operations

1.1. Elimination of redundant attributes and composite attributes: As shown in Table 3, attributes are classified as determinants (A1), dependent on determinants (A2), or other (A3), whereas operations are classified as dependent on determinants (O1) or other (O2).

Table 3. Attributes and operations

Determinant attribute A1	Attribute dependent on determinant A2	Operation dependent on determinant O1	Attribute, other A3	Operation, other O2

1.2. Elimination of composite operations: This stage looks for overlaps between O1 operations that are dependent on the same determinant. The overlap is made into a new operation.

Step 2: Reorganisation of Multiple-Dependent Attributes and Operations

2.1. Elimination of partial dependent attributes and shared operations: This stage determines the attributes and operations in A3 and O2 that can be decided concurrently, and by which determinants. Elimination of transitive dependency:

2.2. Attributes in A3 that are not dependent on any determinants are redefined as new determinants, and then added to A1.

Step 3: Encapsulation of Inherited Classes

3.1. Elimination of homogeneous operations: If two classes have attributes of the same function, or two or more operations that access such attributes, these attributes and operations should be removed from the original classes and encapsulated as a new parent class.

3.2. If there are two subclasses that inherit the polymorphic operations of a parent class, whether either of the subclass operations needs to override the parent class operation is determined depending on the polymorphic operation of the parent class.

Step 4: Encapsulation of Concrete Classes

4.1. Individual concrete classes are encapsulated according to the determinants, and named. Visibility and data type is clearly marked for each attribute, and visibility is also clearly marked for each operation.

Step 5: Encapsulation of Control Classes

5.1. Operations in O2 that access attributes across classes are encapsulated as control classes, and named. Visibility is also clearly marked for each operation.

Step 6: Plotting the Class Diagram

6.1. The relationships between concrete classes are plotted based on the logical relationships derived from use cases.

6.2. The relationships between control classes are plotted based on the accessed attributes of concrete classes.

Step 7: Adjustments for Class Relationships

7.1. If two classes are in a one-to-one relationship, the determinant of the class with more operations is placed in the attribute compartment of another class. The attributes and parameters of the target class that access the determinant operation are also adjusted.

7.2. If two classes are in a one-to-many relationship, the determinant of the class on the "one" side is placed in the attribute compartment of the class on the "many" side. The attributes and parameters of the target class that access the determinant operation are also adjusted.

7.3. If two classes are in a many-to-many relationship, the determinants of the two classes are used as codeterminants to form a new class. The attributes and operations in the original classes and those dependent on those determinants are also moved to the new class.

7.4. If there is only one attribute for the determinant in a class, the class is deleted.

4 Class Diagram Examples of the Seven Steps for Object-Oriented Normalisation

4.1 Case of a Course Selection System

To verify the seven steps, the course selection system in [8] was adapted for a simulation test as follows.

To integrate its information systems, a school decided to combine its student database, faculty database, and the course selection system into a unified system that is accessible to the faculty, administrative staff, and students. The following user requirements were collected through various interviews and meetings:

1. Administrative staff should be granted access to the students' personal information, because it is their duty to input such information at the beginning of each semester.
2. Students should be allowed to file for minors or double majors. Upon receiving an application, the administrative staff should be able to change the applicant's status accordingly.
3. The course selection system should be open at the beginning of each semester to allow students to select their courses. It should be open again in the middle of the semester to allow courses to be added or dropped.
4. At the end of each semester, each professor has to submit a list of courses for the following semester. There should be an upper limit for the number of courses for each professor.
5. Each professor is entitled to a personal office, and has the right to apply for a change of location.
6. A student has the right to join multiple clubs, and participation in club activities should be entered into the student's extracurricular performance record.
7. At the end of the semester, students are required to evaluate the courses taken that semester as feedback for course instructors.
8. Course instructors should submit student grades before the end of the semester. Upon submission, the students concerned should be notified by email and allowed to check their grades online.
9. Students should have access to their own grade report online, and be able to print it out.

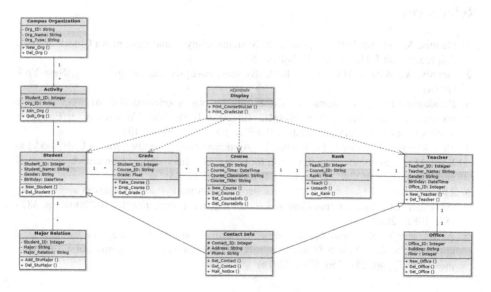

Fig. 2. Class diagram at the end of Step 7

10. The system should be capable of issuing invitations for special campus events to faculty and students, either by email or by hard copies of the invitation, with the capability to send invitations to faculty that are distinct from those sent to students.
11. At the end of the semester, the system should automatically send hard copies of students' grade reports to their contact addresses (Fig. 2).

5 Discussion and Conclusion

Object normalisation encompasses attribute and operation normalisation. While attribute normalisation can be considered the normalisation of the database in accordance with the actual class diagram to prevent discrepancies and duplication in the data, operation normalisation still lacks a comprehensive discourse. On the basis of previously mentioned principles for object-oriented normalisation, the present study has proposed a comprehensive method and steps for object-oriented normalisation that can be applied in the process of object modelling, to maintain the object-oriented characteristics in the resultant class diagram, and also facilitate the consistency, completeness, and accuracy of data.

To remain focused on the theoretical core of object-oriented normalisation, the present study used a simplified and fictional case study. An actual case would have been massive in scale and required considerable labour, not to mention the interpretation of results would have been limited by the capabilities and experience of the researchers. However, the method proposed here was derived from a prevailing method in past studies, and its central concepts should still be applicable under ordinary situations.

References

1. Dennis, A., Wixom, B.H., Tegarden, D.: Systems Analysis and Design: An Object-Oriented Approach with UML. Wiley, Hoboken (2015)
2. Dennis, A., Wixom, B.H., Roth, R.M.: Systems Analysis and Design. Wiley, New York (2014)
3. Papakonstantinou, N., Sierla, S.: Generating an object oriented IEC 61131-3 software product line architecture from SysML. In: 2013 IEEE 18th Conference on Emerging Technologies & Factory Automation (ETFA), pp. 1–8. IEEE (2013)
4. Vogel-Heuser, B., Witsch, D., Katzke, U.: Automatic code generation from a UML model to IEC 61131-3 and system configuration tools. In: 2005 International Conference on Control and Automation, ICCA 2005, pp. 1034–1039. IEEE (2005)
5. Thramboulidis, K.C.: Using UML in control and automation: a model driven approach. In: 2004 2nd IEEE International Conference on Industrial Informatics, INDIN 2004, pp. 587–593. IEEE (2004)
6. Abdulganiyyi, N., Ibrahim, N.: Semantic abstraction of class diagram using logical approach. In: 2014 Fourth World Congress on Information and Communication Technologies (WICT), pp. 251–256. IEEE (2014)

7. De Lucia, A., Gravino, C., Oliveto, R., Tortora, G.: Data model comprehension: an empirical comparison of ER and UML class diagrams. In: The 16th IEEE International Conference on Program Comprehension, ICPC 2008, pp. 93–102. IEEE (2008)
8. Wu, M.S., Wu, S.-Y.: Systems Analysis and Design. West Publishing Co., St. Paul (1994)
9. Yonghui, W., Wenyun, J., Aoying, Z.: Implementation and proof for normalization design of object-oriented data schemes. In: Proceedings of the 36th International Conference on Technology of Object-Oriented Languages and Systems, TOOLS-Asia 2000, pp. 220–227. IEEE (2000)
10. Ambler, S.: About the third rule of class normalization. Computing Canada, December 1996

Sleeping Customer Detection Using Support Vector Machine

Tsun Ku, Pin-Liang Chen, and Ping-Che Yang[✉]

Institute for Information Industry, Taipei, Taiwan, R.O.C.
{cujing, mileschen, maciaclark}@iii.org.tw

Abstract. Customers are difficult to find and sometimes even more difficult to keep. If a company doesn't pay attention to customer relationship management, it will spend lots of money to acquire them and then let them sleep. The goal of this study was to develop a sleeping customer detection system that collected users' social network information and shopping behavior, and then classified them into 3 categories (sleeping customer, napping customer, and general customer). We collected the user information from January 01, 2015 to December 31, 2016. Support vector machine based classification was used. In this study, the overall accuracy was 81.7%. The results can help companies to reactivate sleeping customers as soon as possible.

Keywords: Classification · Customer segmentation · Sleeping customer

1 Introduction

Customer segmentation is the act of dividing customers into groups of similar individuals based on common characteristics. Companies can use different strategies to different groups effectively and appropriately. Sleeping customer detection is one of the most important issues. A sleeping customer means a customer doesn't buy anything at the store for a long time. Customers are difficult to find and sometimes even more difficult to keep. However, many companies don't pay attention to customer relationship management (CRM) and miss the opportunity to reactivate them. Therefore, it is valuable to design a system to detect sleeping customers, and we can reactivate them as soon as possible.

Kreara, a company providing services in data analytics, provides a framework to find sleeping customers using purchase history. They slice customers into active and passive based on the recency, frequency and monetary (RFM) aspects of their purchase. Then they reactivate them by sending customized messages. They announced that as high as 73% of the sleeping customers got reactivated.

In the past, there were several studies focused on customer segmentation [1–5]. Ding classified the credit card customers into five types: sleeping customer, active customer, common customer, potential customer and churn customer, and chose the target customers [6]. Li *et al.* classified the credit card customers into four classifications to reduce "sleeping cards" and increase credit card usage rates [7]. In our previous studies, we classified followers of Facebook fan pages into 6 clusters based on sentiment analysis [8, 9].

© Springer Nature Singapore Pte Ltd. 2018
J. C. Hung et al. (Eds.): FC 2017, LNEE 464, pp. 326–334, 2018.
https://doi.org/10.1007/978-981-10-7398-4_34

In this study, we developed a sleeping customer detection system. The sleeping customer detection system collected users' social network information and shopping behavior from Facebook and a point earning app. The shopping behavior from point earning app was provided by an online to offline (O2O) corporation we cooperated with. Customers could use the app and get reward points from completing its assignments, for example, spending over NT$100 in the designated store. Its cooperating companies included JPMED, KFC, SK-II, etc. Customers could exchange the reward points for some gifts, for example, donuts, tart or gift certificates. Finally, the customers were classified into 3 categories (sleeping customer, napping customer, and general customer). The combination of customer social network information and shopping behavior provided the power to extract business value, find the target customers and improve the conversion rates.

The goal of this study was to develop a sleeping customer detection system that collected users' social network information and shopping behavior, and then classified them into 3 categories (sleeping customer, napping customer, and general customer). The future goal was to reactivate the napping and sleeping customers.

2 Methods

To avoid the loss of customers, we developed a sleeping customer detection system that collected users' social network information and shopping behavior, and then classified them into 3 categories (sleeping customer, napping customer, and general customer). The sleeping customer detection system predicted a customer whether he/she would become a sleeping/napping customer or not in the next two months. In this section, we describe the sleeping customer detection system we developed and the technologies in each module.

2.1 Data Collection

In this section, we describe the data source we used in the paper. We collected the information of the users who installed the point earning app. The point earning app would give the users three types of assignments. One was to check into the store; another was to scan the barcode of some products; the other was to purchase some products in the store. When a user finished one assignment, he/she would be rewarded some reward points and the record would be saved in the database. Besides, the point earning app used Facebook login and needed the authorization to access user's information on Facebook. Therefore, we had two types of information, the user shopping behavior records on point earning app and their social network information on Facebook.

In the data collection process, we collected the user shopping behavior records of point earning app from October 14, 2014 to December 31, 2016. The user records contained user Facebook ID, invite code, corporation name, tax ID, shop name, assignment name, assignment type, reward points, beacon ID, execution time, app platform, invoice no., invoice information, invoice amount, and status. There were three types of assignments on point earning app according to the assignment name. One was the check-in assignment; another was the scan assignment; the other was the shopping assignment.

Furthermore, we collected the social network information of point earning app users by their Facebook ID. The user information on Facebook contained app ID, user Facebook ID, user name, user email, gender, birthday, location (the city he/she lived in), number of followers, social network preferences (the fan pages' IDs, names and categories he/she followed, and the time he/she followed a fan page).

2.2   Data Extraction

In the study, we extracted user records about the retail store. The dates of those user records were between January 01, 2015 and October 31, 2016. Those records covered more than one hundred stores of the retailer in Taiwan. We tracked each customer in the study samples for 2 months from October 31, 2016 to decide whether he/she became a sleeping customer or not.

We used social network features and shopping behavior features of each sample. The social network features contained the following information: age; gender; location; social network preferences. The shopping behavior features contained the following information: the average periods he/she did the check-in assignments, scan assignments and shopping assignments; the average number of times he/she did the check-in assignments, scan assignments and shopping assignments in one month; the average spend in one month; the average periods he/she did the check-in assignments, scan assignments and shopping assignments in the last two months and last month; the average number of times he/she did the check-in assignments, scan assignments and shopping assignments in the last two months and last month; the average spend in the last two months and last month; the last time he/she did a check-in assignment, scan assignment and shopping assignment.

We defined that a sleeping customer is who didn't buy anything in the retail store 30 days after his/her last shopping time. A napping customer is who isn't a sleeping customer and didn't buy anything in the retail store 15 days after his/her last shopping time. Otherwise, he/she is a general customer.

We labeled all samples into 3 categories according to their status change. If one person became a sleeping/napping customer between November 1, 2016 and December 31, 2016, we labeled him/her as a sleeping/napping customer. Otherwise, we labeled him/her as a general customer.

2.3   Sample Selection

In the sample selection step, we removed the outlier and reduced the bias caused by the different ratio of 3 categories (sleeping customer, napping customer, and general customer). First, to avoid the influence of the error records, we deleted the duplicate records or the records with incomplete status. Second, we excluded the samples that already were napping customers or sleeping customers before October 31, 2016. Third, we randomly selected n samples from each category separately to ensure that each category would be included in the training sample.

2.4   Feature Extraction

In the feature extraction step, we described how we set the features. The gender was separated into 2 features: female and male. The location was separated into 20 features: 20 administrative divisions of Taiwan. The social network preferences were separated into 56 features based on the 56 categories of top 10,000 highest liked Facebook fan pages we labeled manually. Table 1 shows some examples of 56 categories of Facebook fan pages. Categories of top 10 highest liked Facebook fan pages are shown in Table 2.

Table 1. Some examples of 56 categories of Facebook fan pages

Categories			
Retail	Cosmetics	Writer	Relationship
Mart	Costume	Illustrator	News
Mall	Accessories	Movie	Drinks
Discount	Shoes	TV	Performer
Online Shopping	Cooking	Game	Organization
House	Snacks	Animation	Music
Communication	Broadcast	Pet	...

Table 2. Categories of top 10 highest liked Facebook fan pages

Top	Fan Page	Category
1	7-ELEVEN	Retail
2	pairs	Relationship
3	Yahoo! News	News
4	Starbucks	Drinks
5	FamilyMart	Retail
6	May Day	Performer
7	Duncan	Illustrator
8	s3beauty	Cosmetics
9	FanFan	Performer
10	86 Shop	Cosmetics

For the gender and location features, a feature value was set to 1 if the sample belonged to the feature; otherwise, it was set to 0. For the social network preferences features, a feature value was set to 1 if the sample followed more than 3 fan pages belonged to the category; otherwise, it was set to 0. If a sample had no information from Facebook, all the social network features would be set to 0.

For the 3 types of assignments, we calculated the features (the average periods, the average spend, the average periods in the last two months and last month, etc.) of each type of assignments separately. Finally, we got 103 features (79 features about social network information and 24 features about shopping behavior) in the next classification step.

2.5   Classification

In the classification step, we classified users into 3 categories (sleeping customer, napping customer, and general customer). Support vector machine (SVM) based classification was used. SVM is a popular machine learning method for classification, regression, and other learning tasks. Here, we built our classifier by LIBSVM [10], a popular SVM library developed by Chang and Lin. 10-fold cross validation was used. Figure 1 shows the architecture of classification module.

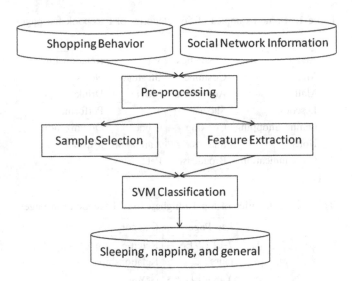

Fig. 1. The architecture of classification module.

3   Experiment Results

In this section, we describe the study sample used in the experiment and report the test results on our sleeping customer detection system.

3.1   Study Sample

In this study, we used the user information between January 01, 2015 and October 31, 2016. There were 20,104 users in total. More than 1 million shopping behavior records were collected. The number of samples and ratio in each category was shown in Table 3.

Table 3. The number of samples and ratio in each cateogry

Category	No. of samples (20,104)	Ratio
Sleeping customer	10153	50.5%
Napping customer	2453	12.2%
General customer	7498	37.3%

The number of napping customers (12.2%) was much fewer than that in other categories. To reduce the sampling bias caused by the different ratio of 3 categories, we applied the sample selection method described above. We randomly selected n samples from each category separately. If the number of samples in one category was less than n, we included all of the samples. Here we set $n = 2000, 3000, 4000$ and 5000.

3.2 Classification Results

Table 4 shows the classification results on different number of samples we selected from each category and the line chart is shown in Fig. 2. The overall accuracy increased when the n increased. However, the accuracy of napping customer group was the highest (73.1%) when $n = 3000$, and it decreased slightly (72.4%) when $n = 4000$. When $n = 5000$ the accuracy of napping customer group decreased a lot (65.3%). The accuracy of sleeping customer group was the most important, and a little accuracy decreased in napping customer group was allowed. Therefore, we finally chose $n = 4000$ and the overall accuracy was 81.7%.

Table 4. The classification results on different number of samples selected from each category

n^*	Category	Correct	Wrong	Accuracy
2000	Sleeping	1446	554	72.3%
	Napping	1364	636	68.2%
	General	1384	616	69.2%
	Total	4194	1806	69.9%
3000	Sleeping	2403	597	80.1%
	Napping	1794	659	73.1%
	General	2337	663	77.9%
	Total	6534	1919	77.3%
4000	Sleeping	3408	592	85.2%
	Napping	1776	677	72.4%
	General	3356	644	83.9%
	Total	8540	1913	81.7%
5000	Sleeping	4425	575	88.5%
	Napping	1602	851	65.3%
	General	4259	741	85.2%
	Total	10286	2167	82.6%

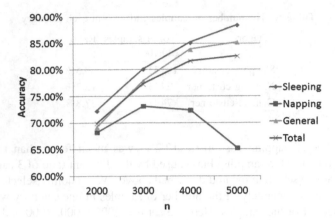

Fig. 2. The line chart of classification results on $n = 2000 \sim 5000$.

Next, we compared the effect of different feature sets. We defined 3 cases: the samples with only social network features, the samples with only shopping behavior features, and the samples with all features. Table 5 shows the classification results in each cases and the bar chart is shown in Fig. 3.

Table 5. The classification results of the samples with different feature sets

Category	Correct	Wrong	Accuracy
Samples with social network features	5592	4861	53.5%
Samples with shopping behavior features	7955	2498	76.1%
Samples with all features	8540	1913	81.7%

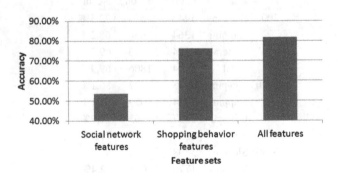

Fig. 3. The bar chart of classification result in each category.

The accuracy of samples with social network features was the lowest (53.5%), followed by the accuracy of samples with shopping behavior features (76.1%). The accuracy of samples with all features was the highest (81.7%). Finally, we used all features to train the prediction model.

4 Discussion

In this study, we developed a sleeping customer detection system that collected users' social network information and shopping behavior, and then classified them into 3 categories (sleeping customer, napping customer, and general customer). The sleeping customer detection system predicted a customer whether he/she would become a sleeping/napping customer or not in the next two months. Finally, the overall accuracy was 81.7%.

We found that the number of sleeping customers was the most, followed by the number of general customers. The number of napping customers was the lowest because many napping customers became sleeping customers soon. That was the reason why we developed a sleeping customer detection system.

The accuracy increased when we selected more customers from each category separately. However, the accuracy of napping customer group decreased when we selected more than 3000 customers from each category. The difference between the numbers of customers in each category was large. Therefore, one possible reason was that once we selected more than 3000 customers, the bias of the training sample increased. Nevertheless, we chose $n = 4000$ finally because the accuracy of sleeping customer group was the most important. A little accuracy decreased in napping customer group was allowed.

The accuracy of samples with only shopping behavior features was 76.1%, which was closed to the accuracy of samples with all features. However, the accuracy of samples with social network features was 53.5%. Besides, the accuracy increased (81.7%) when we added the social network features. Therefore, the social network features were still useful.

The strengths of this study are that we collected a large number of customer social network information and shopping behavior from the point earning app. The size of samples enabled a systematic examination of analysis. There are also some limitations to this study. The number of napping customers is the lowest in 3 categories. The difference between the number of customers in sleeping group and napping group is more than four times. Although we applied the sample selection method to reduce the bias, the bias may still affect the accuracy of classification results.

In conclusion, we developed a sleeping customer detection system that collected users' social network information and shopping behavior, and then classified them into 3 categories. The sleeping customer detection system predicted a customer whether he/she would become a sleeping/napping customer or not in the next two months. The overall accuracy was 81.7%. The results can be used to reactivate the napping and sleeping customers.

Acknowledgment. This study is conducted under the "Online and Offline integrated Smart Commerce Platform (4/4)" of the Institute for Information Industry which is subsidized by the Ministry of Economy Affairs of the Republic of China.

References

1. Alborzi, M.: Using data mining and neural networks techniques to propose a new hybrid customer behaviour analysis and credit scoring model in banking services based on a developed RFM analysis method. Int. J. Bus. Inf. Syst. **23**(1), 1–22 (2016)
2. Shashidhar, H.V., Varadarajan, S.: Customer segmentation of bank based on data mining – security value based heuristic approach as a replacement to K-means segmentation. Int. J. Comput. Appl. **19**, 8–13 (2011)
3. Bošnjak, Z., Grljevic, O.: Credit users segmentation for improved customer relationship management in banking. Presented at the 6th IEEE International Symposium on Applied Computational Intelligence and Informatics (SACI) (2011)
4. Zaza, S., Al-Emran, M.: Mining and exploration of credit cards data in UAE. Presented at the Fifth International Conference on e-Learning (econf) (2015)
5. Sun, N., Morris, J.G., Xu, J., Zhu, X., Xie, M.: iCARE: a framework for big data-based banking customer analytics. IBM **58**(5), 4 (2014)
6. Ding, G.: The research of the China Merchants Bank for applying data mining on the marketing of credit card (2007)
7. Li, W., Wu, X., Sun, Y., Zhang, Q.: Credit card customer segmentation and target marketing based on data mining. Presented at the International Conference on Computational Intelligence and Security, Nanning (2010)
8. Lin, K.-C., Wu, S.-H., Chen, L.-P., Ku, T., Chen, G.-D.: Mining the user clusters on Facebook fan pages based on topic and sentiment analysis. Presented at the IEEE IRI, Redwood City, CA (2014)
9. Lin, K.-C., Wu, S.-H., Chen, L.-P., Yang, P.-C.: Finding the key users in Facebook fan pages via a clustering approach. Presented at the IEEE IRI, San Francisco, CA (2015)
10. Chang, C.-C., Lin, C.-J.: LIBSVM: a library for support vector machines. ACM Trans. Intell. Syst. Technol. **2**(3), 27:1–27:27 (2011)

A New Chaotic Map-Based Authentication and Key Agreement Scheme with User Anonymity for Multi-server Environment

Fan Wu[1], Lili Xu[2], and Xiong Li[3(✉)]

[1] Department of Computer Science and Engineering,
Xiamen Institute of Technology, Xiamen 361021, China
[2] School of Information Science and Technology, Xiamen University,
Xiamen 361005, China
[3] School of Computer Science and Engineering,
Hunan University of Science and Technology, Xiangtan 411201, China
lixiongzhq@163.com

Abstract. The explosive usage of Internet and telecommunications makes the remote server access mechanism necessary. Multi-server architecture mixes various services into one system. A user can get different services from different providers after registering on one registration center. To protect the information transmitted in the sessions, authentication is naturally considered. In the past decades, many multi-server authentication schemes have been presented. But unfortunately, various sorts of attacks are presented to prove the past schemes insecure. To prevent the common attacks, we give a new two-factor authentication scheme for multi-server systems. Through the analysis of security properties and performance, all can see that the proposed scheme is against the common attacks, such as off-line guessing attacks, tracking attacks, etc. And it is suitable for application in real circumstance. ...

Keywords: Multi-server environment · Two-factor authentication
Forgery attacks · Chaotic maps

1 Introduction

One of the advantages of Internet is to make communications between numerous of computers distributed in different regions to be applicable. Scalable service

---

X. Li—This research is supported by Fujian Education and Scientific Research Program for Young and Middle-aged Teachers under Grant No. JA14369, University Distinguished Young Research Talent Training Program of Fujian Province (Year 2016), the National Natural Science Foundation of China under Grant No. 61300220, and the Scientific Research Fund of Hunan Provincial Education Department under Grant no. 16B089. It is also supported by PAPD and CICAEET.

© Springer Nature Singapore Pte Ltd. 2018
J. C. Hung et al. (Eds.): FC 2017, LNEE 464, pp. 335–344, 2018.
https://doi.org/10.1007/978-981-10-7398-4_35

platform with various services, such as e-health and e-commerce, is supported on the public networks. In past time, users must register on the special server to share special service, which is on the server. Multi-server architecture changes the exhaustive state. Users need not register on every sever to get the corresponding service and memorize different pairs of identity and password. Only one registration on one server is enough. Here we put our focus on the multi-servers to server (MS-S) architecture, which is discussed in [5,23]. The user could register on any one server once, which provides at least one service and is in the server group. If the user need access one of the services provided from server group, he can make his identity to be trusted via the registered server when communicating with the target server. In other words, there is no center server, and the model is different from the architecture which has the fixed one for registration.

It is well-known that there is full of danger in the common channels. Authentication is a strong way for getting rid of general attacks and it makes the entities in the session to be trusted for each other [3,4,7–9,20]. For the MS-S architecture mentioned above, three entities are referred in one session: a user U, the server which the user has registered, and another server with the special service which the user need. Here we call the two servers S_y and S_x, and there is a shared secret key between them when the whole system is constructed. When U initializes the communication, he uses his mobile device to contact S_x. Then S_x sends messages to S_y, and S_y verifies the identities of U and S_x. The verifications are sent to S_x and U. Finally U and S_x send encrypted messages to each other with the session key.

Until recently, many authentication schemes based on multi-server environment have appeared [2,5,11–13,16,23]. In 2011, Lee et al. [11] presented an authentication scheme for multi-server systems with dynamic identity. Unfortunately, Li et al. [12] showed that the scheme in [11] could not reach authentication and was vulnerable to forgery attacks, and an improved one was given. But Shunmuganathan et al. [15] claimed that Li et al.'s scheme could not resist attacks such as forgery attacks and off-line password guessing attacks. Their scheme was then criticized by Jangirala et al. [6] due to failing to resist password guessing attacks, forgery attacks, replay attacks and so forth. In 2015, Zhu et al. [23] proposed an authentication scheme for MS-S architecture. But it was pointed out that dangers such as insider attack, stolen verifier attack, tracking attack and lack of user anonymity had effects on Zhu et al.'s scheme by Irshad et al. [5]. Unluckily, we find that the scheme in [5] has weaknesses including insider attack, and tracking attack due to leakage of secret key.

To avoid the common attacks appearing above, we propose a new authentication scheme for the MS-S architecture. Through the informal analysis, our scheme is far away from the attacks and owns necessary security properties. From the performance comparison, our scheme is suitable for practice.

The reminder of the paper is arranged as follows: some preliminary knowledge is shown in Sect. 2. The proposed scheme and the informal analysis are in Sects. 3 and 4, respectively. The comparison of performance is in Sect. 5 and the conclusion comes in Sect. 6.

2 Preliminaries

2.1 Notations

Table 1 shows the symbols employed in the paper.

Table 1. Notations

Symbol	Meaning
U, ID_U, PW_U	The user with his identity and password
S_x, ID_x	The $x - th$ server and its identity
S_y, ID_y	The $y - th$ server and its identity
k_y	The secret key of S_y
k_{xy}	The secret key shared between S_x and S_y
$h(\cdot)$	Hash function
$\mathscr{A}$	The attacker
sk_u, sk_{S_x}	Session keys formed by U and S_x, respectively
$s_1 \oplus s_2$	The exclusive-or operation with s_1 and s_2
$s_1 \| s_2$	The concatenation of s_1 and s_2

2.2 Basic Knowledge of Chebyshev Chaotic Maps

Definition 1. *We use the enhanced Chebyshev polynomial cryptosystem. Based on [10], an integer $s \in (-\infty, \infty)$ and a large prime p are chosen to calculate the following equation:*

$$T_n(s) = \begin{cases} 1 & n = 0 \\ s \mod p & n = 1 \\ 2sT_{n-1}(s) - T_{n-2}(s) \mod p & n \geq 2 \end{cases}$$

According to [22], the semi-group character is feasible for the above equation. And we omit *mod p* below. For example, given positive integers u, v, $T_u(T_v(s)) = T_v(T_u(s))$.

2.3 Assumptions for Scheme Analysis

Assumption 1. *Each secret key in server is secure. $\mathscr{A}$ cannot guess random numbers or find the collision of hash results in polynomial time.*

Assumption 2. *Based on [14,18,19], $\mathscr{A}$ could retrieve data from U's mobile device, and $\mathscr{A}$ could eavesdrop, forge and replay messages from the public channel until the two-factor circumstance is broken. According to [17], in polynomial time, $\mathscr{A}$ can retrieve the user' identity and password from respective finite sets for trial, in order to get the correct strings.*

Assumption 3. *Chaotic-map Discrete Logarithm (CDL) Problem: Given a number y, it is hard to calculate the number u where $T_u(s) = y$ in polynomial time.*

Assumption 4. *Chaotic-map Computational Diffie-Hellman (CCDH) Problem: Given $T_u(s)$ and $T_v(s)$, it is hard to calculate $T_{uv}(s)$ in polynomial time.*

3 Skeleton of Our Scheme

Our scheme includes four phases: initialization, user registration, authentication and password renewal. In scheme [5], there is also a phase for updating the shared key between S_x and S_x via public channel. But it is unnecessary to do that with the trusted servers. So we have not designed that phase.

3.1 Initialization

For n users in the system, each pair of servers (S_x, S_y) $(x \neq y)$ share a common secret key k_{xy} $(1 \leq x, y \leq n, x \neq y)$. And every server has its own secret key and the common parameter s, e.g., k_y is the secret key for S_y.

3.2 User Registration

- Step 1: U selects ID_U, PW_U and a random number b_U, calculates $HPW_U = h(PW_U \| b_U)$ and sends $\{ID_U, HPW_U\}$ to S_y via a secure way.
- Step 2: S_y selects PID_U as U's pseudo-identity, calculates $B_{01} = h(PID_U \| k_y \| ID_y)$, $B_1 = B_0 \oplus HPW_U$, $B_{02} = h(ID_U \| k_y \| ID_y)$ and $B_2 = B_{02} \oplus h(ID_U \| HPW_U)$, stores ID_U in its database, and finally sends $\{PID_U, B_1, B_2, s, h(\cdot)\}$ to U via a secure way.
- Step 3: U calculates $B_3 = b_U \oplus h(ID_U \| PW_U)$ and stores $(PID_U, B_1, B_2, B_3, s, h(\cdot))$ into his own mobile device.

3.3 Authentication

- Step 1: U inputs ID_U and PW_U into his own mobile device. Then the device calculates $b_U = B_3 \oplus h(ID_U \| PW_U)$ and $HPW_U = h(PW_U \| b_U)$. With two random nonces r_U and N_U generated, the mobile device computes $C_1 = T_{r_U}(s)$, $C_2 = B_1 \oplus HPW_U \oplus N_U$, $C_3 = h(N_U) \oplus ID_U$, $C_4 = B_2 \oplus h(ID_U \| HPW_U)$ and $C_5 = h(C_1 \| N_U \| C_4)$. After the computations, U sends $M_1 = \{PID_U, C_1, C_2, C_3, C_5, ID_x, ID_y\}$ to S_x.
- Step 2: S_x selects the nonce r_{Sx}, computes $C_6 = T_{r_{Sx}}(s)$ and $C_7 = h(C_6 \| k_{xy} \| ID_x)$, and finally sends $M_2 = \{PID_U, C_1, C_2, C_3, C_5, C_6, C_7, ID_x\}$ to S_y.
- Step 3: S_y calculates $N_U = C_2 \oplus h(PID_U \| k_y \| ID_y)$, $ID_U = C_3 \oplus h(N_U)$ and searches ID_U in database. If it is found, S_y calculates $B_{02} = h(ID_U \| k_y \| ID_y)$ and checks $C_5? = h(C_1 \| N_U \| B_{02})$ and $C_7? = h(C_6 \| k_{xy} \| ID_x)$. S_y generates PID_U^{new} as the new pseudo-identity for U,

calculates $C_8 = h(ID_x||k_{xy}||C_1||C_6)$, $C_9 = h(ID_U||N_U||PID_U) \oplus PID_U^{new}$, $B_{01}^{new} = h(PID_U^{new}||k_y||ID_y)$, $C_{10} = B_{01}^{new} \oplus h(PID_U^{new}||N_U||ID_U)$ and $C_{11} = h(PID_U^{new}||B_{01}^{new}||N_U||C_1||B_{02}||C_6||ID_x||ID_y)$. At last it sends $M_3 = \{C_8, C_9, C_{10}, C_{11}\}$ to S_x.

- Step 4: S_x checks $C_8? = h(ID_x||k_{xy}||C_1||C_6)$. If it is true, S_x computes the session key $sk_{S_x} = Tr_{S_x}(C_1)$ and $C_{12} = h(C_1||C_6||C_9||C_{10}||sk_{S_x})$, and finally sends $M_4 = \{C_6, C_9, C_{10}, C_{11}, C_{12}\}$ to U.

- Step 5: After U receives M_4, the mobile device calculates the session key $sk_U = Tr_U(C_6)$ and verifies $C_{12}? = h(C_1||C_6||C_9||C_{10}||sk_U)$. If it is true, U calculates $PID_U^{new} = C_9 \oplus h(ID_U||N_U||PID_U)$, and $B_{01}^{new} = C_{10} \oplus h(PID_U^{new}||N_U||ID_U)$, and checks $C_{11}? = h(PID_U^{new}||B_{01}^{new}||N_U||C_1||C_4||C_6||ID_x||ID_y)$. If it is correct, U calculates $B_1^{new} = B_{01}^{new} \oplus HPW_U$ and replaces (B_1, PID_U) with (B_1^{new}, PID_U^{new}).

3.4 Password Renewal

- Step 1: U inputs ID_U and PW_U into his mobile device. The mobile device calculates b_U and HPW_U as in Sect. 3.3. Then it produces N_U and calculates C_2, C_3, C_4 and $C_{13} = h(N_U||C_4||ID_U||PID_U)$. After the operations U sends $M_5 = \{PID_U, C_2, C_3, C_{13}\}$ with a password renewal request to S_y.

- Step 2: S_y calculates N_U, ID_U and searches ID_U as in Sect. 3.3. Then it computes B_{02} and checks $C_{13}? = h(N_U||B_{02}||ID_U||PID_U)$. If it is true, S_y then generates PID_U^{new}, calculates B_{01}^{new}, C_9, C_{10} and $C_{14} = h(N_U||B_{02}||ID_U||PID_U)$. Finally it sends $M_6 = \{C_9, C_{10}, C_{14}\}$ with a permission for password changing.

- Step 3: When U receives M_6, the mobile device calculates PID_U^{new}, B_{01}^{new} and checks $C_{14}? = h(PID_U^{new}||B_{01}^{new}||N_U||ID_U||ID_y)$. If it is correct, U is asked to enter a new password PW_U^{new}. Then the mobile device generates b_U^{new} and calculates $HPW_U^{new} = h(PW_U^{new}||b_U^{new})$, $B_1^{new2} = B_{01}^{new} \oplus HPW_U^{new}$, $B_2^{new} = C_4 \oplus h(ID_U||HPW_U^{new})$ and $B_3^{new} = b_U^{new} \oplus h(ID_U||PW_U^{new})$. Finally it replaces (B_1, B_2, B_3, PID_U) with $(B_1^{new2}, B_2^{new}, B_3^{new}, PID_U^{new})$.

4 Security Characters Analysis

In this section we demonstrate the analysis of security characters of our scheme. The results are in Table 2 with the comparison of same sort schemes in [5, 23]. We use ✓ to express the scheme meets the requirement, or × appears. Readers can read concrete details for the problems mentioned in [5, 23].

4.1 Resistance to Insider Attack

U submits HPW_U to S_y for registration in our scheme. We see that the malicious administrator of S_y cannot get PW_U since it is protected by the random number b_U with hash function. So our scheme resists insider attack.

But we see that S_y and U shares PW_U as a secret in [5]. It is obvious that the administrator can get the password directly. Thus we use × for that blank.

Table 2. Security properties comparison

	[23]	[5]	Ours
Resistance to insider attack	×	×	✓
Resistance to off-line guessing attack	✓	✓	✓
Resistance to tracking attack	×	×	✓
Resistance to forgery attack	✓	✓	✓
Resistance to stolen-verifier attack	×	✓	✓
Resistance to de-synchronization attack	×	✓	✓
Resistance to replay attack	✓	✓	✓
User anonymity	×	×	✓
Strong forward security	✓	✓	✓

4.2   Resistance to Off-Line Guessing Attack

Suppose $\mathscr{A}$ gets $\{M_1^{old}, M_2^{old}, M_3^{old}, M_4^{old}\}$ in the a session and then obtains $(PID_U, B_1, B_2, B_3, s)$ in U's device. Then he guesses (ID^*, PW^*) and uses the following equations: $b^* = B_3 \oplus h(ID^*\|PW^*)$, $HPW^* = h(PW^*\|b^*)$, $C_9^{old} = h(ID^*\|N_U^{old}\|PID_U^{old})$, $C_{10}^{old} = h(PID_U\|N_U^{old}\|ID^*) = B_1 \oplus HPW^*$ and $N_U^{old} = B_1^{old} \oplus HPW^* \oplus C_2^{old}$. Unluckily B_1^{old} disappeared when the last session finished, and $\mathscr{A}$ could not get it. So our scheme can resist this attack.

4.3   Resistance to Tracking Attack

As the pseudo-identity, PID_U always changes in different sessions. And there is no relation between any two of them. So $\mathscr{A}$ cannot track the special user by eavesdropping such information.

But in [5], if $\mathscr{A}$ registers on S_y, he can retrieve $PID_{\mathscr{A}}$ from his own mobile device and calculate $k_y^{part} = PID_{\mathscr{A}} \oplus (ID_{\mathscr{A}}\|PW_{\mathscr{A}})$. Although there is a random number q next to $(ID_{\mathscr{A}}\|PW_{\mathscr{A}})$, it is impossible to block the danger. $\mathscr{A}$ can eavesdrop PID_U and calculate $ID_U\|PW_U = k_y^{part} \oplus PID_U$. The real identity and password are exposed by $\mathscr{A}$.

4.4   Resistance to Forgery Attack

If $\mathscr{A}$ wants to forge any of the message in the session, he must get k_y or k_{xy}. We illustrate every case below.

1. $\mathscr{A}$ must know k_y to calculate C_2 and C_4 at least, or M_1 cannot be forged.
2. $\mathscr{A}$ must know k_y and k_{xy} to calculate C_2, C_4 and C_7 at least, or M_2 cannot be forged.
3. $\mathscr{A}$ must know k_{xy} and k_y to calculate C_8, C_{10} and C_{11} at least, or M_3 cannot be forged.
4. $\mathscr{A}$ must know k_y to calculate C_{10} and C_{11} at least, or M_4 cannot be forged.

4.5   Resistance to Stolen-Verifier Attack

A verifier means that the string which is calculated with the user's password on the server side. There is no verifier stored in the server for our scheme. So this attack is invalid.

4.6   Resistance to De-synchronization Attack

There is no information which needs to change on at least two sides during the whole protocol, e.g., U and S_y. If any message is blocked or lost in public channel, no inconsistent strings will appear, even in the password renewal phase.

4.7   Resistance to Replay Attack

If some message is replayed, the session will lead to be meaningless. For example, if M_1 is replayed, S_x should generate another random number $r_{S_x}^{new}$ to continue the session. However, the final session key should be different and $T_{r_u^{old}}(T_{r_{S_x}^{old}}(s))$ for the last session cannot be reused.

4.8   User Anonymity

In our scheme, every time U sends different PID_U as the pseudo-identity, and the real identity ID_U does not appear in the message. Since the scheme is against tracking attack, according to Sect. 4.3, it is fit for user anonymity without doubt.

4.9   Strong Forward Security

Strong forward security means that even if all the secret keys in the entities are mastered by $\mathscr{A}$, $\mathscr{A}$ still cannot calculate the past session keys [21]. Since the session key in our scheme is constructed on CCDH Problem in Assumption 4, which has little relation to calculate the session key, the scheme keeps this property.

5   Performance Comparison

We list the result comparison among the schemes mentioned in Table 2. First the lengths of relative parameters are demonstrated in Table 3. To keep the robustness of parameters, we define that the identities of servers and random numbers have 160 bits, and parameters s and p have 1024 bits. Also, Sha2-256 is the employed hash function and the result is 256 bits. We use Sha2-256 simply because it is the trend that Sha2 family will replace Sha1 in short future time. For example, Google claimed that Sha1 would not be supported in Chrome from 2017 [1]. The time cost of mentioned cryptographic calculations is in Table 4. The platform is same as [19]. Moreover, the symmetric encryption/decryption is used in [23], and we employ AES, where the encryption blocks are all 128 bits.

Table 3. Lengths of parameters (bits)

$ID_x, ID_y,$ nonces	Hash result	s, p
160	256	1024

Table 4. Time cost of cryptographic operations

Symbol	Meaning	Time (ms)
T_c	Time of one Chebyshev chaotic map	127.042
T_s	Time of one symmetric encryption/decryption	00214835 [19]
T_h	Time of one Sha2-256 calculation	0.005174

We calculate the Chebyshev chaotic maps with the "left to right" way to deal with the exponent n in [10].

The results of performance comparison with [5,23] are in Table 5. We demonstrate six aspects, including time cost on three entities, communication cost in authentication phase, number of messages and security. The analysis is below:

- Our scheme wins in the aspect of the time cost on user side, since in schemes [5,23], three chaotic maps are used, while ours only has two.
- Our scheme also wins in the aspect of S_x time cost. In scheme [23], there are three chaotic maps, while in our scheme and scheme [5], there are two chaotic maps. Moreover, four hash functions are used in [5], while ours employs only three.
- The difference among the S_y time is very clear. Our scheme wins without doubt. There are two, one and zero chaotic maps in [5,23] and our scheme, respectively. And the quantities is the most critical factor for winning.

Table 5. Performance comparison

	[23]	[5]	Our scheme
Time cost of U (ms)	$3T_c + 2T_s + 3T_h$ $= 381.1458187$	$3T_c + 4T_h$ $= 381.146696$	$2T_c + 8T_h$ $= 254.125392$
Time cost of S_x (ms)	$3T_c + 2T_s + 3T_h$ $= 381.1458187$	$2T_c + 4T_h$ $= 254.104696$	$2T_c + 3T_h$ $= 254.130566$
Time cost of S_y (ms)	$2T_c + 4T_s + 4T_h$ $= 254.1390696$	$T_c + 6T_h$ $= 127.073044$	$10T_h$ $= 0.05174$
Communication cost (bits)	9568	8032	8736
Messages	4	5	4
Security	No	No	Yes

- For the aspect of communication cost in authentication, our scheme is in the middle. That is because user's pseudo-identity needs to be replaced. Calculations appear on both user and S_y and the communication cost is higher than scheme in [5].
- There are four messages in scheme [23] and ours, while five messages are included in [5].
- Our scheme satisfies all security requirements in Sect. 4, while the other two both have disadvantages.

6   Conclusion

Multi-server service environment is a common structure in temporary networks and security communication is very important. However, past authentication schemes for multi-server environment have many kinds of security weaknesses. To prevent the problems, we present our authentication scheme. Through the security property analysis, our scheme is robust against various attacks including off-line guessing attack, tracking attack, etc. The performance comparison also proves that our scheme is suitable for applications due to its good results.

References

1. Palmer, C., Sleevi, R.: Gradually sunsetting sha-1 (2017). https://blog.chromium.org/2014/09/gradually-sunsetting-sha-1.html
2. Chaudhry, S.A.: A secure biometric based multi-server authentication scheme for social multimedia networks. Multimed. Tools Appl. **75**, 12705–12725 (2015)
3. He, D., Kumar, N., Shen, H., Lee, J.H.: One-to-many authentication for access control in mobile pay-TV systems. Sci. China Inf. Sci. **59**(5), 1–14 (2015)
4. He, D., Zeadally, S., Kumar, N., Lee, J.H.: Anonymous authentication for wireless body area networks with provable security. IEEE Syst. J. (2016). https://doi.org/10.1109/JSYST.2016.2544805
5. Irshad, A., Ahmad, H.F., Alzahrani, B.A., Sher, M., Chaudhry, S.A.: An efficient and anonymous chaotic map based authenticated key agreement for multi-server architecture. KSII Trans. Internet Inf. Syst. (TIIS) **10**(12), 5572–5595 (2016)
6. Jangirala, S., Mukhopadhyay, S., Das, A.K.: A multi-server environment with secure and efficient remote user authentication scheme based on dynamic id using smart cards. Wirel. Pers. Commun. (2017). https://doi.org/10.1007/s11277-017-3956-2
7. Jiang, Q., Ma, J., Yang, C., Ma, X., Shen, J., Chaudhry, S.A.: Efficient end-to-end authentication protocol for wearable health monitoring systems. Comput. Electr. Eng. (2015). https://doi.org/10.1016/j.compeleceng.2017.03.016
8. Jiang, Q., Ma, J., Wei, F., Tian, Y., Shen, J., Yang, Y.: An untraceable temporal-credential-based two-factor authentication scheme using ECC for wireless sensor networks. J. Netw. Comput. Appl. **76**, 37–48 (2016)
9. Jiang, Q., Zeadally, S., Ma, J., He, D.: Lightweight three-factor authentication and key agreement protocol for internet-integrated wireless sensor networks. IEEE Access **5**, 3376–3392 (2017)

10. Kocarev, L., Lian, S.: Chaos-Based Cryptography: Theory, Algorithms and Applications, vol. 354. Springer (2011)
11. Lee, C.C., Lin, T.H., Chang, R.X.: A secure dynamic id based remote user authentication scheme for multi-server environment using smart cards. Expert Syst. Appl. **38**(11), 13863–13870 (2011)
12. Li, X., Xiong, Y., Ma, J., Wang, W.: An efficient and security dynamic identity based authentication protocol for multi-server architecture using smart cards. J. Netw. Comput. Appl. **35**(2), 763–769 (2012)
13. Li, X., Ma, J., Wang, W., Xiong, Y., Zhang, J.: A novel smart card and dynamic id based remote user authentication scheme for multi-server environments. Math. Comput. Model. **58**(1), 85–95 (2013)
14. Mangard, S., Oswald, E., Standaert, F.X.: One for allcall for one: unifying standard differential power analysis attacks. IET Inf. Secur. **5**(2), 100–110 (2011)
15. Shunmuganathan, S., Saravanan, R.D., Palanichamy, Y.: Secure and efficient smart-card-based remote user authentication scheme for multiserver environment. Canad. J. Electr. Comput. Eng. **38**(1), 20–30 (2015)
16. Tsai, J.L.: Efficient multi-server authentication scheme based on one-way hash function without verification table. Comput. Secur. **27**(3), 115–121 (2008)
17. Wang, D., Wang, P.: On the anonymity of two-factor authentication schemes for wireless sensor networks: attacks, principle and solutions. Comput. Netw. **73**, 41–57 (2014)
18. Wu, F., Xu, L., Kumari, S., Li, X.: A new and secure authentication scheme for wireless sensor networks with formal proof. Peer-to-Peer Netw. Appl. (2015). https://doi.org/10.1007/s12083-015-0404-5
19. Wu, F., Xu, L., Kumari, S., Li, X., Das, A.K., Khan, M.K., Karuppiah, M., Baliyan, R.: A novel and provably secure authentication and key agreement scheme with user anonymity for global mobility networks. Secur. Commun. Netw. **9**, 3527–3542 (2016)
20. Wu, F., Xu, L., Kumari, S., Li, X.: A privacy-preserving and provable user authentication scheme for wireless sensor networks based on internet of things security. J. Ambient Intell. Human. Comput. **8**, 101–116 (2017)
21. Xu, L., Wu, F.: An improved and provable remote user authentication scheme based on elliptic curve cryptosystem with user anonymity. Secur. Commun. Netw. **8**(2), 245–260 (2015)
22. Zhang, L.: Cryptanalysis of the public key encryption based on multiple chaotic systems. Chaos Solitons Fractals **37**(3), 669–674 (2008)
23. Zhu, H.: Flexible and password-authenticated key agreement scheme based on chaotic maps for multiple servers to server architecture. Wirel. Pers. Commun. **82**(3), 1697–1718 (2015)

A Novel Lightweight PUF-Based RFID Mutual Authentication Protocol

Wei Liang[1(✉)], Songyou Xie[1], Xiong Li[1], Jing Long[1], Yong Xie[2], and Kuan-Ching Li[3]

[1] School of Computer Science and Engineering,
Hunan University of Science and Technology, Xiangtan 411201, China
idlink@163.com, songy@hnust.edu.cn, lixiongzhq@163.com
[2] Department of Software and Engineering, Xiamen University of Technology,
Xiamen 361024, Fujian, China
yxie@xmut.edu.cn
[3] Department of Information Science and Engineering, Providence University,
Taichung City 43301, Taiwan
Kuan-ChingLi@gm.pu.edu.tw

Abstract. The widespread use of radio frequency identification (RFID) in IoTs makes authentication of RFID systems be widely concerned. In existing encryption schemes (e.g., Hash function) in electronic products, secure chip is hard to be used in high performance RFID system due to high computation complexity and cost. In this work, we consider problems in these performances and propose a PUF-GIMAP protocol by combining GIMAP protocol and physically unclonable functions (PUFs). The response of PUF is added into protocol. The mutual authentication between label and reader is realized by transmitting information such as secret key and dynamical pseudonym, which greatly ensures data security in transmission. After authentication, the reserved data is updated in time. The protocol analysis shows that the proposed scheme has the advantages of high security and high reliability.

Keywords: Physically unclonable functions · RFID · Low-cost authentication
GIMAP

1 Introduction

Radio frequency identification (RFID) is a novel technique for identity authentication. Traditional RFID system consists tag, reader and database, as shown in Fig. 1. Tag always includes chip and antenna transmitter. Based on the way to acquire energy, RFID is divided into active tag and passive tag [1]. A unique identity code *ID* is embedded into the tag. When a tag sensor enters read/write area of the reader, electromagnetic wave around reader-writer will provide energy for the tag. After processing data in the tag, the data will be sent for authentication. Reader always includes antenna, radio frequency module, and read-write module. When RFID system receives information from a tag, radio frequency module will perform demodulation and the generated data will be sent to backend database. In this way, the corresponding secret key

© Springer Nature Singapore Pte Ltd. 2018
J. C. Hung et al. (Eds.): FC 2017, LNEE 464, pp. 345–355, 2018.
https://doi.org/10.1007/978-981-10-7398-4_36

will be found in the database. At present, there are lots of RFID standards, such as ISO/IEC, EPC global and AIM global. EPC global designed EPC Class1 Generation-2 for electronical products encoding in 2004. The standard can be divided into four categories [2]. (1) non-overridable passive tag. It cannot provide energy itself. Energy during work comes from electromagnetic wave sent by reader. It is suitable to design a low-cost tag. (2) overridable and identifiable passive tag. (3) semi-passive tag. It can provide part of energy during work period. (4) active tag. Some complex algorithm can be computed.

Fig. 1. Simple RFID system

At present, hash function needs lots of hardware resources to support. For example, MD4, MD5 and SHA-256 need 7350–10868 logical gates to support [3]. 3400 logical gates are required to implement AES [4]. Hash cannot resist tracing attacks and has heavy computation and storage. HB family protocol mainly includes HB, HB+, HB#, etc. These protocols have some potential security hazards. Although HB protocol [5] has good performance in terms of computation, cost and overhead, it cannot resist counterfeit and no synchronous attacks. In recent years, there are lots of attacks aiming at RFID system, such as man-in-the-middle, replay, tracing, and no synchronous attacks. These attacks make the communication in RFID device unsafe. Attackers may steal and destroy completeness of confidential data. To address this issue, researchers have proposed many solutions. These solutions can be divided into three types [6]. (1) Physical encryption scheme, such as fire extinguishing method [7], faraday cage [8]. (2) Cryptography based secure protocol, such as Hash function. It has high requirements on computation ability and storage space. (3) Lightweight authentication protocol based on bit operation, such as LMAP + protocol and UMAP protocol.

Existing RFID protocols depend on encryption primitives. Large hardware cost may be caused in manufacture process [9] (e.g., thousands of logical gates are needed to realize a tag), which greatly hinders rapid development of RFID technique. In this work, we propose a secure RFID protocol based on physically unclonable functions (PUFs). It meets the security demands of RFID system and greatly reduces the hardware cost.

This paper is organized as follows. The PUF technique is introduced in Sect. 2. In Sect. 3, we introduce some related work in PUF based RFID protocols. Section 4 introduces the GIMAP protocol and Sect. 5 improves it and describes the new protocol. The protocol analysis is illustrated in Sect. 6. Section 7 summarizes the paper.

2 PUF Technique

PUF is regarded as central module in various cryptographic protocols and secure structure. Many related techniques, such as PUF based intellectual property (IP) protection [10], security authentication [11, 12], signature, are widely reported in academia. It is inevitable to use PUF in low-cost, high-security and high reliability authentication due to its features. Currently, internet of things and cloud computation are rapidly developed. It is more important to realize secure and sensitive task in integrated circuit. PUF is a novel technique to realize hardware security. In 2001, Pappu [13] designed a optic PUF. After that, more and more PUFs are proposed. Based on their implementation method, PUF can be divided into three categories [14], non-electronical PUF, artificial circuit PUF and digital circuit PUF. Digital circuit PUF is the most important and widely used. A complete PUF should be embedded into circuit. Although this PUF has better security, it may cause some extra overhead by comparing to other PUFs.

The circuit is shown in Fig. 2 which consists of a signal transmission delay circuit and an arbiter. There are 64 switches in the circuit. Each switch can change the path of the signal transmission delay path. Each switch has Two input terminals, one control signal terminal and two output terminals. When the control signal is 0, the signal in the two pass directly through the 1, when the cross for the same signal in the two channels on the competition through the arbiter to choose the output.

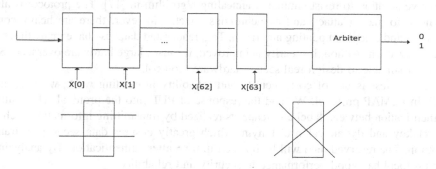

Fig. 2. Arbiter PUF circuit diagram

PUF utilizes random variation in chip manufacturing. Different PUFs with the same challenges will cause various response due to intrinsic wire delay or logical gates. These delays are generally unclonable, such as manufacturing process, fluctuation of quantum mechanics, temperature gradient, electron transfer, parasitic effect, noise [15]. PUF is unpredictable and hard to extract. The behavior of PUF does not exist as digits. And PUF itself contains many built-in nonlinear delay components, making it resilient against modeling attacks. So, it is more difficult for attackers to crack a PUF than an encryption algorithm. The use of PUF can be regarded as adding an unclonable and unpredictable ID for a target device. An adversary cannot crack the regularity of response. PUF also has good resistance against replay attack and memory reading attack [5]. It can be utilized to confidential data protection and security authentication in integrated circuit. Integration of PUF into RFID tagsnot only makes it possess the same nature and uncloning properties of PUF, but also can greatly reduce the manufacturing cost. Besides, the use of PUF in RFID authentication can greatly improve the security and reliability.

3 Related Work

Many reported references have combined PUF and RFID and implemented one-way authentication of reader and tag. The response of PUF is stored into database, which will be extracted in authentication. The authentication is successful when the generated response is consistent with that extracted in database. In 2013, Li et al. [5] combined PUF with HB# protocol. A PUF-HB# based lightweight mutual RFID authentication protocol was proposed. The proposed protocol has advantages of both protocols, which can effectively resist replay attacks and nonsynchronous attacks. It has low false rejection rate and false acceptation rate. Meanwhile, the computation and communication traffic satisfy the standard of RFID. But the protocol will directly expose the tag ID, causing it being captured. In 2012, Gurubani [16] proposed LAMP+ protocol by using some simple bit operations. It updates data expression and secret key K_3 and increases system security. Tracing attack and complete leak attack [17] cannot be resisted. In 2016, Zhu et al. [18] presented a novel PUF based LMAP+ protocol. It is an effective solution to resist simulated annealing algorithm in [17]. The protocol is also immune to tracing attack and asynchronous attack. However, there are heavy computation and complex updating algorithm. Tag, reader and database have many times of data interaction. Although security is improved, it causes large hardware overhead. So, it is not suitable to design real secure hardware protocol.

To address issues of cost, security and reliability in existing work, we introduce PUF in GIMAP protocol. We add the response of PUF into the protocol. The mutual authentication between label and reader is realized by transmitting information such as secret key and dynamical pseudonym, which greatly ensures data security in transmission. The reserved data will be updated in time after authentication. By analyzing, the protocol has good performance in security and reliability.

4 Lightweight RFID Authentication Protocol: GIMAP

Zheng et al. [19] proposed a novel lightweight authentication protocol to strengthen security in existing RFID protocol, namely GIMAP. It satisfies EPC Class Gen2 standard. Mutual authentication between tag and backend database is implemented in this protocol. Furthermore, it utilizes CRC correction and random number generator PRNG. Good performance and low hardware complexity are achieved. Some parameters of GIMAP protocol are listed in Table 1.

Table 1. Parameters of GIMAP protocol

Symbol	Definition
μ	Column offset of matrix
l	Bit length of RFID tag
S	n × m matrix, satisfying $K \cdot S = 0$
K	m × matrix
K^+	Generalized inverse of matrix K
ξ	The minimum values of row number and column number in matrix K
$\lambda_i(X)$	Matrix generated by right shifting i columns of matrix X
$\lambda_{-i}(X)$	Matrix generated by left shifting i columns of matrix X
$\delta_i(X)$	Matrix generated by up shifting i rows of matrix X
$\delta_{-i}(X)$	Matrix generated by down shifting i rows of matrix X

In GIMAP, RFID reader sends a random number R_r to a tag. Tag calculates $M_1 = (R_r \oplus R_t) \cdot K^+ + (R_r \oplus ID) \cdot S$ and $M_2 = (R_t \oplus ID) \cdot K^+ + (R_r \oplus R_t) \cdot S$ with the received R_r, the generated random number R_t and other stored data ID, matrix S and matrix K^+. M_1 and M_2 are sent to backend database via reader. Based on definition of generalized matrix, $K^+ K K^+ = K^+$, both sides of M_1 equation multiplies by $(K^+ K)$. So, we have $(R_r \oplus ID) \cdot S = M_1 \cdot (E - K^+ K)$. It is possible to determine whether the generated equation is equal to $(R_r \oplus ID) \cdot \lambda_{-\mu}(S) = M_1 \cdot (E - K^+ K)$. The result can determine whether the tag is legal. After successfully authenticated, the values of i and $\lambda_i(S)$ are regarded as new values of μ and S. The new values will be stored in database.

After authenticating the tag, both sides of M_2 equation multiplies by $(K^+ K)$ and $(R_r \oplus R_t) \cdot S = M_2 \cdot (E - K^+ K)$ is generated. With known conditions, we can get the value of R_t. It is utilized to perform a modulus on the minimum value ξ of matrix row number m and matrix column number n. Then i is calculated and used as row offset and column offset of K and K^+ to generate new matrix. Moreover, $M_3 = \delta_i(K)\lambda_i(K^+)$ $CRC(ID \oplus R_r)$ is computed and sent to tag with other related information via reader-writer. With the received M_3 and R_t, it performs a modulus on the minimum value ξ of matrix row number m and matrix column number n to get the value of i. Generalized matrix has the features of $\lambda_i(K^+) = \delta_i^+(K)$. $\lambda_i(K^+)$ is the generalized inverse of $\delta_i(K)$, satisfying $\lambda_i(K^+)\delta_i^+(K)\lambda_i(K^+) = \lambda_i(K^+)$. The computation result can be used to determine whether $\lambda_i(K^+) \cdot M_3$ is equal to $\lambda_i(K^+) \cdot CRC(ID \oplus R_r)$ in order to verify validity of backend database. If successfully verified, the value of S will be updated as $\lambda_i(S)$. The protocol implementation is described as Fig. 3.

The protocol implementation has good performance. It satisfies requirements of lightweight RFID systems and resists against replay attack and asynchronous attack. But there is heavy computation between database and tag. Some information should be found in verification by traversing the database. This protocol does not verify the reader-writer and not update ID stored in tag and system. If ID is captured, it is easy to be traced. To address this issue, we propose a lightweight PUF-GIMAP authentication protocol in this work. It can resist against above illegal attacks and have good improvements on security, reliability and cost.

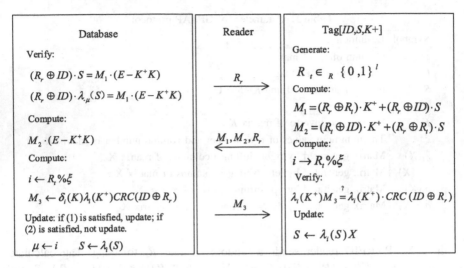

Fig. 3. The implementation of GIMAP protocol

5 PUF-GIMAP Protocol

In this section, we give a description about the proposed PUF-GIMAP protocol. Firstly, some parameters in PUF-GIMAP protocol are listed in Table 2.

Table 2. Parameters of PUF-GIMAP protocol

Symbol	Definition
K_r	Output of PUF
μ	Column offset of matrix
S	n × m matrix
K	m × n matrix
K_p	System defined value, to protect K_r
$\lambda_i(X)$	Matrix generated by right shifting i columns of matrix X
$\lambda_{-i}(X)$	Matrix generated by left shifting i columns of matrix X
$\delta_i(X)$	Matrix generated by up shifting i rows of matrix X
$\delta_{-i}(X)$	Matrix generated by down shifting i rows of matrix X
D^n	Dynamical pseudonym in each round of authentication

At initial stage, a tag stores values of $K^+, K, \mu, S, ID, K_p, ID_0, ID, D^n$ and reader stores $K^+, K, S, ID, K_p, K_r, R_r$. Here, $K_r = PUF(D^n)$. The implementation is shown in Fig. 4.

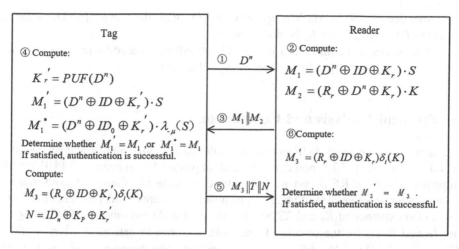

Fig. 4. Implementation of PUF-GIMAP protocol

Detailed steps of protocol implementation are described as follows.

1. When a tag enters sensing area of a reader, the reader sends authentication command *HELLO* to the tag.
2. After receiving *HELLO*, the tag sends a dynamic pseudonym D^n to the reader.
3. With the received D^n, reader generates random number R_r and finds related secret key for D^n, such as S, ID, K_r. Then it computes $M_1 = (D^n \oplus ID \oplus K_r) \cdot S$ and $M_2 = (R_r \oplus PID^n \oplus K_r) \cdot K$. The result of $M_1 \| M_2$ is sent to the tag.
4. The tag receives M_1 and M_2. It computes $K'_r = PUF(D^n)$, $M'_1 = (D^n \oplus ID \oplus K'_r) \cdot S$ and $M^*_1 = (D^n \oplus ID_0 \oplus K'_r) \cdot \lambda_{-\mu}(S)$. Then whether M'_1 is equal to M_1, or M^*_1 is equal to M_1 should be determined. Any one success proves the authentication is successful.

   - In (4), if $M'_1 = M_1$ is satisfied, we can calculate R_r by $M_2 = (R_r \oplus D^n \oplus K_r) \cdot K$ due to the known values of M_2, PID^n, K_r, K. After that, we perform modulus on ξ to get i. i is regarded as row offset of matrix K to get new matrix. The tag calculates $M_3 = (R_r \oplus ID \oplus K'_r)\delta_i(K)$ and uses ID as input of PUF to compute $PUF(ID) = ID_n$ and $N = ID_n \oplus K_P \oplus K'_r$. $M_3 \| N$ is sent to the reader. After that, the tag updates ID to ID_n. The old ID is stored as ID_o. $\lambda_i(S)$ is stored as new value of S. μ is instead by i, namely, $\mu = i$.
   - In (4), if $M^*_1 = M_1$, it is caused by asynchronous data update between tag and reader. Tag ID and matrix S are updated. The reader cannot update data correctly if being attacked or tampered. To avoid a successful authentication of an attacker with illegal ID and S, the tag will not update data when $M^*_1 = M_1$.

5. The reader receives $M_3 \| N$ and performs modulus on ξ to get i. M'_3 can be calculated by $M'_3 = (R_r \oplus ID \oplus K_r)\delta_i(K)$. If $M'_3 = M_3$ is satisfied, the reader successfully authenticates a tag. Otherwise, the authentication is failed. When successfully

authenticated, the matrix S is updated by $\lambda_i(S)$ with the value of i. The value of $ID = ID_n = N \oplus K_p \oplus K_r$ is also updated.

6. When mutual authentication is successful, D^n will be updated by $D^{n+1} = D^n \oplus K'_r$. It is stored in the tag. The authentication is finished.

6 Protocol Analysis and Comparison

In the proposed protocol, PUF technique is utilized to realize mutual authentication in RFID system [15]. The non-clonable and unpredictable features of PUF greatly improve security of RFID system. In two times of authentications, it doesn't transmit ID and other sensitive data of tag. Even an attacker captures $T\|N$, he cannot obtain ID due to the existence of K_p and K'_r. So, the original tag ID and output of PUF in tag are not leaked to ensure the security. In step (4), we need to determine whether $(D^n \oplus ID_o \oplus K_r) \cdot \lambda_{-\mu}(S) = M_1 \cdot (E - K^+ K)$ is satisfied. The purpose is to resist illegal tampering attack. If the legal authentication message is tampered, reader will fail to authenticate the legal tag. Thus, the values of ID and S cannot be updated in time. So, to determine the above equation could resist asynchronous attack.

In reference [20], the probability for a PUF generating error output is 0.7. With a very low probability, the stored K_r and K_r generated by tag are different. At initial stage, it can repeatedly input challenge D^n. PUF will generate several outputs. To choose a main output can ensure stability and reliability of the system. There is another way. A tag utilizes several PUFs concurrently. Each PUF generates its own combination chain independently. It improves the probability of successful authentication. Furthermore, PUF consumes less resources than that of digital cipher. So, it is unnecessary to consider hardware cost and computation complexity. We can combine both methods to ensure the consistence between stored K_r and that generated by tag. The stability of PUF based system is also improved. In Table 3, we have analyzed several protocols and compared our protocol to them.

Table 3. Security analysis of various authentication protocols

	PUF-GIMAP	GIMAP	HB	LAMP+	Hash-lock
Cost	Low	High	Low	Low	High
Modeling attack	Safe	Low	Dangerous	Dangerous	Dangerous
Tracing attack	Safe	Low	Dangerous	Dangerous	Safe
Replay attack	Safe	Safe	Dangerous	Safe	Safe
Asynchronous attack	Safe	Safe	Dangerous	Dangerous	Safe
Hardware tampering attack	Safe	Dangerous	Dangerous	Dangerous	Dangerous
Computation complexity	Low	High	Low	Low	High

- Modeling attack. An attacker may attempt to build a model for a PUF, observe or learn the behavior of the PUF. In the proposed PUF-GIMAP protocol, the behavior of PUF is hard to predict. If the tag is cracked by brute force and delay of inner physical components is measured, the circuit component will be destroyed.
- Tracing attack. Data is transmitted between tag and reader. If some information is regular or not updated, it may be used in illegal request after being captured. The tag will be traced based on response of tag or reader. In the proposed protocol, the stored values of ID, S, PID^n, K_r will be also updated after each successful update. Due to the unpredictability of PUF, it is hard for an attacker to find regularity of these information. If the tag uses equation of $(D^n \oplus ID_o \oplus K_r) \cdot \lambda_{-\mu}(S) = M_1 \cdot (E - K^+K)$ to authenticate reader, the updated data is stored in the tag in the last authentication. With this equation, asynchronous legal reader will pass authentication of tag.
- Replay attack. If the transmitted data between tag and reader is the same at each round of authentication, it will be possible for an attacker to capture and replay information at last authentication. In this case, he will illegally pass the authentication. In this protocol, PID^n, $M_1\|M_2$ and $M_3\|T\|N$ are transmitted. R_r will participate in each round of authentication. The data will be updated in time after successfully authenticated. So, replay attacked will be resisted.
- Hardware tampering attack. It is an attack by physically tampering or destroying behavior of PUF. Attackers can inject faults by making glitches of power supply or clock signal, destroying RFID circuit and analyze result of fault circuit to get secret key. The difficulty, strength and method to perform hardware tampering attack will be related to control ability of attacker for fault position and time.

7 Conclusions

This work combines PUF and GIMAP and proposes a novel lightweight PUF based RFID mutual authentication protocol. The unclonable and unpredictable features of PUF are utilized. Challenge-response pair of PUF is encrypted to ensure data confidentiality and security. After each round of authentication, response data of the system will be updated in time. It makes up for the shortage of low ability against tracing attacks. In addition, dynamical pseudonym is used to look up related information. It is more effective and correct by comparing to the method of traversing database in GIMAP protocol. The proposed protocol can be used in RFID system to address security threat in existing RFID system. It protects RFID system from being attacked by clone attack, memory read attack, etc., which is more suitable for complex application environment of RFID system. In future, we will further consider structural difference between multiple PUFs to address the limitations of RFID system in power and cost. Meanwhile, the impaction of environment on PUF authentication will be concentrated. Thus, stability and security can be ensured in authentication.

Funding. The author(s) disclosed receipt of the following financial support for the research, authorship, and/or publication of this article: This work was supported by the National Science Foundation of China (grant no. 61572188, 61502405), Hunan Provincial Natural Science Foundation of China (grant no. 2016jj2058), Scientific Research Project of Hunan University of Science and Technology (grant no. E51697), Xiamen science and technology Foundation (grant no. 3502Z20173035).

References

1. Pagnin, E., Yang, A., Hu, Q., et al.: HB + DB math container loading Mathjax: distance bounding meets human based authentication. Future Gener. Comput. Syst. **6**(11), 6–21 (2016)
2. Liang, W., Liao, B., Jiang, Y., Long, J., Peng, L.: Study on PUF based secure protection for IC design. Microprocess. Microsyst. **8**(45), 56–66 (2016)
3. Manifavas, C., Hatzivasilis, G., Fysarakis, K., et al.: Lightweight cryptography for embedded systems–a comparative analysis. In: Data Privacy Management and Autonomous Spontaneous Security, pp. 333–349. Springer, Heidelberg (2014)
4. Arbit, A., Livne, Y., Oren, Y., et al.: Implementing public-key cryptography on passive RFID tags is practical. Int. J. Inf. Secur. **14**(1), 85–99 (2015)
5. Li, H., Deng, G.: PUF-HB#: a lightweight RFID mutual authentication protocol. J. Beijing Univ. Posts Telecommun. **36**(6), 13–17 (2013)
6. Rührmair, U., Sölter, J., Sehnke, F., et al.: PUF modeling attacks on simulated and silicon data. IEEE Trans. Inf. Forensics Secur. **8**(11), 1876–1891 (2013)
7. Nilsson, E., Svensson, C.: Ultra-low power wake-up radio using envelope detector and transmission line voltage transformer. IEEE J. Emerg. Sel. Topics Circ. Syst. **3**(1), 5–12 (2013)
8. Wikipedia. The Free Encyclopedia. http://en.wikipedia.org/wiki/FaradayCage
9. Ding, Z., Li, J., Feng, B.: Research on hash-based RFID security authentication protocol. J. Comput. Res. Dev. **46**(4), 583–592 (2009)
10. Maiti, A., Gunreddy, V., Schaumont, P.: A systematic method to evaluate and compare the performance of physical unclonable functions. In: Embedded Systems Design with FPGAs, pp. 245–267. Springer, New York (2013)
11. Li, X., Ma, J., Wang, W., Xiong, Y., Zhang, J.: A novel smart card and dynamic ID based remote user authentication scheme for multi-server environment. Math. Comput. Model. **58**(12), 85–95 (2013)
12. Li, X., Niu, J., Khan, M.K., Liao, J.: An enhanced smart card based remote user password authentication scheme. J. Netw. Comput. Appl. **36**(5), 1365–1371 (2013)
13. Pappu, R.S.: Physical One-way Functions. Massachusetts Institute of Technology, Boston (2001)
14. Zhang, Z., Guo, Y.: Survey of physically unclonable function. J. Comput. Appl. **32**(1), 3115–3120 (2012)
15. Igier, M., Vaudenay, S.: Distance bounding based on PUF. In: International Conference on Cryptology and Network Security, pp. 701–710. Springer International Publishing (2016)
16. Gurubani, J.B., Thakkar, H., Patel, D.R.: Improvements over extended LMAP+: RFID authentication protocol. In: IFIP Advances in Information and Communication Technology, pp. 225–231 (2012)
17. Wang, C.: Heuristic attack strategy for improving LMAP+ protocol. Comput. Sci. **41**(5), 143–149 (2014)

18. Zhu, F.: New lightweight RFID bidirectional authentication protocol: PUF-LMAP+. Microcomput. Appl. **35**(1), 1–4 (2016)
19. Zheng, J., Chen, B., Zhou, Y.: A RFID lightweight authentication protocol GIMAP. J. Chin. Comput. Syst. **34**(3), 530–534 (2013)
20. Herder, C., Yu, M.D., Koushanfar, F., et al.: Physical unclonable functions and applications: a tutorial. Proc. IEEE **102**(8), 1126–1141 (2014)

Comparison of Similarity Measures in Collaborative Filtering Algorithm

Jing Wang[✉]

Neusoft Institute, Guangdong, Foshan, China
jingyun_wj@163.com

Abstract. Collaborative filtering algorithms help people make choices based on the opinions of other people. User-based and item-based collaborative filtering algorithms predict new ratings by using ratings of similar users or items. Similarity calculation is the key step in the algorithms. This paper compares the prediction quality of four commonly used similarity measures on different datasets. Experimental results show that Adjusted Cosine similarity consistently achieves best prediction accuracy.

Keywords: Similarity measure · Prediction quality · Collaborative filtering
Recommendation system

1 Introduction

With the growing popularity of Internet, some commercial websites are changing our life way. We enjoy the convenience and rapidity of on-line consumption and relaxation. However, we often feel lost in so much information. Frequently, before we find what we want, we must spend much time and energy in browsing them. To solve such problems, some famous commercial websites such as Amazon[1], Netflix[2], have adopted the recommendation systems. On one hand, recommendation system can help customers find their valuable information. On the other hand, information will be shown to those who are interested in, so as to realize the win-win of information providers and customers. Recommendation system is now widely used in many fields, such as e-commerce, movies, music, social networks online systems, and so on. Recommendation algorithm is the core and key part of the system, commonly used algorithms include: content-based recommendation, collaborative filtering recommendation, knowledge-based recommendation and combination recommendation [1]. At present Collaborative filtering algorithm is one of most successful recommendation techniques [2].

---

[1] http://www.amazon.com.
[2] http://www.netflix.com.

© Springer Nature Singapore Pte Ltd. 2018
J. C. Hung et al. (Eds.): FC 2017, LNEE 464, pp. 356–365, 2018.
https://doi.org/10.1007/978-981-10-7398-4_37

2 Collaborative Filtering Algorithm

2.1 Description of Ratings

Collaborative filtering algorithm works by building a database of ratings for items by users. Assuming that there are m users $U = \{u1, u2, ... um\}$ and n items $I = \{i_1, i_2, ... i_m\}$ in the database. Collaborative filtering algorithm represents the entire m × n user-item data as a ratings matrix R(m, n) in Table 1:

Table 1. User-item ratings matrix

	i_1	...	i_t	...	i_n
u_1	$R_{1,1}$	...	$R_{1,t}$	...	$R_{1,m}$
...					
u_a	$R_{a,1}$	...	$R_{a,t} = ?$	...	$R_{a,n}$
...					
u_m	$R_{m,1}$	...	$R_{m,t}$	...	$R_{m,n}$

Here, R(m, n) represents that user m rates an item n. The user has expressed his/her preference through ratings, which can be a certain numerical scale. The higher the rating, the more that user likes the item. If a user m does not rate an item n, then R(m, n) = 0.

2.2 User-Based Collaborative Filtering Algorithm

User-based collaborative filtering algorithm is based on the assumption that in order to find the items which a user is interested in, the better way is to find other users with similar interests, then recommend other users' interested items to him. The basic idea is very easy to understand, in daily life, we tend to accept friends' recommendations to make some choices. Tapestry [3] is one of the earliest implementations of user-based collaborative filtering. Later GroupLens research system [4] adopts it to recommend news and movies for users. Obviously, finding similar users is the key step in the algorithm. Then, algorithm can predict a target user's preference on an unrated item based on rating information from similar users, as follows:

$$P_{user}(R_{u,i}) = \overline{R_u} + \frac{\sum\limits_{a \in S(u)} Sim(a, u) \cdot (R_{a,i} - \overline{R_a})}{\sum\limits_{a \in S(u)} |Sim(a, u)|} \tag{1}$$

Here, $\overline{R_u}$ and $\overline{R_a}$ is the average rating of user u and a respectively. $R_{a,i}$ is the ratings of user a on item i, $R_{u,i}$ is the ratings of user u on item i. $Sim(a, u)$ denotes similarity between user a and u. After the ratings of unrated items are predicted, the next task is to recommend a list of items to the object user. The vast majority of current systems recommend the top-K items with the highest-predicted ratings [6].

2.3   Item-Based Collaborative Filtering Algorithm

Item-based collaborative filtering algorithm is first brought forward by Sarwar et al. [5]. The basic principle of the algorithm is similar, and the key point is also to find the similar items. Then the prediction is computed by the target users ratings on these similar items, as follows:

$$P_{item}(R_{u,i}) = \overline{R_i} + \frac{\sum\limits_{j \in S(i)} Sim(i,j) \cdot (R_{u,j} - \overline{R_j})}{\sum\limits_{j \in S(i)} |Sim(i,j)|} \tag{2}$$

Here, $\overline{R_i}$ and $\overline{R_j}$ is the average rating of item i and j respectively. $R_{u,i}$ is the ratings of user u on item i, $R_{u,j}$ is the ratings of user u on item j. $Sim(i,j)$ denotes similarity between item i and j. Likewise, the algorithm is also recommend top-K items to the object users.

3   Similarity Measures

Computing the similarity between the users and items is an important step in the collaborative filtering algorithm. The commonly used similarity measures include Pearson Correlation Coefficient (PCC), Cosine (COS), Adjusted Cosine (ACOS), and Euclidean Distance (ED). All signs in the following formulas are described in Table 2.

Table 2. Description of signs in the formulas

Sign	Description
$R_{u,i}$	The rating of user u has assigned to item i
$R_{a,i}$	The rating of user a has assigned to item i
$R_{u,j}$	The rating of user u has assigned to item j
$\overline{R_u}$	The average rating of user u
$\overline{R_u}'$	The average rating of user u on co-rating pair
$\overline{R_a}$	The average rating of user a
$\overline{R_a}'$	The average rating of user a on co-rating pair
I_{ua}	$I_{ua} = I_u \cap I_a$ means the item set rated simultaneously by user u and a
I_u	The items rated by user u
I_a	The items rated by user a
$\overline{R_i}$	The average rating of item i
$\overline{R_i}'$	The average rating of item i on co-rated pair
$\overline{R_j}$	The average rating of item j
$\overline{R_j}'$	The average rating of item j on co-rated pair
I_{ij}	$I_{ij} = I_i \cap I_j$ means the users who have both rated item i and j
I_i	The users who has rated item i
I_j	The users who has rated item j

3.1   Pearson Correlation Coefficient Similarity

Pearson Correlation Coefficient is also used in statistics to evaluate the degree of linear relationship between two variables. It ranges from -1 (a perfect negative relationship) to $+1$ (a perfect positive relationship), with 0 stating that there is no relationship. This is the most commonly used similarity measures in collaborative filtering algorithm. The Pearson Correlation Coefficient similarity between user u and a is defined as Eq. (3):

The Pearson Correlation Coefficient similarity between user u and a is defined as follows:

$$Sim(u,a) = \frac{\sum\limits_{i \in I_{ua}} (R_{u,i} - \overline{R_u'})(R_{a,i} - \overline{R_a'})}{\sqrt{\sum\limits_{i \in I_{ua}} (R_{u,i} - \overline{R_u'})^2} \sqrt{\sum\limits_{i \in I_{ua}} (R_{a,i} - \overline{R_a'})^2}} \tag{3}$$

The Pearson Correlation Coefficient similarity between item i and j is defined as follows:

$$Sim(i,j) = \frac{\sum\limits_{u \in I_{ij}} (R_{u,i} - \overline{R_i'})(R_{u,j} - \overline{R_j'})}{\sqrt{\sum\limits_{u \in I_{ij}} (R_{u,i} - \overline{R_i'})^2} \sqrt{\sum\limits_{u \in I_{ij}} (R_{u,j} - \overline{R_j'})^2}} \tag{4}$$

3.2   Cosine Similarity

Cosine similarity is a measure of similarity between two non-zero vectors of an inner product space that measures the cosine of the angle between them. This measure is widely used in information retrieval and text mining, which is used to compare two text document, and the document vector can be expressed by words [8, 9]. Cosine similarity (CS) between two vectors x and y is defined as:

$$COS(x,y) = \frac{x^T y}{\|x\|\|y\|} \tag{5}$$

The resulting similarity measure is always within the range of -1 and $+1$. So, in the collaborative filtering algorithm, we may look on co-ratings as user or item vectors, and calculate the angle of two vectors to measure similarity between two users or two items.

The Cosine similarity between user u and a is defined as follows:

$$Sim(u,a) = \frac{\sum\limits_{i \in I_{ua}} (R_{u,i} \times R_{a,i})}{\sqrt{\sum\limits_{i \in I_{ua}} (R_{u,i})^2} \sqrt{\sum\limits_{i \in I_{ua}} (R_{a,i})^2}} \tag{6}$$

The Cosine similarity between item i and j is defined as follows:

$$Sim(i,j) = \frac{\sum\limits_{u \in I_{ij}} (R_{u,i} \times R_{a,i})}{\sqrt{\sum\limits_{u \in I_{ij}} (R_{u,i})^2} \sqrt{\sum\limits_{u \in I_{ij}} (R_{u,j})^2}} \tag{7}$$

3.3 Adjusted Cosine Similarity

Cosine similarity measure doesn't take into account the differences in rating scale between different users. The adjusted cosine similarity offsets this drawback by subtracting the corresponding user's or item' s average rating from co-rating or co-rated pairs [10, 11].

The Adjusted Cosine similarity between user u and a is defined as follows:

$$Sim(u,a) = \frac{\sum\limits_{i \in I_{ua}} (R_{u,i} - \overline{R_u'})(R_{a,i} - \overline{R_a'})}{\sqrt{\sum\limits_{i \in I_u} (R_{u,i} - \overline{R_u})^2} \sqrt{\sum\limits_{i \in I_a} (R_{a,i} - \overline{R_a})^2}} \tag{8}$$

The Adjusted Cosine similarity between item i and j is defined as follows:

$$Sim(i,j) = \frac{\sum\limits_{u \in I_{ij}} (R_{u,i} - \overline{R_i'})(R_{u,j} - \overline{R_j'})}{\sqrt{\sum\limits_{u \in I_j} (R_{u,i} - \overline{R_i})^2} \sqrt{\sum\limits_{u \in I_j} (R_{a,i} - \overline{R_j})^2}} \tag{9}$$

3.4 Euclidean Distance Similarity

Euclidean distance is the "ordinary" (i.e. straight-line) distance between two points in Euclidean space, which is usually used to measure the similarity in clustering and classification algorithms [12, 13]. In Cartesian coordinates, if $x = (x_1, x_2, ..., x_n)$ and $y = (y_1, y_2, ..., y_n)$ are two points in Euclidean n-space, then the distance between x and y is given by the formula:

$$dist = \sqrt{\sum_{i=1}^{n} (x_i - y_i)^2} \tag{10}$$

If x and y are completely similar, distance is 0. If they are far away, distance will be large. In order to guarantee distance between 0 and 1, we usually take the reciprocal of distance, as follows: $1/(1 + dist(X, Y))$. Since, most ratings are 0s, we just calculate the distance of co-rating or co-rated pairs.

The Euclidean distance similarity between user u and a is defined as follows:

$$Sim(u,a) = \frac{1}{1 + \sqrt{\sum_{i \in I_{ua}} (R_{u,i} - R_{a,i})^2}} \tag{11}$$

The Euclidean distance similarity between item i and j is defined as follows:

$$Sim(i,j) = \frac{1}{1 + \sqrt{\sum_{i \in I_{ij}} (R_{u,i} - R_{u,j})^2}} \tag{12}$$

4   Empirical Analysis

4.1   Dataset

We experimented with the MovieLens datasets, which are collected by the GroupLens Research Project at the University of Minnesota[3]. We utilize one of the datasets which contains 100,000 ratings (1–5 scales) from 943 users on 1682 movies (items), where each user has rated at least 20 items. The density of the user-item matrix is: $100000/(943 * 1682) = 6.3\%$. We use 5-fold cross validation, which means we repeat experiments 5 times with each different 80% training set and 20% test set and average the results.

4.2   Metrics

The Mean Absolute Error (MAE) and RMSE (Root Mean Square Error) are commonly used metrics to evaluate the prediction quality by summarizing the differences between the actual and predicted ratings in collaborative filtering algorithms. MAE gives equal weight to all errors, while RMSE gives extra weight to large errors.

$$MAE = \frac{\sum_{i=1}^{N} |R_{u,i} - \hat{R}_{u,i}|}{N} \tag{13}$$

$$RMSE = \frac{\sum_{i=1}^{N} \sqrt{(R_{u,i} - \hat{R}_{u,i})^2}}{N} \tag{14}$$

Here $R_{u,i}$ denotes the actual rating that user u gave to item i. $\hat{R}_{u,i}$ denotes the prediction that user u gave to item i which is predicted by our algorithms. N denotes the number of the predicted ratings.

---

[3] movielens.umn.edu.

4.3   Experimental Results

First, we compute the similarity between users or items using Pearson Correlation Coefficient (PCC), Cosine (COS), Adjusted Cosine (ACOS), and Euclidean Distance (ED). Then, we choose k nearest neighbors for every user or item. In the experiment, we choose k as 10, 20, 30, 40, 50, 60, 70, 80, 90 and 100. At last, we compute a prediction of the target user's rating to an unrated item from a combination of the selected neighbors' ratings by using User-based Collaboration Filtering (UBCF) as Eq. (1), and Item-based Collaboration Filtering (IBCF), as Eq. (2) respectively. By using 5-fold cross validation, we obtain average MAE and RMSE.

From Fig. 1 to 2, and Fig. 3 to 4, we can see that MAE is less than RMSE. As we has mentioned, RMSE has enlarge punishment to inaccurate ratings, so this metric has more strict evaluation to algorithms. When using Adjusted Cosine similarity, both in UBCF and IBCF algorithms, MAE and RMSE are the lowest, which means Adjusted Cosine similarity (ACOS) achieves the most accurate prediction quality. When using Euclidean Distance (ED) similarity, both in UBCF and IBCF algorithms, MAE and RMSE are the largest, which means Euclidean Distance (ED) similarity has the worst prediction quality. When using Pearson Correlation Coefficient (PCC) similarity and Cosine (COS) similarity, MAE and RMSE are between them. Because of considering the user's rating scale, by subtracting the user's average rating, ACOS and PCC measures both gain better prediction than COS and ED measures. With the increase in the number of neighbors (the parameter k), MAEs and RMSEs are decrease in PCC, COS and ED measures. However, MAEs and RMSEs are also increase in ACOS measure. It is observed that if the number of neighbors is less, PCC, COS and ED measures can't obtain accurate similarity both in UBCF and IBCF algorithms. Inversely, in ACOS measure, more neighbors lead to more errors. Even we select 100 neighbors, MAE and RMSE in ACOS measure are far less than that in another three measures when selecting 10 neighbors. Both in UBCF and IBCF algorithms, it's the best choice to select ACOS measure as similarity.

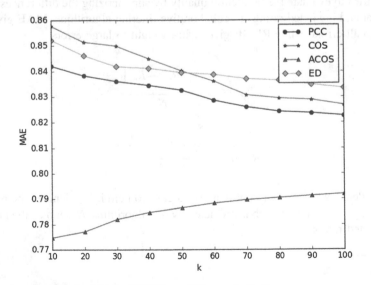

Fig. 1. MAE of UBCF on different similarity measures

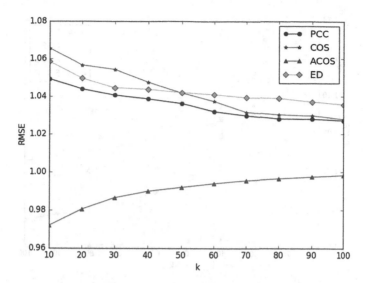

Fig. 2. RMSE of UBCF on different similarity measures

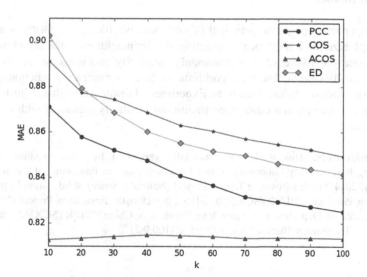

Fig. 3. MAE of IBCF on different similarity measures

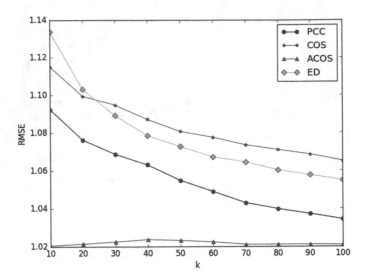

Fig. 4. RMSE of IBCF on different similarity measures

5 Conclusion

In this paper we explain the principal of collaborative filtering algorithms and especially emphasize the importance of selecting similar neighbors in the algorithms. Then we elaborate the usage of four commonly similarity measures in user-based and item-based collaborative filtering algorithms. At last, we compare prediction quality of the four measures on different datasets. Experimental results show that Adjusted Cosine similarity outperforms the other three traditional similarity measures with equal computation cost.

Acknowledgements. This work was financially supported by 2016 Opening Project of Guangdong Province Key Laboratory of Big Data Analysis and Processing at the Sun Yat-sen University, 2014 Youth Innovative Talents Project (Natural Science) of Education Department of Guangdong Province (2014KQNCX248), 2016 Characteristic Innovation Project (Natural Science) of Education Department of Guangdong Province of China (2016KTSCX162), and Foshan Science and Technology Bureau Project (2016AG100382).

References

1. Xu, H.L., Wu, X., Li, X.D., Yan, B.P.: Comparison study of internet recommendation system. J. Softw. **20**(2), 350–362 (2009)
2. Ma, H.W., Zhang, G.W., Li, P.: Survey of collaborative filtering algorithm. J. Chin. Comput. Syst. **30**(7), 1282–1288 (2009)
3. Goldberg, D., Nichols, D., Oki, B., Terry, D.: Using collaborative filtering to weave information tapestry. Commun. ACM **35**(12), 61–70 (1992)

4. Resnick, P., Iacovou, N., Suchak, M., Bergstorm, P., Riedl, J.: GroupLens: an open architecture for collaborative filtering of netnews. In: ACM CSCW, pp. 175–186 (1994)
5. Sarwar, B., Karypis, G., Konstan, J., Riedl, J.: Item-based collaborative filtering recommendation algorithms. In: 10th Proceeding of WWW Conference, Hong Kong, 1–5 May 2001
6. Wang, J., Yin, J.: Enhancing accuracy of user-based collaborative filtering recommendation algorithm in social network. In: International Conference on System Science, Engineering Design and Manufacturing Informatization (ICSEM 2012), pp. 142–145. IEEE Press (2012)
7. Zhou, J.F., Tang, X., Guo, J.F.: An optimized collaborative filtering recommendation algorithm. J. Comput. Res. Dev. **14**(10), 1842–1847 (2004)
8. Tata, S., Patel, J.M.: Estimating the selectivity of tf-idf based cosine similarity predicates. ACM Sigmod Rec. **36**(2), 7–12 (2007)
9. Li, B.: Distance weighted cosine similarity measure for text classification. In: Ideal 2013, vol. 8206, pp. 611–618 (2013)
10. Guo, L., Peng, Q.K.: A combinative similarity computing measure for collaborative filtering. Appl. Mech. Mater. **347–350**, 2919–2925 (2013)
11. Shabanpoor, M., Soofipoor, S., Mahdavi, M.: An E-health recommender system based on adjusted cosine similarity. Int. J. Acad. Res. **7**(5), 84–90 (2015)
12. Deng, S., Cao, H., Pan, Q., Shen, J., Xing, T.: Analysis and improvement of affinity propagation based on Euclidean distance similarity. J. Comput. Inf. Syst. **8**(6), 2293–2300 (2012)
13. Kumar, A., Panigrahi, R.K.: Classification of hybrid-pol data based on Euclidean distance between Stokes vectors. In: IEEE Radar Conference, pp. 422–425 (2015)

Study of CR Based U-LTE Co-existence Under Varying Wi-Fi Standards

A. C. Sumathi[1](✉), M. Akila[2], and Sangaiah Arunkumar[3]

[1] PSG Institute of Technology and Applied Research, Coimbatore, India
sumathi.ac@gmail.com
[2] PGP College of Engineering and Technology, Namakkal 637207, India
akila@nvgroup.in
[3] School of Computing Science and Engineering,
Vellore Institute of Technology (VIT), Vellore 632014, India
sarunkumar@vit.ac.in

Abstract. Long Term Evolution in unlicensed band extends the benefits of Long Term Evolution and Long Term Evolution - Advanced to deploy in 5 GHz unlicensed spectrum, enabling mobile operators to offload data traffic onto unlicensed frequencies. The License Assisted Access with Long Term Evolution allows co-existence with Wi-Fi through carrier aggregation. The recently evolved intelligent technology viz. Cognitive radio which supports the efficient spectrum utilization is applied in the proposed system model to detect the white spaces in 5 GHz band to accomplish Listen-Before-Talk regulatory requirement of radio communication in Long Term Evolution - unlicensed band. Another major goal of Long Term Evolution - unlicensed to co-existence along with Wi-Fi/Internet of Things users in a non-interference style is also accomplished by the use of Cognitive radio. Simulation results demonstrate their coexistence along with effectiveness of resource allocation in a varying 5 GHz compatible 802.11 wireless local area networks environment.

Keywords: LTE · U-LTE · Cognitive radio · Carrier aggregation
Coexistence issues · Wi-Fi/IoT in 5 GHz band

1 Introduction

1.1 LTE in Unlicensed Spectrum

Fourth Generation (4G) Long Term Evolution (LTE) is one of the several competing 4G standards along with Ultra Mobile Broadband and WiMax (IEEE 802.16). The main goal of LTE is to provide a high data rate, low latency and packet optimized radio-access technology, supporting flexible bandwidth deployments. LTE networks carries large amount of data. In spite of efficient cell management, large spectrum is required to handle such huge data. To address these issues, Long Term Evolution in unlicensed (LTE – U) is considered as the best innovations to meet the high performance and seamless user experience. Since LTE radio technology is based on state of the art technology, it can achieve both high data rates and at the same time high spectral efficiency in the unlicensed arena. Known in (3rd Generation Partnership Project) 3GPP

© Springer Nature Singapore Pte Ltd. 2018
J. C. Hung et al. (Eds.): FC 2017, LNEE 464, pp. 366–375, 2018.
https://doi.org/10.1007/978-981-10-7398-4_38

as LTE License Assisted Access (LTE-LAA) or more generally as LTE-U, it enables access to unlicensed spectrum especially in the 5 GHz (Industrial, Scientific and Medical) ISM band.

Qualcomm, Huawei and Ericsson requested the 3GPP standards committee to allow LTE service to run on the 5 GHz band. The 5 GHz spectrum offers a large amount of bandwidth and is one of the two unlicensed bands that are typically used by Wi-Fi service. It has a shorter communication range due to higher path loss but has wider available bandwidth [1].

The 802.11 standard defines 23 20 MHz wide channels in the 5 GHz spectrum. Each channel is spaced 20 MHz apart and separated into three Unlicensed National Information Infrastructure (UNII) bands. Wireless devices specified as 802.11a/n/ac are capable of operating within these bands. In the United States, UNII-1 (5.150 to 5.250 GHz) containing channels 36, 40, 44, and 48 and UNII-3 (5.725–5.825) containing channels 149, 153, 157, 161 are permitted. UNII-2 (5.250–5.350 GHz and 5.470–5.725 GHz) which contains channels 52, 56, 60, 64, 100, 104, 108, 112, 116, 120, 124, 128, 132, 136, and 140 are permitted in the United States, but shared with radar systems.

1.2 Coexistence of U-LTE with Wi-Fi/IoT

Due to non-exclusive usage nature of unlicensed spectrum by U-LTE, there are two main challenges. The foremost challenge of design of U-LTE is its coexistence with Wi-Fi/IoT systems on a fair and friendly basis. The Wi-Fi/IoT systems are the user-deployed systems and they are the incumbent users or primary users of the unlicensed band. The PHY/MAC implementation differences between LTE transmissions and Wi-Fi/IoT, hinders the direct implementation of U-LTE transmissions as it can generate continuous interference to Wi-Fi/IoT systems.

Second is the coexistence of two or more different U-LTE operators in the same unlicensed band. The operation in unlicensed band also needs to factor in the regulatory requirements of a given region. In some markets, like Europe, Japan and India, a specific waveform requirement on supporting Listen-Before-Talk (LBT) at milliseconds scale is required which would need changes in LTE air interface. In other markets, like US, Korea and China, there are no such requirements. In these countries, with carefully designed coexistence mechanisms realizable by software to ensure peaceful coexistence with Wi-Fi/IoT, operators can deploy LTE in unlicensed bands that are compatible with Rel. 10/11 3GPP LTE standards. In markets where LBT is required, LTE in unlicensed operation can be further optimized through air interface enhancement with the introduction of LBT feature potentially in 3GPP Release 13 [2, 3].

1.3 Usage 5 GHz Band and Wi-Fi/IoT Usage

The Internet of Things (IoT) or Machine-to-Machine (M2M) communications is one of the most exciting and fastest growing technologies across the globe today. It employs the embedded technology with sensors and actuators, connects to other devices or to cloud, and automatically transmits information. There are many standards and proprietary solutions used for connecting devices to each other or to cloud such as Wi-Fi, Bluetooth, ZigBee, Active RFID, loWPAN, EtherCAT, NFC, RFID, etc. Wi-Fi has been the most successful among these technologies listed above due its adaptability, scalability, ease of use and cost. As the 2.4 GHz band is meant for unlicensed users, it is presently heavily crowded and hence, the IoT device communication and Wi-Fi are shifting to less crowd 5 GHz band.

In this work, IEEE 802.11a, 802.11n and 802.11ac standards of 5 GHz band are considered. 802.11a uses an orthogonal frequency division multiplexing encoding scheme and provides up to 54-Mbps. With multiple-input multiple-output (MIMO), the real speed of 802.11n would be 100-Mbps. The 802.11ac specification operates only in the 5 GHz frequency range and features support for wider channels (80 MHz and 160 MHz) to deliver data rates of 433Mbps per spatial stream, or 1.3 Gbps in a three-antenna (three streams) design.

1.4 Cognitive Radio and 5 GHz Band

The Federal Communication Commission (FCC) defined CR as the radio that can change its transmission parameters based on interaction with the environment in which it operates [4]. The Wireless communication has been increased along with the increase in high data rate requirements. The licensed spectrum space remains idle at most of the times [5] due to the inefficient allocation of frequencies whereas the cellular bands are overloaded. For effective utilization of spectrum and also to meet the spectrum demands, FCC revisited the problem of spectrum management [6] and this leads to CR invention. The IEEE 802.22 is the standard for cognitive wireless regional area networks (WRANs). The main goal of CR is to identify the unused licensed spectrum for secondary users (SU) without interfering Primary Users (PU) and this method of sharing is often called as Dynamic Spectrum Access (DSA) [7].

The flexibility in the hardware and its feasibility to program for a band or mode enables CR to adapt over to an ISM band or up to an IEEE band and even to unlicensed 5 GHz bands. CR dynamically monitors certain bands of the 5 GHz unlicensed spectrum to identify the idle spectrum and uses them as needed [8].

1.5 CR Applications in U-LTE

In authors' previous work [9], a system was so modeled to utilize the attributes of the CR to optimally operate U-LTE in 5 GHz band. Figure 1 depicts the architecture of U-LTE in 5 GHz band. User Equipment (UEs) and IoT devices, communicates with Wi-Fi Access Point (AP) using the unlicensed spectrum, form a femto cell and become primary users. During their communications with eNodeB (eNB), UEs form a small cell and try to utilize the unlicensed spectrum and become secondary users.

U-LTE operates in two modes: Supplemental Downlink (SDL) and Time Division Duplex (TDD). In SDL mode, the unlicensed spectrum is used only for the downlink traffic, thereby eNB performs most of the necessary operations to ensure the reliable communications, including checking whether the intended unlicensed channel is free from other use [10]. For TDD mode, the unlicensed spectrum is used for uplink and downlink, resulting additional implementation complexity in UEs for LBT feature. Latest release of 3GPP LTE standard Rel.13 supports usage of unlicensed spectrum in both operating modes [3].

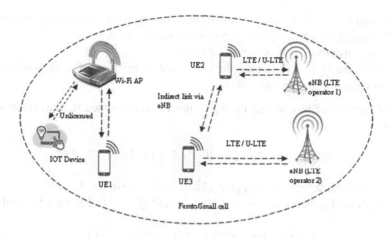

Fig. 1. U-LTE architecture in 5 GHz band

Considering the SDL mode, the eNB equipped with CR realizes the LBT feature with an efficient mechanism to share the unlicensed spectrum with Wi-Fi, the primary users of the specified band in a non-interference basis and also paving the way for the currently evolving IoT or M2M communications.

Channel and traffic model was formulated to optimize the functionality of U-LTE in terms of clean channel searching and co-existence of secondary users with primary users and presented in the previous work [9]. The same work is extended to study the data traffic and resource allocation for eNB/Wi-Fi nodes in SDL mode for IEEE 802.11a and 802.11n standards and presented in this paper.

The mathematical theory formulation for resource allocation in 23 channels of 5 GHz band is explained in Sect. 2. Simulation and performance analysis of U-LTE in 802.11a and 802.11n 5 GHz bands is presented in Sect. 3 and concluded in Sect. 4.

2 Problem Formulation

Considering the U-LTE architecture in 5 GHz band as shown in Fig. 1, the Wi-Fi access points are the primary users and eNBs are secondary users equipped with CRs. The network is considered to be operates in time slots t, t + 1, t + 2 etc. The time slot for U-LTE is defined by the ETSI standards [11] as Channel Occupancy Time (CCT) for 10 ms. The downlink U-LTE traffic (in bps) is enqueued in the eNBs and then transmitted to UEs using packet-scheduling procedure as detailed in LTE standard [12]. Based on the resource allocation procedures [13] and queuing theory [14, 15], the following notations are used in sequel:

1. $Q_i^L(t)$ and $Q_i^w(t)$ denotes the queue size (in bits) at eNB node and Wi-Fi/IoT access point at the beginning of the time slot t at i^{th} channel.
2. $S_i^L(t)$ and $S_i^w(t)$ denotes the service rate (in bps) at the time slot t at i^{th} channel for eNB node and Wi-Fi/IoT access point.

3. $D_i^L(t)$ and $D_i^w(t)$ denotes the amount of data (in bps) generated during the time slot t at i^{th} channel for eNB node and Wi-Fi/IoT access point.
4. C_i is the capacity (data rate) of the channel.
5. $Q_i^L(t+1)$ and $Q_i^w(t+1)$ denotes the queue size (in bits) at eNB node and Wi-Fi/IoT access point at the next time slot.

The total number of bits served by the channel cannot exceed its maximum serving capacity

$$S_i^L(t) \le C_i, \ S_i^w(t) \le C_i \ \forall i \in l, l = \{1, 2, \dots\dots.23\} \tag{1}$$

Here l denotes the 23 non-overlapping channels of 5 GHz band.

By applying Lindley's equation, $Q_i^L(t+1)$ and $Q_i^w(t+1)$ can be estimated as

$$Q_i^L(t+1) = [Q_i^L(t) + D_i^L(t) - S_i^L(t)]^+, \forall i \in l \tag{2}$$

$$Q_i^w(t+1) = [Q_i^w(t) + D_i^w(t) - S_i^w(t)]^+, \forall i \in l \tag{3}$$

Where

$$[x]^+ = \max[0, x] \tag{4}$$

By using Eq. (2), the queue length for n time slots is determined. The resource can be allocated as continuous time slots not exceeding the channel capacity can be shown as

$$\sum_{t=1}^{n} S_i^L(t) \le C_i, \forall i \in l, \ l = \{1, 2, \dots\dots.23\} \tag{5}$$

The application of Eq. (5) is considered for U-LTE (secondary users) alone whereas the resource allocation and data transmission for primary or the Wi-Fi/IoT users are assumed to be taken care by its own MAC protocol.

3   Simulation

3.1   System Model

The system model is simulated in compliance to Rel. 13 3GPP LTE standards using MATLAB showing the coexistence of U-LTE and Wi-Fi/IoT in 5 GHz band ranging from 5.0 to 5.8 GHz (U-NII 1/2/2e), 23 channels each with assumed bandwidth of 20/40/80 MHz's. The general simulation parameters are listed in Table 1. Table 2 lists the simulation parameters of U-LTE under different Wi-Fi standards.

The block diagram for simulating the coexistence of U-LTE with Wi-Fi/IoT with LBT using CR is shown in Fig. 2. The simulation system consists of two separate

Table 1. General simulation parameters

Parameters	Values
No. of channels (l)	23
Time slot (t)	10 ms
Aggregated time slots	10/20/50/100 ms
Wi-Fi band	5 GHz (UNII –1,2 & 2e)
No. of users	10 to 500
Energy of Wi-Fi signal	–60 dBm to –30 dBm
Energy of LTE signal	–80 dBm to –65 dBm
CCA threshold	20 µs and < –80 dbm

Table 2. U-LTE simulation parameters under different Wi-Fi standards

Type of Wi-Fi	Bandwidth	$S_i^L(t)$	C_i	$D_i^L(t)$	$Q_i^L(t)$
Case 1: 802.11a	20 MHz	540 Kbits	54 Mbps (max)	100/200/500/1000 Mbps	50/100/200/500 Mbps
Case 2: 802.11n	20 MHz	700 Kbits	72 Mbps (max)	100/200/500/1000 Mbps	50/100/200/500 Mbps
Case 3: 802.11n	40 MHz	30 Mbits	600 Mbps (max) 300 Mbps (avg)	100/200/500/1000 Mbps	50/100/200/500 Mbps

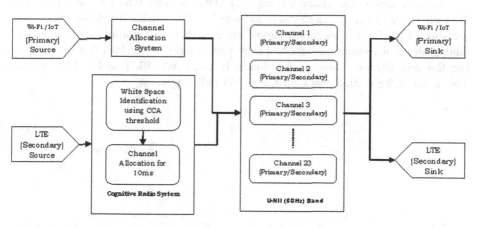

Fig. 2. Block diagram of channel allocation in LBT scenario with application of cognitive radio

sources, one for LTE and other for Wi-Fi/IoT generating multiple signals in random fashion with respective energy and transmission time. Due to the application of CR in this system model, Wi-Fi/IoT systems are considered as the primary users and U-LTE systems as secondary users. Incoming signals from Wi-Fi/IoT sources occupy the channels by following their own MAC protocol by means of Channel allocation system. Whereas, U-LTE signals, in order to accomplish LBT feature, they follows the ETSI standards [10] for channel occupancy through CR system. The channel and the traffic model are formulated based on the state transition model and queuing theory [16] in our previous work and the same is utilized in this system.

The CR system at the LTE eNB node, on getting a transmission request from LTE source, identifies the white space (free channel) in U-NII band and estimates the Clear Channel Assessment (CCA) threshold. CCA threshold is estimated by measuring the energy level of the free channel for a listening period of 20 μs.

If the energy level in the channel is below −80 dBm for the listening period of 20 μs, considering it as low interference level in the channel and assumed to be free. The Channel is then allocated for the U-LTE signal transmissions for the duration equal to Channel Occupancy Time of 10 ms. Later, if LTE source wishes to continue its transmission, it repeats the white space detection and CCA process for channel allocation. On every channel request and allocation process, the Channel Occupancy time for LTE source is maximum 10 ms. Therefore, the above process is repeated until LTE source completes its transmission. On every cycle, the channel allocated for LTE transmissions is different or same based on the free channel availability. The two scenarios depicting coexistence of U-LTE with Wi-Fi/IoT with LBT using CR at different continuous instances of time for 100 users are shown in Figs. 3 and 4.

In Fig. 3, the channel occupancy status during the initial time span of 100 ms is shown. During that time, 13 channels out of 23 are occupied by Wi-Fi/IoT signals and the remaining 10 channels are occupied by U-LTE signals. Therefore, channels of U-NII band have been occupied almost fairly by U-LTE and Wi-Fi/IoT systems during the initial phase of channel allocation.

Figure 4 depicts the channel occupancy status at later time after 200 ms. Here, Wi-Fi/IoT signals occupy 15 channels and 8 by U-LTE signals. The channel occupancy measure of Wi-Fi/IoT systems is almost same as that of in earlier time span. Maintaining almost the same channel occupancy status even after the later time span depicts the fair and friendly sharing of channels between Wi-Fi/IoT and U-LTE signals. This leads to the reduced back off rate of Wi-Fi/IoT systems.

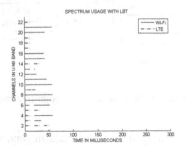 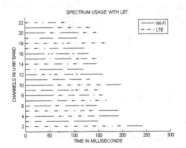

Fig. 3. Scenario 1: initial channel occupancy

Fig. 4. Scenario 2: channel occupancy status after 200 ms

3.2   Performance Analysis

Considering the channel capacity C_i (data rate) of the Wi-Fi standards, timeslots are aggregated as 10, 20, 50 and 100 ms to service the U-LTE signals. For every cases depicted in Table 2, the queue size $Q_i^L(t)$ at the eNB node is varied as 50/100/200/500 Mbps at the beginning of the time slot for every channel and the behavior of the system is analyzed. Similarly, the system is analyzed for various amounts of data $D_i^L(t)$ such as 100/200/500/1000 Mbps generated during the time slot at every channel.

In case 1 (802.11a) and case 2 (802.11n 20 MHz) as shown in Figs. 5a and 6a, Wi-Fi back off rate increases. The increase in back off rate is due to the dominance usage of medium by LTE-U users due to time slot aggregation with increased data traffic. Increase in back off rate results in throughput deficiency. Figures 5b and 6b compares the throughput of Wi-Fi and LTE-U. As the medium is most utilized by LTE-U, its throughput ratio is more than Wi-Fi. The imbalance utilization of medium by LTE-U and Wi-Fi indicates the reduced chance of fair and friendly co-existence between them.

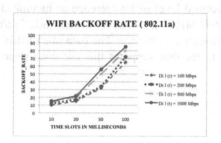

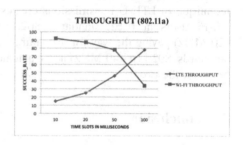

Fig. 5a. Wi-Fi Back off rate for IEEE802.11a   **Fig. 5b.** Throughput for IEEE802.11a

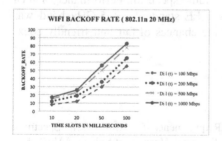

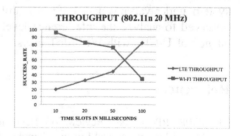

Fig. 6a. Wi-Fi Back off rate for IEEE802.11n (20 MHz)

Fig. 6b. Throughput for IEEE 802.11n (20 MHz)

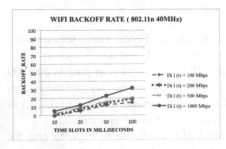
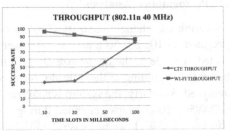

Fig. 7a. Wi-Fi Back off rate for IEEE802.11n (40 MHz)

Fig. 7b. Throughput for IEEE802.11n (40 MHz)

In case 3 (802.11n 40 MHz) as shown in Fig. 7a, the Wi-Fi back off rate reduced to 50% of case1 and 2. Though the throughput of LTE-U increases with increased data traffic and time slot aggregation, Wi-Fi throughput does not falls below 80% as shown in Fig. 7b.

Based on the above observations, it can be summarized that, time slot aggregation technique of LTE-U transmissions creates reduced level of interference to incumbent Wi-Fi users in IEEE standards 802.11n (40 MHz) and 802.11a than 802.11a/n (20 MHz) paving the way for their fair and friendly coexistence. The data rate of IEEE standards 802.11n (40 MHz)/ac is a major factor that supports the fair coexistence between the two.

4   Conclusion

The spectrum utilization of radio frequencies is gaining momentum due to the invasion of wireless equipment in every field of human life. In this regard, there is a change over from licensed LTE to U-LTE in view of the evident advantages of later in terms of speed, cost etc. An improved method is proposed in this work to include CR in U-LTE for effective utilization of the white spaces in the radio spectrum. Performance of such system under varying data traffic for different IEEE standards of 5 GHz band was observed to analyze the interference level and the chances of fair and friendly coexistence of U-LTE with Wi-Fi/IoT users.

References

1. 3GPP RP-140808: Review of Regulatory Requirements for Unlicensed Spectrum. Alcatel-Lucent, Alcatel-Lucent Shanghai Bell, Ericcson, Huawei, HiSilicon, IASEI, LG, Nokia, NSN, Qualcomm, NTT Docomo, June 2014
2. Qualcomm, Qualcomm Research LTE in Unlicensed Spectrum: Harmonious Coexistence with Wi-Fi, June 2014
3. 3GPP TS 36.201 version 13.0.0 LTE Evolved Universal Terrestrial Radio Access (E-UTRA) LTE physical layer, General description Release 13 (2016)

4. Haykin, S.: Cognitive radio: brain-empowered wireless communications. IEEE J. Select. Areas Commun. **23**(2), 201–220 (2005)
5. Das, D., Das, S.: Primary user emulation attack in cognitive radio networks: a survey. IRACST – Int. J. Comput. Netw. Wirel. Commun. **3**(3), 312–318 (2013)
6. Stevenson, C.R., Chouinard, G.: IEEE 802.22: the first cognitive radio wireless regional area network standard. IEEE Commun. Mag. **47**, 130–138 (2009)
7. Clancy, T.C.: Dynamic Spectrum Access in Cognitive Radio Networks. Ph. D. thesis, University of Maryland, College Park, MD (2006)
8. Zamblé, R., et al.: Peaceful coexistence of IEEE 802.11 and IEEE 802.16 standards in 5 GHz unlicensed bands. World Acad. Sci. Eng. Technol. Int. J. Electr. Comput. Energ. Electron. Commun. Eng. **4**(7), 1054–1059 (2010)
9. Sumathi, A.C., Priya, M., Vidhyapriya, R.: Realization of LBT for co-existence of U-LTEwith Wi-Fi using cognitive radio. In: International Conference on Innovative Trends in Electronics Communication and Applications, pp. 153–158 (2015). Print
10. Nokia, Nokia LTE for Unlicensed Spectrum, June 2014
11. ETSI, ERM TG28: Electromagnetic compatibility and Radio spectrum Matters (ERM); Short Range Devices (SRD); Radio equipment to be used in the 25 MHz to 1000 MHz frequency range with power levels ranging up to 500 mW. European harmonized standard EN 300.220: v2
12. Paul, K., et al.: Spectrum policy task force. Federal Communications Commission, Washington, DC, Rep. ET Docket, 02-135 (2002)
13. Asheralieva, A., Mahata, K.: A two-step resource allocation procedure for LTE-based cognitive radio network. Comput. Netw. **59**(11), 137–152 (2014)
14. Karpelevich, F.I., Kelbert, M.Ya., Suhov, Yu.M.: Higher-order Lindley equations. Stoch. Proc. Appl. **53**(1), 65–96 (1994)
15. Gass, S.I., Fu, M.C. (eds.): Lindley's Equation. In: Encyclopaedia of Operations Research and Management Science. Springer US, Boston, pp. 880–881 (2013)
16. Jin, Z., Anand, S., Subbalakshmi, K.P.: Performance analysis of dynamic spectrum access networks under primary user emulation attacks. In: Global Telecommunications Conference (GLOBECOM 2010), pp. 1–7. IEEE (2010)

Hybrid Intelligent Bayesian Model for Analyzing Spatial Data

J. Velmurugan[1]([✉]) and M. Venkatesan[2]

[1] School of Computing Science and Engineering, VIT University, Vellore, India
velmurugan85@gmail.com
[2] Department of Computer Science and Engineering,
National Institute of Technology Karnataka, Mangalore, India
venkisakthi77@gmail.com

Abstract. Spatial data mining refers to the extraction of Geo Spatial Knowledge, maintaining their spatial relationships, along with other interesting patterns not explicitly stored in spatial datasets. The overall objective of this research work is to apply GIS based data mining classification modeling techniques to assess the spatial landslide risk analysis in Nilgris district, Tamilnadu, India. Landslide is one of the most important hazards that affect different parts of India in the every year. Landslides cover broad range impact on the people of the affected area in terms of the devastation caused to material and human resources. Landslide is generated by various factors such as rainfall, soil, slope, land use and land covers, geology, etc. Each landslide factor has a different level of values. The ranking of values and assignment of weight to the landslide factor gives good classification of landslide risk level. Data science and soft computing play major role in landslide risk analysis. The rank and weight are assigned to the landslide factor and its different levels using classification data science techniques. In this paper, we proposed a new model with integration of rough set and Bayesian classification called Hybrid Intelligent Bayesian Model (HIBM) to analyze the possibilities of various landslide risk level. The proposed model is compared with real-time data, and performance is validated with other data science models.

Keywords: GIS · Rough set · Bayesian · Landslide · Disaster

1 Introduction

Environmental disasters like cyclone, earthquakes, rainfall, tsunamis and landslides cause incalculable deaths and fearful damage in the world and the changes in environment infrastructure. Landslides are the main disaster which causes huge damages in the infrastructure and incalculable of deaths are happening in the every year.

Landslide is a frequently occurring natural hazard in the hilly terrains of India. It is found that the preponderance of such activity happens during monsoon period from July to September which is rainy period and after the snow fall during January to March. Earthquake activity which triggers Landslide between 5.0 ritcher scales to 8.0 ritcher scales also causes landslide, particularly in regions marked by critically disposed and unstable slopes.

© Springer Nature Singapore Pte Ltd. 2018
J. C. Hung et al. (Eds.): FC 2017, LNEE 464, pp. 376–391, 2018.
https://doi.org/10.1007/978-981-10-7398-4_39

Recently many landslides happened in India. On 30 July 2014 Morning a very large landslide occurred in the village of Malin in Pune, western India. Figure 1 shows the landslide in Malin, where the mud landslide struck and effectively wiped out the village of Malin, which is located close to Bhimashankar in the Western Ghats. It is found that about 40 houses were buried – the timing of the landslide meant that most would have been fully occupied and of course the darkness would have impeded any escape.

Fig. 1. Landslides at Malin, 2014 (Pune)

The Himalayan State of India, Uttarakhand on 16 June 2013 faced one of the toughest situations of the century in form of a natural disaster with landslides and flash floods. Figure 2 shows various landslides in Uttarakhand.

Fig. 2. Landslides in Uttarakhand

Landslide predictions are to be recognized using various statistical methods based on the Geographical Information System (GIS) technology. In the past few years, many research works are carried out to identify the landslides in order to improve the efficiency of landslide prediction. In these cases few works may success in the prediction of landslide using data mining technologies. In this paper, we consider the Coonoor District, Tamilnadu because the huge landslides are happening in the every year due to heavy rainfall.

There are so any factors are involving in the landslides but the effects of the landslides are different area to area. The statistical bond between the factors which impact landslide and the landslide strength prediction is solid to obtain. However, it is a comparatively exact method to get a mathematical investigation model with the past data. To get an exact prediction result of landslides, we have to evaluate the geological and environmental circumstances and consider positive factors are soil, slope, rainfall, land use and land cover, geomorphology and geology.

There are so many components are involved to identify the landslides. In these few, components may fully depend for the prediction of landslides. The importance of the attributes is identified by using the soft computing approach is Rough Set theory. To get an exact and consistent prediction result of landslides, we consider the few of environmental factors such as the Soil, Slope, Rainfall, Land use and Land Cover, Geomorphology & Geology. In the existing studies, Statistical models, Neural Networks, Fuzzy based Neural Networks and Data Mining classification techniques are applied to solve these problems. To improve the ability of the existing works the above said factors are considered to construct a new model called Hybrid Intelligent Bayesian Model (HIBM) to predict the landslides, which may give accurate results.

2   Literature Survey

The primary work concerned on analysis of landslide monitor and control alongside the approaches to determine the occurrence of landslide or abnormal phenomenon leads to occurrence of landslide is a major concern, due to its threat or hazardous nature among human life. Hence, the natural hazard has been expressed as "an element in the physical environment harmful to man", or as "a harmful interaction of people and nature" (White 1973), or defined as "the probability of occurrence of a potentially damaging phenomenon" (UNDRO 1982), or "a physical event which has a harmful impact on human beings and their environment" (Alexander 1993).

Researchers had suggested multiple versions for landslide and its management, where the popular definitions of landslide can be judged as follows: "The movement of a mass of rock, debris, or earth down a slope" (Cruden 1991). "The product of local geomorphic, hydrologic and geologic conditions, the modification of these by geodynamic processes, vegetation, land use practices and human activities; and the frequency and intensity of precipitation and seismicity" (Soeters and Van Westen 1996).

Factors behind occurrence of Landslide: Landslides occur repeatedly primarily due to low concentration of the plan used for the land and its slope stability analysis. Due to the abundant tropical rainfall major landslides occur (Pradhan and Lee 2010), while the ground cause of natural landslide is absenteeism of erect vegetative structure, soil nutrients & soil structure. The slope destruction occurs because of the groundwater pressure, while the river or ocean wave's does not support on land erosion or an abrasion of the land slope due to snow melt, glaciers melting, or heavy rain saturation, which penetrate through the slope. Earthquake is another major cause for the landslide, which results in condensation and destruction of the slope and an increase in volcanic eruptions.

Landslides are occurred due to heavy rainfall, earthquake ground motion and other relative environmental factors, In addition to that, the environmental factors are also considered for the landslide incident, such as Soil, Slope, Land use and Land Cover, Slope Geology and Morphological parameters. The existing studies had focus on mathematical methods, including Analytical hierarchy method, Data Mining Approaches, Soft Computing Techniques. Likelihood ratio and artificial neural networks (e.g., Melchiorre et al. 2008; Yilmaz 2009; Chung 2006; Nefeslioglu et al. 2008; Wu and Chen 2009). On the other hand, accuracy of these methods is calculated by manually using mathematical methods. The disadvantages of these methods are the value and relative weightings are manually assigned the professed influence on the occurrence of landslides. In addition to that, these methods are integrated with ecological factors. In authenticity, only some factors or combination of factors had an enormous input to identify the landslides.

To differentiate these factors are very difficult, so that landslide susceptibility can be exactly mapped. The present research works are carried out in the data mining area has created more awareness in the data discovery for landslides prediction ("Gorsevski and Jankowski 2008; Saitoet al. 2009; Wan 2009; Wan et al. 2009"). For example, the environmental factors are categorized and evaluated systematically by using data mining classification method i.e. decision tree (entropy based) to constructing the knowledge based rules. Naturally, data for landslide forecast can be supposed as a body of data, which constitutes our domain of interest. The knowledge is represented in the form of rows and columns. The columns are labeled by attributes name, and tuples (rows) by environmental conditional factor values are presence. The Data mining decision tree classification problems for the collection of environmental factors can be formulated using the above decision table formalism to predict the landslides susceptibility. In the existing study, the rough set theory proposed by Pawlak (1982) was considered for core and reduct of a knowledge table in such a manner that probability can be considered with a less number of attributes. Also in the rough set processing, the landslide susceptibility database are regarded as a decision knowledge table, which consist of condition attributes (factors which affecting landslides) and decision attributes (landslide occur or not).

The Rough set theory dealing with totally different from the conventional mathematical analyses that understand distributions in the independent variables ("Arciszewski and Ziarko 1990; Pawlak and Slowinski 1994; Nguyen and Slezak 1999; Beynon 2001; Wan et al. 2009; Slowinski et al. 2009"). The Bayesian classification rough sets are invented for making classification decisions based on available knowledge information. Bayesian authentication rough sets are proposed for weighting pieces of evidence given by correspondence classes. The models can be studied with respect to three basic issues. For calculating the thresholds, we have a systematic method for a Bayesian classification rough set model according to Bayesian decision theory (Yao and Zhou 2015). Rough set theory is one of the mathematical model that provides a various statistical concept to extract the feature knowledge from real data involving ambiguity, uncertainty and impreciseness and the extracted knowledge will be supplied successfully in the field of machine learning, pattern recognition and knowledge discovery (Saxena et al. 2014).

The analyzed landslide risk by weighted decision tree prediction model. The four important landslide induced factors such as rainfall, land use/land cover, slope and geology are considered for the analysis. The Study Contains Remote sensing images and field data are used to prepare various thematic maps. The performance of the weighted decision tree prediction model is compared with existing classification approaches. Weighted decision tree prediction model is more suitable and accurate than decision tree classifier (Anbalagan and Chandrasekaran 2015a; Venkatesan et al. 2013; Venkatesan and Thangavelu 2015). The weighted decision tree prediction modeling approach, combined with the use of remote sensing and Geographical Information Systems (GIS) spatial data, yields a reasonable accuracy in the landslide risk analysis.

Analytical Hierarchy Process Frequency Ratio (FR-AHP) method, it performs better than the conventional Frequency Ratio (FR) and Analytical Hierarchy Process (AHP) methods. In addition, the individual weights of factors' giving to landslide occurrence ranked by the experts efficiently incorporate the factors and historical landslides in the study area (Zhou et al. 2016).

3   Hybrid Intelligent Bayesian Model (HIBM)

3.1   Rough Set Approach

Rough set theory was developed by the Pawlak (1982). It has created the awareness too many research scholars all over the countries, which contributed fundamentally to its development and applications.

It also found many attractive applications. It seems that the rough set approaches are involved in the fundamental importance of Artificial Intelligence, Soft Computing and Cognitive Sciences, especially in the areas of machine learning, knowledge mining, decision analysis and pattern recognition.

Definition 1: $RS = (A, X, V, f)$ are set to an information system. Among them $A = \{A_1, A_2, A_3, A_n\}$ is non empty finite sets which is called the domain space, $X = \{x_1, x_2,, x_n\}$ is non empty finite attribute set, which is called the attribute set, $V = \bigcup X_a$, $x \in X$, V_x is attribute's domain range, $f: A \times X \rightarrow V_x$ is the information function. When x is a, x has unique value in V_a. On the side, for sequence $C(c_1(x), c_2(x) \ldots c_n(x))$ and sequence $D(d_1(x), d_2(x) \ldots d_n(x))$, $B = C \cap D$, $C \cup D = \varnothing$, $S = (A, X, V, f)$ is called as decision table of the information system. $c_1(x), c_2(x) \ldots c_n(x)$ is called as the condition attribute set.

Definition 2: For the given knowledge representation system $S = (A, X, V, f)$, the in-discernable relationship of any attribute is as follows:

$$RIND(R) =: \{(x,y) \in AX \ A : \forall a \in B(f(x,a) = f(y,a)\} \qquad (1)$$

Definition 3: For the given knowledge representation system $S = (A, X, V, f)$, $P \subseteq X$, $V \subseteq$, $x \in A$ the lower and upper approximation set for A with regard to $RIND(B)$ is as below respectively:

$$\underline{R(B)} = \cup \{a \in A : RIND(X) \subseteq B\} \qquad (2)$$

$$\overline{R(B)} = \cup \left\{a \in A : RIND(X) \bigcap B = \emptyset\right\} \qquad (3)$$

Definition 4: For the given knowledge representation system $S = (A, X, V, f)$, if P, $Q \subseteq X$, the positive domain POSp(Z) is defined as:

$$POS_p(Z) = \cup \frac{R(A)}{x \subseteq U/P} \qquad (4)$$

Among them, R(A) is the lower approximation of A.

Let P, $Z \subseteq C \cup D$, the given the upper and lower approximations $\underline{P(B)}$ and $\overline{P(B)}$, the P-positive region of A can be defined as

$$POS_p(Z) = \cup \{\underline{PB} : B \in A/RIND(Z)\} \qquad (5)$$

The positive region $POS_P(Z)$ contains all the object sin A that can be classified into one class without an error defined by RIND(Z). The bound region can be defined as $\underline{PB} - \overline{PB}$, and the negative region as A-$\underline{PB}$. The dependency of Z on P is defined as

$$\Upsilon_p(Z) = (card(POS_P(Z)))/(card(A)) \qquad (6)$$

A measure of significance of the attribute $a \in P$ from the set P with respect to the classification A/RIND(Z) generated by a set Q is

$$\mu_{P,Z}(a) = (card(POS_P(Z)) - card(POS_{p-(a)}(Z))/(card(A)) \qquad (7)$$

A measure of the accuracy of an approximation of a set in the space P is defined as

$$\mu(B) = (\underline{R(B)})/(\overline{R(B)}) \qquad (8)$$

3.2 Naive Bayes Classifier

The most popular mathematical theorem called Bayes theorem which was defined by mathematician Naïve Bayesian. Bayesian classification can be depends on the Bayes theorem. Bayesian classifier assumes that the effect of an attribute value on a given class is independent of the values of the other attributes. This condition is called conditional independence. By using the Bayesian classifiers it can produces the high accuracy and speed when applied to large data bases. In naïve Bayesian classification will be like in this way. X is considered as evidence described of measurements and it can be made on a set of attributes. c_i be hypothesis and data tuple belongs to specified class C. The simple Bayesian classifier can be works as follows.

$$p(c_i/x) = p(x/c_i)p(c_i)/p(x) \tag{9}$$

There are i layers of spatial map data containing "causal" factors, which are known to associate with the occurrences of future landslides in the study area. The significant landslide factors from the study area are extracted from the map sources and represented in Table 1. Assume that we have Five classes of landslide susceptibility D1 = High, D2 = Low, D3 = medium, D4 = Very High, D5 = Very Low.

Let A be a training set of tuples and their connected with class labels. Each tuple is mentioned by an n dimensional attribute vector, $C = (c_1, c_2, \ldots ,c_i)$, depicting m measurements made on the tuple from j attributes, respectively $A_1, A_2, \ldots ,A_j$. Suppose that there are i classes $C_1, C_2, \ldots ,C_i$.

3.3   Weighted Naive Bayesian Approach in Landslide Prediction

Weighted Bayesian classifiers are statistical classifiers that predict class membership probabilities by training the past data. This algorithm obtain the probabilities that a given data instance belongs to particular class. This works according to Bayes' theorem.

The basic thing of Bayes's theorem is that the outcome of a hypothesis or an event (H) can be predicted based on some evidences (X) that can be observed.

A priori probability of H or P(H): This is the probability of an event before the evidence is observed.

A posterior probability of H or P(H/X): This is the probability of an event after the evidence is observed.

$$p(H/X) = (p(X/H)W_X p(H))/P(X)$$

Weighted Bayesian Classifier works according to following series of steps as mentioned below precisely.

Firstly D to be taken as training set of tuples or data instances and their associated class labels, tuple X = (x1, x2, ... xn), depicts n measurements made on the tuple from n attributes, respectively, A1, A2, ..., An.

If there are m classes, C1, C2, ..., Cm. For a given tuple X, the classifier will predict that X belongs to the class having the highest posterior probability, conditioned on X.

3.4   Proposed Architecture

The new approach is to classify and predict the possibilities of landslides by using a novel Hybrid Intelligent Bayesian Model (HIBM). The innovative data mining

Table 1. Training landslide data set

S. No.	Soil	Slope	Rainfall	Land use & Land cover	Geomorphology	Geology	Zone
1	CLAYEY	0–8.029984	109.397533–124.594088	Agriculture	Ridge type structural hills (large)	Charnockite group	Moderate
2	CLAYEY	0–8.029984	135.63122–150.827775	Scrub forest	Ridge type structural hills (large)	Charnockite group	Moderate
3	CLAYEY	0–8.029984	135.63122–150.827775	Scrub forest	Ridge type structural hills (large)	Charnockite group	Moderate
4	CLAYEY	0–8.029984	150.827775–171.75126	Forest plantations	Ridge type structural hills (large)	Charnockite group	High
5	CLAYEY	0–8.029984	150.827775–171.75126	Land without scrub	Ridge type structural hills (large)	Charnockite group	High
6	CLAYEY	0–8.029984	150.827775–171.75126	Scrub forest	Ridge type structural hills (large)	Charnockite group	High
7	CLAYEY	0–8.029984	150.827775–171.75126	Scrub forest	Ridge type Structural hills (large)	Charnockite group	Low
8	CLAYEY	0–8.029984	150.827775–171.75126	Scrub forest	Ridge type structural hills (large)	Charnockite group	Low
9	CLAYEY	0–8.029984	171.75126–200.559911	Forest plantations	Ridge type structural hills (large)	Charnockite group	High
10	CLAYEY	0–8.029984	59.665397–88.474048	Dense forest	Ridge type structural hills (large)	Gneiss	Low
11	CLAYEY	0–8.029984	59.665397–88.474048	Dense forest	Ridge type structural hills (large)	Gneiss	Very low
12	CLAYEY	0–8.029984	59.665397–88.474048	Land with scrub	Ridge type structural hills (large)	Gneiss	Low

(continued)

Table 1. (*continued*)

S. No.	Soil	Slope	Rainfall	Land use & Land cover	Geomorphology	Geology	Zone
13	CLAYEY	0–8.029984	59.665397–88.474048	Land with scrub	Ridge type structural hills (large)	Gneiss	Moderate
14	CLAYEY	0–8.029984	59.665397–88.474048	Scrub forest	Ridge type structural hills (large)	Gneiss	Low
15	CLAYEY	0–8.029984	59.665397–88.474048	Waterbodies	Water body mask	Gneiss	Very low
16	CLAYEY	0–8.029984	59.665397–88.474048	Waterbodies	Water body mask	Charnockite group	Very low
17	CLAYEY	0–8.029984	59.665397–88.474048	Waterbodies	Water body mask	Charnockite group	Very low
18	CLAYEY	11.7–19.798347	150.827775–171.75126	Residential	Ridge type structural hills (large)	Charnockite Group	Very high
19	CLAYEY	11.7–19.798347	150.827775–171.75126	Residential	Ridge type structural hills (large)	Charnockite group	Very high
20	CLAYEY	11.7–19.798347	171.75126–200.559911	Agriculture	Ridge type structural hills (large)	Charnockite group	Very high

classification approach is applied to predict the occurrence of the landslide using the detected landslide locations.

To predict the possibilities of landslides in hills areas by using various systematic models are used. For these predict analysis models are used so many environmental geological attributes. In that some of the attributes are fully involved to predict class labels but few attributes are partially involved. To overcome this disadvantage we developed a novel Hybrid Intelligent Bayesian Model (HIBM) to predict the incidence of the landslide using the detected landslide locations and the constructed spatial database. The predicted landslide analysis results are verified using the landslide location test data for each studied area. The results obtain are verified for accuracy and further fine tuned using hybrid soft computing techniques (Fig. 3).

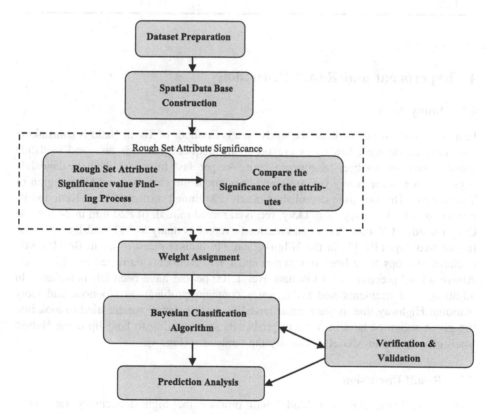

Fig. 3. Proposed frame work

3.5   Proposed HIBM Algorithm

Input Att: set of attributes, n: number of attributes
Output HIBM classifier
BEGIN
1. Calculate Significance of all attributes using eq.

$$\mu_{P_Z}(a) = \frac{card\left(POSp(Z)\right) - card\left(POS_{p_{-(a)}}(Z)\right)}{card(U)}$$

2. Compare the Significance values of each attribute aj,
3. Assign the weight Wi based on the Significance values of attributes aj,
4. For each j= i+1 To n DO

    4.1 calculate $P(Y|\mathbf{X}) = \dfrac{P(\mathbf{X}|Y)P(Y)}{P(\mathbf{X})}$ * Wj value for the all class labels.
    4.2 Compare the class labels value and predict the landslide occurrence.
5. NEXT j
END

---

4   Experiment and Result Discussion

4.1   Study Area

In this research, we considered the study area is Coonoor of Tamil Nadu is painstaking the analysis. Because landslides are occurring frequently in this area and landslide analysis has always been a concern here. As per fast historical data the landslide triggered due to the heavy rain occurred throughout the Coonoor and Ooty region of Tamilnadu. "The landslide demolished nearly 300 tinned roof mud huts. Ketti and its border, about 7 km away from Ooty, received record rainfall of 820 mm in 24 h while Ooty recorded 170 mm. As per another media report as many as 543 landslips occurred in just two days (10–11) in the Nilgiris, and 816 houses razed to trash. Besides, 600 hectares of crops have been devastated and road revetments damaged in 145 places. Above all, 43 precious lives lost and over 1,100 people have been left homeless". In addition, road accidents and traffic are a common problem on Coonoor and Ooty National Highway due to the regular landslips. In this paper, we are tried to look into the given region of land in order to predict its susceptibility to landslip using Hybrid Intelligent Bayesian Model to classify the landslides (Fig. 4).

4.2   Result Discussion

Hybrid Intelligent Bayesian Model will produce the highest accuracy rate when compared to other data mining classification algorithms. This classifier will accept different types of continuous or categorical variable values. The Hybrid Intelligent Bayesian Model classifier technique is also suitable for high dimension data. The Hybrid Intelligent Bayesian Model educates to improve the probability in predicting accurate landslide prediction results.

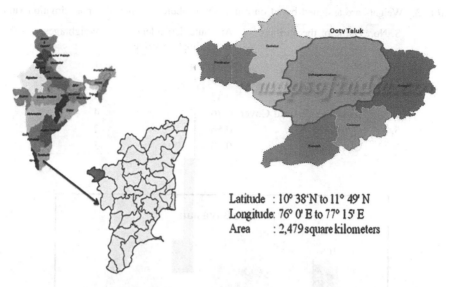

Latitude : 10° 38'N to 11° 49' N
Longitude: 76° 0' E to 77° 15' E
Area : 2,479 square kilometers

Fig. 4. Study Area Coonoor Dist of Tamil Nadu

There are six landslide factors considered in this analysis such as Soil, Slope, Rainfall, Land Use & Land Cover, Geomorphology, & Geology. The training dataset contains 445 record samples; each has 6 conditional attributes and 1 decision attribute. We applied the Soft computing Techniques called rough set method to find the dependency of the landslide attributes. To find the attribute significance using rough set method we have developed java code. The attributes are to be classified based on the attribute significant value which we obtained through the computational program as shown in the Table 2.

Table 2. Rough set attribute significance approximation values

S. No.	Name of the attribute	Attribute dependency approximation value
1.	Soil	0.51
2.	Slope	0.99
3.	Rainfall	0.87
4.	Land Use & Land Cover	0.76
5.	Geomorphology	0.56
6.	Geology	0.68

Depending upon the attribute dependency value we assigned weights to the different landslide attributes as show in the Table 3.

According to the Table 3, the land sliding attribute values and the corresponding weight are supplied to the weighted Bayesian algorithm. The accuracy of Hybrid Intelligent Bayesian Model is 72.03% but when we applied the original data set to the Naïve Bayesian without weight the accuracy of the result is 66.52%. The results have been shown in the graphs as follows (Figs. 5, 6 and 7).

Table 3. Weights are assigned based on rough set attribute significance approximation values

S. No.	Name of the attribute	Attribute dependency approximation value	Weightages
1.	Soil	0.51	1
2.	Slope	0.99	6
3.	Rainfall	0.87	5
4.	Land Use & Land Cover	0.76	4
5.	Geomorphology	0.56	2
6.	Geology	0.68	3

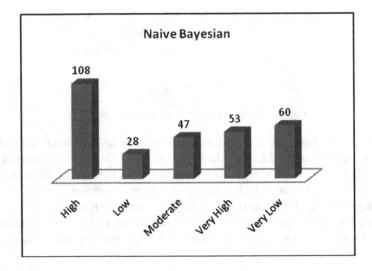

Fig. 5. Naïve Bayesian without Weight Assignment

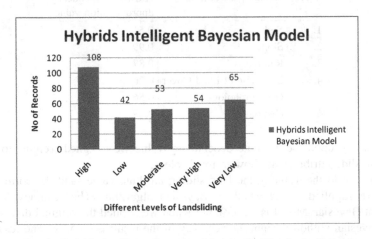

Fig. 6. Hybrids Intelligent Bayesian Model with Weight Assignment

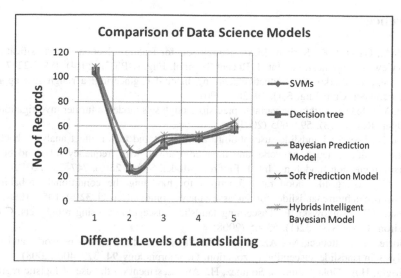

Fig. 7. Analysis of accuracy

Hybrids Intelligent Bayesian Model compared with other data mining classification algorithms like SVM, DTs. The accuracy of the Hybrids Intelligent Bayesian Model is more when compared to the other models as shown in Table 4.

Table 4. Accuracy comparisons with other data mining techniques

S. no.	Classifier	Accuracy %
1	SVMs	62.23
2	Decision tree	64.02
3	Bayesian classification	66.21
4	Soft Bayesian classification	71.23
5	Hybrids Intelligent Bayesian Model	72.03

5 Conclusion

In this research, Rough Set & Bayesian classification approaches are used to build novel Hybrid Intelligent Bayesian Model (HIBM) to classify the landslides with the help of spatial knowledge database and geographic information system. And also compared the accuracy of the Hybrid Intelligent Bayesian Model approach with the Naïve Bayesian Classification. Hybrid Intelligent Bayesian Model approach is more suitable and accurate than Naïve Bayesian model. The advantage of the proposed Model (HIBM) is that it handles the ambiguity and vagueness present in the large amount of data. In future work, more environmental factors will be considered for landslide prediction and also covering based rough set techniques can be incorporated to improve the performance of proposed model.

References

Saxena, A., Gavel, L.K., Shrivas, M.M.: Rough sets for feature selection and classification: an overview with applications. Int. J. Recent Technol. Eng. (IJRTE) (2014). ISSN 2277-3878

Arciszewski, T., Ziarko, W.: Inductive learning in civil engineering: a rough sets approach. Microcomput. Civil Eng. 5(1), 19–28 (1990)

Beynon, M.: Reducts within the variable precision rough sets model: a further investigation. Eur. J. Oper. Res. 134(3), 592–605 (2001)

Pradhan, B., Lee, S.: Landslide susceptibility assessment and factor effect analysis: backpropagation artificial neural networks and their comparison with frequency ratio and bivariate logistic regression modelling. Int. J. Environ. Model. Softw. 25(6), 747–759 (2010)

Chung, C.J.: Using likelihood ratio functions for modeling the conditional probability of occurrence of future landslides for risk assessment. Comput. Geosci. 32(8), 1052–1068 (2006)

Gorsevski, P.V., Jankowski, P.: Discerning landslide susceptibility using rough sets. Comput. Environ. Urban Syst. 32(1), 53–65 (2008)

Melchiorre, C., Matteucci, M., Azzoni, A., Zanchi, A.: Artificial neural networks and cluster analysis in landslide susceptibility zonation. Geomorphology 94, 379–400 (2008)

Nefeslioglu, H.A., Gokceoglu, C., Sonmez, H.: An assessment on the use of logistic regression and artificial neural networks with different sampling strategies for the preparation of landslide susceptibility maps. Eng. Geol. 97(3–4), 171–191 (2008)

Anbalagan, P., Chandrasekaran, R.M.: A novel weighted decision tree pre diction model for landslide risk analysis. Adv. Nat. Appl. Sci. 9(8), 22–28 (2015a)

Pawlak, Z.: Rough sets. Int. J. Comput. Inf. Sci. 11, 341–356 (1982)

Pawlak, Z.: Rough Sets: Theoretical Aspects of Reasoning about Data. Kluwer Academic Publishers, Dordrecht (1991)

Pawlak, Z., Slowinski, R.: Rough set approach to multi-attribute decision analysis. Eur. J. Oper. Res. 72, 443–459 (1994)

Anbalagan, P., Chandrasekaran, R.M.: A novel weighted decision tree prediction model for landslide risk analysis. In: Advances in Natural and Applied Sciences, 14 July 2015, pp. 22–28 (2015b)

Pawlak, Z.: Rough set approach to knowledge-based decision support. Eur. J. Oper. Res. 99(1), 48–57 (1997)

Rouse, J.W., Haas, R.H., Deering, D.W., Schell, J.A.: Monitoring the Vernal Advancement and Retrogradation (Green Wave Effect) of Natural Vegetation. Prog. Rep. RSC 1978-1. Remote Sensing Center, Texas A&M University, College Station (1973)

Saito, H., Nakayama, D., Matsuyama, H.: Comparison of landslide Susceptibility based on a decision-tree model and actual landslide occurrence: the Akaishi Mountains, Japan. Geomorphology 109, 108–121 (2009)

Zhou, S., Chen, G., Fang, L., Nie, Y.: GIS-based integration of subjective and objective weighting methods for regional landslides susceptibility mapping. Sustainability 8(4), 1–15 (2016)

Slowinski, R., Greco, S., Matarazzo, B.: Rough sets in decision making. In: Meyers, R.A. (ed.) Encyclopedia of Complexity and Systems Science, pp. 7753–7786. Springer, New York (2009)

Swiniarski, R.W., Skowron, A.: Rough set methods in feature selection and recognition. Pattern Recogn. Lett. 24(6), 833–849 (2003)

Venkatesan, M., Thangavelu, A.: A Delaunay diagram-based Min-Max CP-Tree algorithm for spatial data analysis. WIREs Data Min. Knowl. Discov. 50(3), 142–154 (2015)

Venkatesan, M., Thangavelu, A., Prabhavathy, P.: An improved Bayesian classification data mining method for early warning landslide susceptibility model using GIS. In: Bansal, J., Singh, P., Deep, K., Pant, M., Nagar, A. (eds.) Proceedings of Seventh International Conference on Bio-Inspired Computing: Theories and Applications (BIC-TA 2012). Advances in Intelligent Systems and Computing, vol. 202. Springer, India (2013)

Wan, S.: A spatial decision support system for extracting the core factors and thresholds for landslide susceptibility map. Eng. Geol. 108(3–4), 237–251 (2009)

Wan, S., Lei, T.C., Chou, T.Y.: A novel data mining technique of analysis and classification for landslide problems. Nat. Hazards 52(1), 211–230 (2009)

Wang, F.W., Zhang, Y.M., Hu, Z.J., Matsumoto, T., Huang, B.L.: The July 14, 2003 Qianjiangping landslide, Three Gorges Reservoir, China. Landslides 1, 157–192 (2004)

Wu, C.H., Chen, S.C.: Determining landslide susceptibility in Central Taiwan from rainfall and six site factors using the analytical hierarchy process method. Geomorphology 112(3–4), 190–204 (2009)

Yilmaz, I.: Landslide susceptibility mapping using frequency ratio, logistic regression, artificial neural networks and their comparison: a case study from Kat landslides (Tokat–Turkey). Comput. Geosci. 35, 1125–1138 (2009)

Yao, Y., Zhou, B.: Two Bayesian approaches to rough sets. Eur. J. Oper. Res. 251(3), 904–917 (2015)

Zeng, Z.P., Wang, H.B., Zhang, Z., Xue, C.S.: GIS/RS-based landslide susceptibility assessment in the Qingganhe River of Three Gorges Area. Chin. J. Rock Mech. Eng. 25(Suppl), 2777–2784 (2006)

Zhang, J., Jiao, J.J., Yang, J.: In site rainfall infiltration studies at a hillside in Hubei Province, China. Eng. Geol. 57(1–2), 31–38 (2000)

Design and Performance Characterization of Practically Realizable Graph-Based Security Aware Algorithms for Hierarchical and Non-hierarchical Cloud Architectures

Rahul Vishwanath Kale[1(✉)], Bharadwaj Veeravalli[1], and Xiaoli Wang[2]

[1] Department of Electrical and Computer Engineering,
National University of Singapore,
4 Engineering Drive 3, Singapore 117576, Singapore
{elekrv,elebv}@nus.edu.sg
[2] School of Computer Science and Technology, Xidian University,
Xi'an 710071, Shaanxi, China
wangxiaoli@mail.xidian.edu.cn

Abstract. Applications processing massive amount of data demand superior time-performance and data-storage capabilities. Many organizations are exploring Cloud Computing to manage such applications because of scalability and convenience of access from different geographic locations, whereas data security and privacy are few of the major concerns preventing them from fully embracing it. Data security while maintaining time-performance becomes an important consideration for designing data placement strategies. We first present the reader with a quick survey on the recent approaches and solutions in data placement oriented problems that address security concern and then characterize the performance for graph based algorithms that are practically realizable. We evaluated the performance of conventional strategies such as, Random, T-coloring and revisited these algorithms in the view of providing maximum security. We refer to our strategy as Data Security Preferential (DSP) data placement strategy and evaluated via rigorous performance evaluation tests to identify which strategy will best suit the requirements of the user on three state-of-the-art hierarchical cloud platforms viz., FatTree, ThreeTier and DCell and non-hierarchical cloud platforms.

Keywords: Data placement strategy · Data center networks
Cloud Computing · Security · Retrieval time

1 Introduction

With the rapid growth in technology, immense amount of data is getting generated through smartphones, wearable sensors, Internet of Things, Smart City

© Springer Nature Singapore Pte Ltd. 2018
J. C. Hung et al. (Eds.): FC 2017, LNEE 464, pp. 392–402, 2018.
https://doi.org/10.1007/978-981-10-7398-4_40

based applications every day. Many organizations are exploring Cloud Computing to manage these applications because of the scalability and convenience of access from different geographic locations. Since the cloud systems can provide resources on demand, the applications can be made scalable and can potentially be used for increasing demand without major tweaks. However, data storage on remote servers is not without its disadvantages, data retrieval time and data privacy being the two most concerning ones.

For cloud based storage systems, the data stored should be readily accessible, otherwise it may impact the overall performance adversely. For large-scale cloud systems, however, providing an efficient retrieval time is a huge concern as storage nodes are distributed geographically and connected through links with varying bandwidths. By increasing the number of copies, faster response time can be achieved as more storage nodes are available for data retrieval service, however, even the attack probabilities are also increased as more potential target sites are available for attack. Suppose a cloud platform has N storage nodes, data file to be stored is split in n chunks and each chunk is replicated m times, then the probability of one chunk of data getting compromised is $m \times n/N$, compared to n/N sans replication.

Different types of cyberattacks, such as Trojans, Denial of Service (DoS), Distributed Denial of Service (DDoS), Packet Forging Attacks, Application Layer Attacks, Fingerprinting Attacks etc. [1], are major causes of cloud security concerns. In [2,3], different encryption techniques and protocols were proposed for cloud platforms. Since encryption involves performance overhead in terms of encrypting and decrypting data for each file access, time performance becomes an important consideration.

An example of Intelligent Transportation Systems (ITS) can be used to explain scenarios where both retrieval time and security play a crucial part in decision-making simultaneously with Cloud Computing being the preferred platform to implement ITS. ITS can be described as a set of advanced applications providing traffic management using intelligent and innovative technologies in information and communication [4]. In ITS, a constant information exchange between the vehicles and the cloud is used to improve road safety, travel productivity and reliability; therefore to avoid any potential accidents, the decisions need to be taken promptly, for which quick data retrieval holds significant importance. At the same time, ITS needs to be protected from attackers aiming to compromise data confidentiality, integrity or availability [5] and to gain unauthorized control of the systems and data for malicious purposes. From this example, we can conclude that both security and time performance are crucial for cloud based systems such as ITS.

The major contributions and organization of this paper are summarized as follows. In Sect. 2, we present the reader with a quick survey on recent approaches and solutions in data placement oriented problems that address security concern. In Sect. 3, we formulate the problem to be addressed and introduce retrieval time and security factor metric In Sect. 4, we design and propose a new strategy for data placement – DSP Strategy to maximize security by maximizing the geodesic distance between any two data chunks to minimize the amount of compromised data. In Sect. 5, we compare the performance of the proposed strategy with conventional placement strategies such as Random and T-Coloring and also evaluate the impact of the number of backups on data security through extensive simulations on different non-hierarchical cloud platforms and three popular Data Center Network (DCN) hierarchical topologies viz. ThreeTier [6], FatTree [7] and DCell [7]. We conclude the paper in Sect. 6.

2   Survey on Data Placement Models and Solutions

Optimal strategies for data placement on cloud platforms aiming at high performance and security have been extensively investigated in the literature. Below we report only the most recent papers that have relevance to the context of the work presented in this paper. In [2], cloud services are assigned the responsibility of handling user data, fine-grained access control, and encryption/decryption files to guarantee data security. However, these security techniques based on encryption affect the overall performance by introducing computational overheads in encryption and decryption operations. Moreover, an additional effort is required from users to take responsibility for the safety of encryption keys. Authors in [3] proposed an authorization architecture for object-based file systems. It is a combination of customer namespace isolation and native access control. This architecture is based on the idea that in the multi-tenant virtualization environment if storage consolidation is performed at the file system level, it will have benefits such as data sharing, optimized performance, and administration efficiency. Individual tenants and services are uniquely identified through a set of cryptographic keys. However, improper information scrubbing before the reallocation may result in data privacy breach. Authors in [8] designed and implemented a secure overlay cloud storage system that achieves secure access control and data assured deletion. It associated outsourced data with data access policies and assuredly deleted data to make them unrecoverable to anyone upon revocations of data access policies. To achieve such security goals, the authors leveraged existing cryptographic techniques by maintaining a set of cryptographic key operations by a quorum of key managers that are independent of third-party clouds. On the similar lines, in [9], authors suggested managing the authentication, integrity and confidentiality of data, and communications between concerned entities by cryptography based Public Key Infrastructure, and suggested using a Trusted Third Party to take the responsibility of maintaining necessary security requirements. However, in both [8,9], the security level of the data is limited by safety of the encryption keys. If encryption keys have been tampered with, then the security level can not be guaranteed. For data security in scientific-cloud workflows, security model to quantitatively measure the security services provided by data center was introduced in [10]. Data security is compromised when sensitive user data is uploaded on the untrusted cloud servers and it is further complicated when constraints are imposed regarding the nature of data for scientific workflows that can be uploaded to the cloud.

In this work, an Ant Colony Optimization (ACO) based algorithm was then used to select ideal data center for intermediate data storage such that data security can be improved while maintaining data transfer time for scientific workflows. The data placement problem on cloud storage systems was formulated as linear programming model in [11] with the objective of minimizing the retrieval time with the security constraint, for data divided and distributed over storage nodes. Then Security-aware data placement heuristic algorithm, SEDuLOUS, based on graph coloring and greedy algorithm for data chunk assignment was used to achieve secure data placement. In [12], data placement with focus on data Security along with access performance was explored for P2P Data Grids. Data fragmentation and data replication was combined to achieve the desired data confidentiality, load balancing as well as efficient access. The problem was modeled as multi-objective Pareto-optimal problem and genetic algorithm was developed to obtain better solutions than conventional randomized algorithms. SecHDFS, a secure data allocation scheme without compromising performance for heterogeneous Hadoop systems was proposed in [13]. To achieve security, the DataNodes are classified according to their security vulnerabilities. Maximum different groups of DataNodes are utilized and secret sharing scheme is also implemented within the allocation mechanism. For secure and efficient data placement in Mobile Healthcare Services based on cloud computing, a mechanism based on fragmentation and caching was used in [14]. Fragmentation was used to split the data based on confidentiality and affinity constraints, and hence data security was achieved by enforcing confidentiality constraints to control data sharing.

3 Data Placement Optimization Model

3.1 Problem Description

We consider a cloud storage system consisting of N storage nodes, denoted as $V = \{v_1, v_2, \cdots, v_N\}$. The connections among N nodes are represented by symmetric matrix $E = (e_{ij})_{N \times N}$, where $e_{ij} = e_{ji} = 1$ indicates there exists a physical link connecting nodes v_i and v_j, while the case of $e_{ij} = e_{ji} = 0$ means nodes v_i and v_j are not connected directly. The topology of non-hierarchical/hierarchical cloud system can be denoted as a graph $G(V, E)$, where V and E are the sets of vertices (nodes) and edges (connection links), respectively.

The volume of data D to be stored on the system is partitioned into n independent chunks and will be stored on the cloud system. Let α_i with $i = 1, 2, \cdots, n$ denotes the size of i-th chunk. Each chunk contains partial sensitive information of the entire data D and each chunk will be stored on a separate node. Hence, even if a node in the system is subjected to cyberattacks, a compromised chunk of data stored on this node will be insufficient to extract the entire information. To prevent any loss of chunks, we attempt to store m number of backup copies for each chunk in the storage system.

Based on actual large-scale cloud platforms, the access points/nodes and storage nodes in the system are part of different levels of architecture. For a data placement solution H, we will first formulate the objective functions of retrieval time $T(H)$ and security factor $S(H)$.

3.2   Retrieval Time

For data retrieval, it may be noted that in a real-life scenario when a read request comes from any one of the access nodes, all the chunks of the data need to be retrieved for entire file retrieval. We incorporate this in our retrieval process. Therefore, for optimal data retrieval, the data chunks have to be placed as close as possible to the access nodes. Further, as with any real-time application that attempts to retrieve contents from a network based system, here too, we will retrieve all the relevant chunks from the respective nodes in parallel and hence, retrieval time for the entire file will be equal to the maximum of the retrieval time for the individual chunks. We assume that the transmission time between access node and storage node, the access path, equals to the sum of time required to traverse all of the links within the shortest path between the two, where shortest path also takes into account the varying link bandwidths that exist between different layers of the architecture. We can derive the retrieval time T_i for data request from access node v_i as,

$$T_i(H) = \max(\alpha_1 \times z_{i1}, \alpha_2 \times z_{i2} \cdots, \alpha_n \times z_{in}). \tag{1}$$

Where α_N is a data chunk and z_{iN} is the transmission time between access node and that chunk. The average retrieval time for the given placement with N access nodes can be represented as $T(H) = \sum_{i=1}^{N} T_i(H)/N$.

3.3   Security Factor

From CSP's perspective, it is critical to identify most vulnerable nodes and avoid allocating data on them to prevent a huge loss of information. Centrality analysis provides measures for defining the importance of nodes in a system. Different centrality measures, such as degree centrality, closeness centrality, betweenness centrality, and eigenvector centrality, capture the importance of nodes from different perspectives. As for betweenness centrality, the importance of a node is determined by the number of shortest paths in the network that will pass this node and more the number of shortest paths passing through the node, higher is the betweenness centrality. Nodes with higher betweenness centrality may favor faster retrieval times, however, such nodes will also be susceptible to larger data loss if attacked, thus, avoiding data placement on them seems to be the best choice for sensitive data. For the above reason, we elaborate only on the betweenness centrality in this paper.

Besides the guidance from centrality, we make further effort to improve data security. We identify a pair of nodes to store chunks in a manner that they are at the largest geodesic distance from each other. This ensures that even in a case of successful attacking to a particular chunk at a node, attackers have no idea of the location of other chunks, thereby further minimizing the probability of compromising data. By "geodesic distance" we mean the number of hops within the shortest path between two nodes. Denote d_{ij} as the geodesic distance between nodes v_i and v_j.

From the above analysis, we conclude that the probability of data being compromised at a node is largely influenced by two factors – first is the probability of a single node that stores data being attacked; the other is the geodesic distance between any two nodes that both storing data chunks. Based on this fact, we design a security measurement that quantifies the security level of a data placement solution H, denoted as $S(H)$, as follows.

$$S(H) = Ad(H) \times \frac{1}{Ac(H)}, \qquad (2)$$

where $Ad(H)$ indicates the average geodesic distance between all pairs of nodes storing data chunks. Larger the geodesic distance between two nodes that store sensitive chunks of data, less amount of data would be compromised. $Ac(H)$ represents the average centrality of nodes storing data chunks. Contrary to geodesic distance, larger the centralities of nodes selected to store data chunks, higher the data loss if attacked, thereby more attack threat for the data. $Q = \{q_i \mid i = 1, 2, \cdots, n \times m\}$ be the set of nodes that are selected to store data chunks according to H, where n and m are the numbers of chunks and backup copies, respectively. Thus $n \times m$ indicates the total number of nodes required for data storage. Let $q_k = h_{ij}$, where $k = (i - 1) \times m + j$ with $i = 1, 2, \cdots, n$ and $j = 1, 2, \cdots, m$. The average geodesic distance of H can be computed as,

$$Ad(H) = \frac{\sum_{i=1}^{n \times m} \sum_{j=1}^{n \times m} d_{q_i, q_j}}{(n \times m)(n \times m - 1)/2}, \qquad (3)$$

The average centrality $Ac(H)$ is given by,

$$Ac(H) = \frac{\sum_{i=1}^{n} \sum_{j=1}^{m} c_{h_{ij}}}{n \times m} = \frac{\sum_{k=1}^{n \times m} c_{q_k}}{n \times m}. \qquad (4)$$

For DCN topologies considered in this paper, all the storage nodes have equal betweenness centralities and hence Eq. 2 can be simplified to $S(H) = Ad(H)$. This equation will be used for calculation of $S(H)$ for hierarchical cloud networks.

4 Design of Data Security Preferential (DSP) Strategy

We design the strategy to maximize the geodesic distance between the location of data chunks to ensure that even in a case of successful attack on a single chunk at a node, attackers are not aware of the location of other chunks, thereby minimizing the probability of compromising larger data. The pseudo code for the strategy is presented in Algorithm 1. This strategy can be used to achieve desired or maximum security factor. The initial part of the DSP Algorithm is designed to avoid placement of chunks on highly central nodes that is to minimize the average centrality of the placement nodes. Second half of the algorithm is designed to maximize the separation among the placement nodes. Both these parts together contribute to achieve overall more secure placement. As established earlier, corresponding expression for security factor will be used in case of hierarchical and non-hierarchical cloud networks. In the next section, we will evaluate the performance of this DSP algorithm on both hierarchical and non-hierarchical cloud networks.

Algorithm 1. Data Security Preferential (DSP) Strategy

Require: An initial random placement $H = \{h_{ij}\}_{n \times m}$.
Ensure: An improved placement $H' = \{h'_{ij}\}_{n \times m}$.
1: Let $H' = H$. Compute the security level $S(H')$.
   //**Decreasing the average centrality.**
2: Select 10% of nodes from $\{h_{ij}\}_{n \times m}$ with the largest betweenness centralities.
3: **for** each selected h_{ij} **do**
4:     Randomly generate an integer k satisfying that $k \notin H$ and $c_k < c_{h_{ij}}$. Let $h_{ij} = k$ and compute the security level $S(H)$.
5:     If $S(H) > S(H')$, let $H' = H$; otherwise let $H = H'$.
6: **end for**
   //**Increasing the average geodesic distance.**
7: **for** $i = 1, \cdots, n$ and $j = 1, \cdots, m$ **do**
8:     Compute the geodesic distances between h_{ij} with all of the rest elements in H to find $h_{pq} \in H$ that has the minimum geodesic distance with h_{ij}.
9:     Let $s = h_{ij}$. Sort $\{d_{sk} \mid k = 1, 2, \cdots, N\}$ in an decreasing order and denote the sorted set as DE.
10:    **while** $DE \neq \emptyset$ **do**
11:        Take an element d_{sk} from DE in order without replacement.
12:        If $k \notin H$, let $h_{pq} = k$ and compute $S(H)$.
13:        If $S(H) > S(H')$, let $H' = H$ and go to **Step 14**; otherwise let $H = H'$.
14:    **end while**
15: **end for**

---

5   Performance Evaluation and Analysis

We conducted extensive performance evaluation studies on three popular DCN topologies and non-hierarchical cloud platforms. The sizes of data chunks were randomly generated from [100, 300] units, where we have normalized the data size values (original data size expressed in GBs). The transmission bandwidth between a pair of nodes for a unit size of data was set according to the network specification in [6]. It may be noted that as in a real-life scenario, the retrieval time is dependent on the largest sized chunk because link bandwidths remain constant within a layer. Thus, the retrieval time is the characteristic of the underlying network architecture. Hence, in our experiments, security factor is the deciding metric as it depends upon the design of the placement strategy.

5.1 Effectiveness of the Proposed Strategies

We evaluate the performance of the proposed DSP strategy for increasing network size on each of the three topologies and non-hierarchical cloud platforms. The number of file fragments to be stored is generated as per the request. For FatTree (Hierarchical) Network, it can be seen from Fig. 1 that as the number of storage nodes increases, the security factor tends to increase since the number of potential placement positions with maximum possible separation also increases. As DSP is designed to maximize the separation among the data chunks, the security factor of the placements suggested by DSP shows a continuous increase for increasing number of storage nodes. However, the behavior of t-coloring algorithm becomes unpredictable with increasing number of nodes as it tends to behave like a random placement. This is because, to place same number of chunks, more potential nodes are available for placement and hence the trend becomes unpredictable. In case of hierarchical three-tier DCN, it can be observed from Fig. 2 that security factor for t-coloring $(2 < t < 5)$ and DSP strategy remains identical for different network sizes. In case of ThreeTier architecture, the separation between two storage nodes can either be 2 or 4. Therefore, for $2 < t < 5$, t can only take value of 4 which makes the behavior t-coloring and DSP identical as value of t coincides with the maximum separation. For t = 2, t-coloring will reduce to random placement as minimum separation between any two storage nodes is 2 and hence all the storage nodes can be colored with a single color. For $t >= 5$, each of the storage node will require different color since t is greater than maximum possible separation among storage nodes. For DCell architecture, as seen in Fig. 3, the trend is not as smooth as previous DCN trends for t-coloring. However, for DSP, with increase in number of nodes, the security factor increases almost uniformly. The success of t-coloring algorithm is sensitive to the t-value and availability of correspondingly separated nodes in the given architecture. However, DSP algorithm is designed to perform independent of network properties or limitations. This fact is further underlined in Fig. 4. In case of non-hierarchical cloud platforms, the DSP strategy performs exceptionally well compared to other strategies as shown in Fig. 4. In case of DCN topologies, the architecture is well defined and hence increase in storage nodes usually results in more potential storage sites. However, in case of non-hierarchical networks, increase in number of nodes does not necessarily mean an increase in the potential storage sites.

5.2 Impact of Replication on Data Security

Even though data replication is a useful way to increase and maintain data availability, however, as shown in Fig. 5 for non-hierarchical platforms, data replication reduces the data security. This is because, with increase in the number of data chunks, the separation among the chunks may decrease, resulting in a decrease in the security factor. Indirectly, increasing the number of copies of the data will also make more nodes vulnerable to attack, thereby reducing data security. Finally, it may be noted that a similar trend is expected even for the case of hierarchical networks.

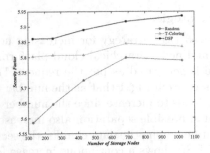

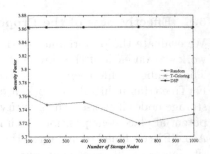

Fig. 1. Security factor vs Network size: FatTree (Hierarchical)

Fig. 2. Security factor vs Network size: ThreeTier (Hierarchical)

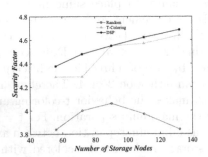

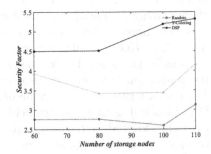

Fig. 3. Security factor vs Network size: DCell (Hierarchical)

Fig. 4. Security factor vs Network size: non-hierarchical cloud platform

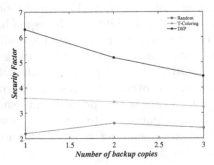

Fig. 5. Impact of replication on security: non-hierarchical cloud platform

6   Conclusions

Data security and time performance are among the top concerns for potential cloud users. We designed the Data Security Preferential (DSP) Strategy with objective of obtaining the most secure data placement by maximizing the separation among the data chunks. We compared the performance of our placement strategy DSP against conventional strategies such as t-coloring and random placement. We evaluated the behavior of our strategy on three popular DCN architectures and non-hierarchical cloud architectures. It has conclusively shown that our strategy outperformed the other two strategies, except in the case where the maximum node separation property of the cloud architecture is used. One of the strengths of this paper is in attempting to incorporate real-life models and scenarios for data placement, retrieval and in providing required security. Also, it may be noted that for practical Cloud architectures considered in this paper, all the storage nodes have equal betweenness centralities, however, for non-hierarchical Cloud platforms one needs to compute and bias the storage decisions using centrality measures. There were number of studies in this direction that exploit centrality measures however, none of them consider in simultaneous optimization of time and security metrics. This would be an interesting direct extension to the study presented in this paper.

Acknowledgment. The NUS authors would like to thank the funding support by MOE Tier-1 grant No. R-263-000-C14-112, entitled "Security-Aware Data Protection and Availability Maximization for Cloud Platforms", in carrying out this project. The third author would like to thank the funding support by National Natural Science Foundation of China (No. 61402350, No. 61472297, and No. 61572391) and China Scholarship Council.

References

1. Hoque, N., et al.: Network attacks: taxonomy, tools and systems. J. Netw. Comput. Appl. **40**, 307–324 (2014)
2. Li, Y., Dai, W., Ming, Z., Qiu, M.: Privacy protection for preventing data over-collection in smart city. IEEE Trans. Comput. **65**(5), 1339–1350 (2016)
3. Kappes, G., Hatzieleftheriou, A., Anastasiadis, S.V.: Dike: virtualization-aware access control for multitenant filesystems. University of Ioannina, Greece, Technical report No. DCS2013-1 (2013)
4. Bitam, S., Mellouk, A.: ITS-cloud: cloud computing for Intelligent transportation system. In: 2012 IEEE GLOBECOM, pp. 2054–2059 (2012)
5. Yan, G., Wen, D., Olariu, S., Weigle, M.C.: Security challenges in vehicular cloud computing. IEEE Trans. Intell. Transp. Syst. **14**(1), 284–294 (2012)
6. Cisco Data Center Infrastructure 2.5 Design Guide (2014)
7. Liu, Y., Muppala, J., Veeraraghavan, M., Lin, D., Hamdi, M.: Data Center Networks. Springer, Berlin (2013)
8. Tang, Y., Lee, P., Lui, J., Perlman, R.: Secure overlay cloud storage with access control and assured deletion. IEEE Trans. Dependable Secure Comput. **9**(6), 903–916 (2012)
9. Zissis, D., Lekkas, D.: New approaches to security and availability for cloud data. Future Gener. Comput. Syst. **28**(3), 583–592 (2012)
10. Liu, W., et al.: Security-aware intermediate data placement strategy in scientific cloud workflows. Knowl. Inf. Syst. **41**(2), 423–447 (2014)
11. Kang, S., Veeravalli, B., Aung, K.M.M.: A security-aware data placement mechanism for big data cloud storage systems. In: 2016 IEEE 2nd International Conference on Big Data Security on Cloud (BigDataSecurity), IEEE HPSC, and IEEE IDS, pp. 327–332 (2016)

12. Tu, M., et al.: Data placement in P2P data grids considering the availability, security, access performance and load balancing. J. Grid Comput. **11**(1), 103–127 (2013)
13. Tian, B., et al.: SecHDFS: a secure data allocation scheme for heterogenous Hadoop systems. In: 2016 IEEE International Conference on Networking, Architecture and Storage (NAS), pp. 1–2 (2016)
14. Kayem, A.V.D.M., Elgazzar, K., Martin, P.: Secure and efficient data placement in mobile healthcare services. In: Proceedings of DEXA 2014, Munich, Germany, 1–4 September 2014, Part I, pp. 352–361 (2014)

Modeling and Interpreting User Navigation Patterns in MOOCs

Xiangyu Zhang and Huiping Lin[⊠]

School of Software and Microelectronics, Peking University,
No. 5 Yiheyuan Road, Haidian District, Beijing, China
pkusszxy@pku.edu.cn, linhp@ss.pku.edu.cn

Abstract. Over the past few years, MOOCs have trigged an education revolution. Clickstream data of user were recorded by MOOCs platform, providing valuable insights about the way user interact with the MOOCs. In this paper, we study user navigation patterns in MOOCs. We propose a metric to measure the similarity between user session-level navigation path, and build an unsupervised clustering model to capture user navigation patterns in MOOCs. Based on the user behavior clustering result, we further explore engagement of each navigation pattern from the perspective of dropout. To measure the effectiveness of our model, we conduct experiment on real world dataset with five weeks of interaction logs of 3,914 users. Through our analysis, clustering model proposed effectively identifies 13 types of user navigation patterns, which help us understand user behavior in MOOCs.

Keywords: Clickstream · User behavior analysis · Learning analysis
MOOC

1 Introduction

MOOCs have experienced a rapid growth in recent years by leveraging the open and highly connective nature of the Internet. MOOCs platforms, such as Coursera, edX and Udacity, provide people around the world the opportunities to receive high quality education at low cost. Without any doubt, MOOCs are leading an educational revolution by gathering global education resources and reshaping the learning environment. Although thousands of people enrolled in MOOCs, the dropout rate is extremely higher than we can expect. According to some estimates, the dropout rates of MOOCs may reach as high as 90% on average [1], fewer than 7% of users actually completed the courses [2]. Therefore, there is an emergent need to understand user behavior in MOOCs.

To survey users on how they participate and interact with MOOCs is one intuitive solution, but this approach is limited by two factors. First, survey is limited in scale because the cost is always huge. Second, survey heavily depend on hypotheses, new style of user behavior patterns always cannot be found in survey.

This problem can be addressed in a more intelligent way using clickstream analysis, i.e., build user behavior models using clickstream data. Clickstream data is generated by users during their web browsing "sessions" or interactions with mobile apps [3, 4], provides information about the sequence of pages viewed by users when they navigate

© Springer Nature Singapore Pte Ltd. 2018
J. C. Hung et al. (Eds.): FC 2017, LNEE 464, pp. 403–413, 2018.
https://doi.org/10.1007/978-981-10-7398-4_41

a website. Almost all records of user online behavior are collected in MOOCs, providing us an unprecedented chance to take a closer look at user behavior. Comparing with survey, clickstream analysis can not only scale to thousands of users easily, but also help identify previous unknown behavior patterns.

However, identifying user behavior patterns using clickstream data is a challenging task. Most previous studies [5, 6] only focused on navigation path of user within a website, ignored the time spent in each webpage. In fact, time spent in a particular page is a strong indication of the user's intent. Studies [7, 8] on clustering user sessions considered both webpage visited and the time spent in corresponding webpage, but the similarity metrics of session are not easy to scale. As we know, describing an appropriate abstraction of user's intent is difficult due to discrete and complex web events, we need a more sophisticated method to better understand user behavior, it must satisfy two requirements. First, it should be able to be scale and work well on large and noisy clickstream dataset. Second, it can help others understand user behavior in an easy way.

In this paper, we investigate the navigation patterns of users within MOOCs and propose a clickstream model to characterize user behavior in MOOCs. We seek to achieve two goals by analyzing clickstream data. (1) Capture: to identify similar session-level navigation path to detect user intent and interactive patterns. (2) Understand: to find correlation between user's navigation patterns and dropout behavior to better understand user's engagement in MOOCs.

To the best of our knowledge, we are the first to analyze patterns of user's session-level navigation path in MOOCs, and the first to study user's session-level navigation path from the perspective of user's dropout behavior. In summary, our contributions are as follows:

- We proposed an unsupervised model to characterize user session-level clickstream. To validate our model, we perform real-world case studies on KDD CUP 2015 dataset, we can capture the typical user navigation paths in MOOCs, which help interpret user intent.
- We further explore engagement of different navigation patterns based on navigation patterns found, we find there exists correlation between navigation patterns and dropout behavior.

The rest of this paper is organized as follows. Section 2 presents related works. Section 3 provides unsupervised user behavior model. Section 4 introduces our experiment. In Sect. 5, we conclude our work and indicate its future directions.

2   Related Work

2.1   User Session Clustering

Previous research used clickstream data for web usage mining, Studies [5–8, 17] related to user sessions clustering focus on mining useful information concerning user behavior patterns, the purpose is to find groups of users with similar preferences within a specific website. Actually, the knowledge of user groups with similar behavior patterns is extremely valuable for web applications.

Diversity of similarity metrics have been discussed for measuring user's session similarity. Nasraoui et al. [5] assigned binary values based on user access a webpage or not, then use a cosine similarity to compute the similarity between any two user sessions, but this method didn't consider navigation sequence of user. Fu et al. [6] proposed an approach that pages accessed in each user session were substituted by a generalization-based page hierarchy scheme, then clustered the generalized sessions using a hierarchical clustering method. Forsati et al. [7] proposed an approach that assigns a quantitative weight to each page, taking into account of time spent by each user on each page and the visiting frequency of each page. Banerjee et al. [8] considered both webpage visited and the corresponding time spent on a webpage, proposed a similarity metric using Longest Common Subsequences (LCS) to cluster user sessions, similarities between LCS paths were computed as a function of the time spent on the corresponding pages in the paths weighted by a certain factor. Wang et al. [17] proposed a similarity metric considering both type of click event and corresponding time, mapped the time gap into the five discrete time buckets, then extracted *k-grams* subsequences from the clickstreams as features to calculate similarity.

2.2   Clickstream Analysis in MOOCs

There have been many analysis efforts focusing on learning behavior through a variety of features derived from the clickstream. Much of the prior work on clickstream data analysis to understand user behavior has occurred in the context of MOOCs setting.

Most prior studies have focused on using the clickstream data to explore student's engagement in MOOCs and predicting completion or learning outcomes within a MOOC. Anderson et al. [9] divided students into five categories by analyzing how students interact with the course, mainly based on learning activities such as viewing a lecture and attempting an assignment for credit, then analyzed the relationship between engagement and student's grades. Taylor et al. [10] divided learners into four types based on whether they participated in the class forum or helped edit the class wiki pages or not. Breslow et al. [11] focused on the amount of time that students spent in different activities and analyzed demographic information of students. Huang [12] explored super-posters on discussion forum and studied their engagement patterns, finding that super-posters tended to display higher engagement, enroll in more courses and achieve higher academic performance than average forum participants.

Dropout in MOOCs has attracted much attention as the dropout rates is far more than our expectation, many researchers extract features from clickstream data to predict dropout. Wang et al. [13] proposed a nonlinear state space model to show how student's latent states at different time steps can be learned via this model, and demonstrate its outperforming prediction accuracy relative to related methods through experiment. Yang et al. [14] developed a survival model that allows us to measure the influence of factors extracted from learning behavior data on student dropout rate. Bayer et al. [15] described an extraction of new features from both student data and behavioral data represented by a social graph, then introduced a method that employs cost-sensitive learning to lower the number of incorrectly classified unsuccessful students.

3 Unsupervised User Behavior Modeling

In this section, we introduce our unsupervised behavior model built from user's clickstream data. In the following, we first describe the notion of user session-level clickstream. Then we introduce the metric of measuring the similarity between user session-level navigation path. Finally, we describe our unsupervised clustering model.

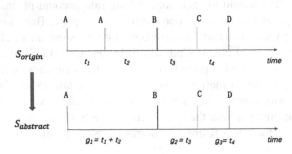

Fig. 1. Convention between original clickstream and abstract clickstream

3.1 Clickstream Model

A session with n pages viewed can be treated as a sequence of pairs consist of webpage and duration, which is given by $S = \{(l_1,t_1),(l_2,t_2),\ldots,(l_n,t_n)\}$, where $L = \{l_1,l_2,\ldots,l_n\}$ is the webpage sequence user clicked, l_i is the ith webpage clicked in a session, $T = \{t_1,t_2,\ldots,t_n\}$ is the corresponding time sequence spent on the webpage, t_i is the ith webpage duration in a session. In order to make the complex and discrete click events understandable, we only consider the click type and combine the consecutive events with the same type together. The abstract clickstream can be described as $S = \{(cat_1,g_1),(cat_2,g_2),\ldots,(cat_n,g_n)\}$, where g_i is the total time of consecutive events spent on the same type cat_i. For example, as shown in Fig. 1, if the original clickstream $S_{origin} = \{(A,t_1),(A,t_2),(B,t_3),(C,t_4)\}$, where A, B or C is the event type, the corresponding abstract clickstream $S_{abstract} = \{(A,g_1),(B,g_2),(C,g_3)\}$, where $g_1 = t_1 + t_2$, $g_2 = t_3, g_3 = t_4$.

3.2 Clickstream Similarity Metric

In order to detect similar session, we need a metric to measure similarity between different user session-level clickstreams. Banerjee [4] was the first to apply LCS algorithm on all pairs of user sessions, but the similarity metric is not easy to scale. We proposed a similarity metric that based on LCS, which is more flexible and can be configurable according to the emphasis of research.

For abstract clickstreams $S_X = \{(cat_{i1},g_{i1}),(cat_{i2},g_{i2}),\ldots,(cat_{im},g_{im})\}$ and $S_Y = \{(cat_{j1},g_{j1}),(cat_{j2},g_{j2}),\ldots,(cat_{jn},g_{jn})\}$, the corresponding click sequence is $L_X = \{cat_{i1},cat_{i2},\ldots,cat_{im}\}$ and $L_Y = \{cat_{j1},cat_{j2},\ldots,cat_{jn}\}$, the corresponding time sequence is $T_X = \{t_{i1}, t_{i2}, \ldots, t_{im}\}$ and $T_y = \{t_{j1}, t_{j2}, \ldots, t_{jn}\}$ respectively. Let $L_K = \{cat_{p1}, cat_{p2}, \ldots, cat_{pk}\}$ be LCS of L_X and L_Y. The click similarity between S_X and S_Y is given by

$$Sim_{click}(S_X, S_Y) = \frac{len(L_K)}{\max(len(L_X), len(L_Y))} \tag{1}$$

Obviously, $Sim_{click}(S_X, S_Y) \in [0, 1]$ and $Sim_{click}(S_X, S_Y) = Sim_{click}(S_Y, S_X)$.

Then we take time spent in the same type event, the element in LCS corresponding indices in L_X and L_Y is represented by $IDX(L_X) = \{a_1, a_2, \ldots, a_k\}$ and $IDX(L_Y) = \{b_1, b_2, \ldots, b_k\}$. We know time spent on the same type sequence depend on LCS (i.e., click similarity), thus the time similarity between S_X and S_Y is given by

$$Sim_{time}(S_X, S_Y) = Sim_{click}(S_X, S_Y) * \frac{1}{k} \sum_{i=1}^{k} \frac{\min(t_{a_i}, t_{b_i})}{\max(t_{a_i}, t_{b_i})} \tag{2}$$

Obviously, $Sim_{time}(S_X, S_Y) \in [0, 1]$ and $Sim_{time}(S_X, S_Y) = Sim_{time}(S_Y, S_X)$.

The metric of clickstream similarity proposed consider both click sequence and duration sequence, the similarity of clickstream S_X and S_Y is given by

$$Sim(S_X, S_Y) = \alpha * Sim_{click}(S_X, S_Y) + (1 - \alpha) * Sim_{time}(S_X, S_Y) \tag{3}$$

where $\alpha(0 \leq \alpha \leq 1)$ is an adjustment factor. Obviously, $Sim(S_X, S_Y) \in [0, 1]$ and $Sim(S_X, S_Y) = Sim(S_Y, S_X)$.

If we emphasize on the click sequence, we can set α a large value, we only consider the common navigation paths without the consideration of duration time when $\alpha = 1$. If we emphasize on the time sequence spent on the type of webpage, we can set α a small value, we only consider the without the consideration of click sequence when $\alpha = 0$.

3.3 Clustering Model

Our goal is to find groups of user sessions with similar navigation path without prior knowledge of user behavior. As user behavior can be change over time, we expect clusters of user sessions in a hierarchy structure instead of flat structure. Moreover, we want to observe the merge process of user clickstreams, thus use Agglomerative Hierarchical Clustering (AHC) [16] as our clustering model. AHC clustering model makes it flexible to analyze fine-grained user behavior without retraining the clustering model. In the hierarchy structure, sessions in higher-level cluster are less similar than lower-level cluster.

When using agglomerative method to find similar sessions, a core need is to measure the distance between two clusters, where each cluster is generally a set of sessions. There are some metrics measuring the distance between clusters, such as single-linkage, complete linkage and average linkage, we choose linkage algorithm based on the result of experiment.

4 Experiment

In the following, we first introduce the real-world dataset we use. Then we describe user behavior patterns captured by our unsupervised model. Finally, we further explore the engagement of each navigation pattern.

4.1 Dataset

We obtain clickstream data from KDD Cup 2015 [18], which was collected from XuetangX[1], which is the largest MOOCs platform of China. The clickstream dataset provides 8,157,277 click events from 79,186 users in 39 courses, which contains information with users engaging in activities such as watching videos, attempting quizzes, accessing course materials, accessing wiki and accessing course forum. The dataset defines a user leaves no activity record more than 10 days at the end of the course as dropout. The dataset contains 7 types of events:

- Problem: Working on course assignments.
- Video: Watching course videos.
- Access: Accessing course materials except videos and quizzes.
- Wiki: Accessing the course wiki.
- Discussion: Accessing the course forum.
- Navigate: Navigating to other part of the course.
- Page_close: Closing the web page.

Among the seven event types, two are meaningless to infer user's learning interest, i.e., navigate and page_close, so we only consider other five types of events. Unfortunately, users do not always explicitly end their sessions, thus we assume that a user's session is over if they do not take any actions for 30 min. Webpage duration is calculated as the time interval between consecutive activities within a session.

Due to the complex and noisy nature of clickstream, we first preprocess the dataset to convert the original clickstream to abstract clickstream. As user may click the webpage by mistake during their learning process, we omit webpage type with duration less than 10 s in the abstract clickstream. The convention process is show in Table 1.

We random pick user behavior data of 3 courses, after preprocessing, we get 3,914 users with 11,885 sessions. To provide contexts for user behavior in MOOCs, we analyze user session-level characteristics. Figure 2 shows different statistics regarding the sessions of users. As shown in Fig. 2a, most sessions contain less than 20 click events, the number of sessions runs down with the rise of click number. In addition, as shown in Fig. 2b, the vast majority of users only participated MOOCs in several sessions.

---

[1] http://www.xuetangx.com.

Table 1. Clickstream convention example

Original clickstream	Abstract clickstream
video1: 1081 ->	video: 1192 ->
video2: 100 ->	~~access: 5 ->~~
video9: 11 ->	problem: 80 ->
access4: 5 ->	access: 115
problem1: 8 ->	
problem2: 72 ->	
access6: 115	

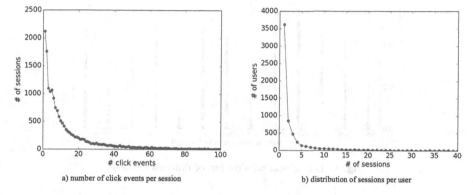

a) number of click events per session   b) distribution of sessions per user

Fig. 2. User session-level characteristics

4.2 Clustering Analysis

In this section, we analyze user behavior patterns captured by our unsupervised model. As we have described extraction of clickstream and calculation of pairwise clickstream similarity in detail. In the following, we show our clustering result and explanation of user behavior.

We first calculate the similarity between clickstreams using proposed method, then clustering the clickstreams using AHC. In the study, we emphasize on the navigation path of user's clickstream, we set $\alpha = 0.8$. To get easy understanding clustering result, single-linkage, complete-linkage and average-linkage are used as metrics of distance between two clusters respectively, we find clustering model with average-linkage is better than models with the other two metrics, it can avoid the outlier sensitivity problem.

Determining the number of sub-clusters is a difficult task. To identify fine-grained behavioral clusters that are easy to interpret, we observe the navigation pattern while continually merge sub-clusters to larger clusters. We first get 6 higher level behavior clusters, which are labeled as $C1 \sim C6$, the navigation patterns are problem centered, video centered, hybrid, access centered, wiki centered and discussion centered respectively. To take a closer look at users' navigation patterns, we further observe the merge process of AHC, we finally find 13 types of navigation patterns after trial and error. As shown in Fig. 3, the vertical axis is the distance between clusters, the black dot means the merge of sub-clusters. We find navigation paths within a cluster have higher similarity, but are very dissimilar to navigation paths in other clusters.

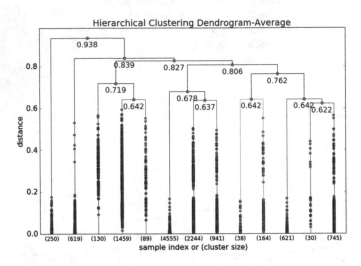

Fig. 3. Hierachical relationship of different clusters

The 13 types of navigation patterns are shown in Table 2, clickstreams in cluster *C1* or in *C2* (i.e., problem centered and video centered) are very similar, the two clusters need not to be divided, but the other clusters are divisible.

- Cluster *C3* contains *C3-1*, *C3-2* and *C3-3* three sub-clusters, the navigation patterns are *access -> video -> problem -> wiki/discussion*, *access -> problem -> video* and *discussion -> access*.
- Cluster *C4* contains *C4-1*, *C4-2* and *C4-3* three sub-clusters, the navigation patterns are *access*, *access -> video* and *access -> problem*.
- Cluster *C5* contains *C5-1* and *C5-2* two sub-clusters, the navigation patterns are wiki and *wiki -> access*.
- Cluster *C6* contains *C6-1*, *C6-2* and *C6-3* three sub-clusters, the navigation patterns are *discussion*, *discussion -> problem* and *access -> discussion*.

4.3   Engagement Analysis

According to Trowler [19], student engagement is a degree of participation in education related activities, the activities happened either in or out of the classroom that is related to measurable learning outcome. In this section, we seek to apply the result of clustering analysis to further explore the relationship between navigation pattern and engagement from the perspective of dropout.

Based on $< enrollment_{id}, session_{id}, cluster_{label} >$ from clustering results and $< enrollment_{id}, dropout_{label} >$ from the dataset, the $dropout_{label}$ is a binary value which means user dropout or not, we get $< session_{id}, cluster_{label}, dropout_{label} >$. Then we use statistical method to analyze the relationship between navigation patterns and dropout behavior.

Table 2. Clickstream navigation patterns and description

Cluster ID	Navigation pattern	Typical navigation paths
C1	problem	(1) problem: 425 (2) problem: 359 -> video: 19
C2	video	(1) video: 186 -> access: 552 (2) video: 46
C3-1	access -> video -> problem -> wiki/discussion	(1) access: 53 -> video: 452 -> access: 90 -> problem: 119 -> video: 68 -> discussion: 536
C3-2	access -> problem -> video	(1) access: 288 -> problem: 22 -> access: 15 -> video: 13
C3-3	discussion -> access	(1) discussion: 32 -> access: 13 -> discussion: 62
C4-1	access	(1) access: 288
C4-2	access -> video	(1) access: 11 -> video: 698
C4-3	access -> problem	(1) access: 77 -> problem: 20 (2) access: 504 -> problem: 10 -> access: 1155
C5-1	wiki	(1) wiki: 113 (2) wiki: 26 -> discussion: 1044
C5-2	wiki -> access	(1) wiki: 25 -> access: 24 -> wiki: 10 (2) wiki: 21 -> access: 13
C6-1	discussion	(1) discussion: 1206 (2) discussion: 83
C6-2	discussion -> problem	(1) discussion: 32 -> problem: 262 (2) access: 1204 -> discussion: 1135 -> problem: 100
C6-3	access -> discussion	(1) access: 291 -> discussion: 660 -> access: 862

The dropout rate of each navigation pattern is shown in Table 3, we find several meaningful information through analysis.

- C1, C2, C4-1, C5-1 and C6-1 are single learning resource centered cluster, among these clusters, dropout rate is increasing in C6-1, C1, C2, C4-1 and C5-1 in turn, which means engagement in navigation patterns of accessing course forum, attempting quizzes, watching videos, accessing webpage materials and accessing wiki is decrease in turn. According to [9, 12], forum is MOOCs plays an important role in online learning, users frequently participate discussion in MOOCs display above-average engagement. In addition, attempting quizzes is also a strong engagement indicator. Whereas watching video, accessing webpage materials and accessing wiki show less engagement is MOOCs.

- Among clusters with accessing multiple types of learning resources, users with clickstream in *C6-2*, *C3-3* and *C3-3* tend to persist to the end of the course. More specifically, users with navigation paths of *discussion -> problem, access -> video -> problem -> wiki/discussion* and *discussion -> access* display higher engagement. Whereas, navigation paths with *wiki* or *access* centered display less engagement.

Table 3. Dropout rate of each navigation pattern

Cluster ID	# session	# dropout	% dropout rate
C1	250	84	33.60
C2	619	340	54.93
C3-1	130	43	33.08
C3-2	1,459	612	41.95
C3-3	89	32	35.96
C4-1	4,555	2,749	60.35
C4-2	2,244	1,345	59.94
C4-3	941	398	42.30
C5-1	38	28	73.68
C5-2	164	110	67.07
C6-1	621	160	25.76
C6-2	30	6	20.00
C6-3	745	359	48.19

5 Conclusion and Future Work

In this work, we describe a clickstream analysis framework to model user behavior in MOOCs. By analyzing tens of thousands sessions in XuetangX, we are able to identify 13 different types of navigation patterns which are fine-grained and easy to understand. We study user's engagement in MOOCs based on the navigation patterns, and find the engagement differs in different kinds of activities. Engagement in activities, such as accessing course forum, attempting quizzes, watching videos, accessing webpage materials and accessing wiki are decreasing in turn.

To extend this study, navigation patterns in different clusters can be explored and analyzed in more detail. In addition, we only explore the path user navigate without considering the content, we hope to use natural language processing techniques to analyze the content user clicked to better understand user interest.

References

1. Rayyan, S., Seaton, D.T., Belcher, J., Pritchard, D.E., Chuang, I.: Participation and performance in 8.02 x electricity and magnetism: the first physics MOOC from MITx. arXiv preprint arXiv:1310.3173 (2013)
2. Parr, C.: New study of low MOOC completion rates| Inside Higher Ed. Inside Higher Ed (2013)

3. Lu, L., Dunham, M., Meng, Y.: Mining significant usage patterns from clickstream data. In: International Workshop on Knowledge Discovery on the Web, pp. 1–17. Springer, Heidelberg (2005)
4. Sadagopan, N., Li, J.: Characterizing typical and atypical user sessions in clickstreams. In: Proceedings of the 17th International Conference on World Wide Web, pp. 885–894. ACM (2008)
5. Nasraoui, O., Petenes, C.: Combining web usage mining and fuzzy inference for website personalization. In: Proceedings of the WebKDD Workshop, pp. 37–46 (2003)
6. Fu, Y., Sandhu, K., Shih, M.-Y.: Clustering of web users based on access patterns. In: Proceedings of the 1999 KDD Workshop on Web Mining, San Diego, CA. Springer (1999)
7. Forsati, R., Meybodi, M.R., Rahbar, A.: An efficient algorithm for web recommendation systems. In: 2009 IEEE/ACS International Conference on Computer Systems and Applications, AICCSA 2009, pp. 579–586. IEEE (2009)
8. Banerjee, A., Ghosh, J.: Clickstream clustering using weighted longest common subsequences. In: Proceedings of the Web Mining Workshop at the 1st SIAM Conference on Data Mining, vol. 143, p. 144 (2001)
9. Anderson, A., Huttenlocher, D., Kleinberg, J., Leskovec, J.: Engaging with massive online courses. In: Proceedings of the 23rd International Conference on World Wide Web, pp. 687–698. ACM (2014)
10. Taylor, C., Veeramachaneni, K., O'Reilly, U.-M.: Likely to stop? Predicting stopout in massive open online courses. arXiv preprint arXiv:1408.3382 (2014)
11. Breslow, L., Pritchard, D.E., DeBoer, J., Stump, G.S., Ho, A.D., Seaton, D.T.: Studying learning in the worldwide classroom: research into edX's first MOOC. Res. Pract. Assess. **8** (2013)
12. Huang, J., Dasgupta, A., Ghosh, A., Manning, J., Sanders, M.: Superposter behavior in MOOC forums. In: Proceedings of the First ACM Conference on Learning@ Scale Conference, pp. 117–126. ACM (2014)
13. Wang, F., Chen, L.: A nonlinear state space model for identifying at-risk students in open online courses. In: Proceedings of the 9th International Conference on Educational Data Mining (2016)
14. Yang, D., Sinha, T., Adamson, D., Rosé, C.P.: Turn on, tune in, drop out: anticipating student dropouts in massive open online courses. In: Proceedings of the 2013 NIPS Data-Driven Education Workshop, vol. 11, p. 14 (2013)
15. Bayer, J., Bydzovská, H., Géryk, J., Obsivac, T., Popelinsky, L.: Predicting drop-out from social behaviour of students. In: International Educational Data Mining Society (2012)
16. Day, W.H.E., Edelsbrunner, H.: Efficient algorithms for agglomerative hierarchical clustering methods. J. Classif. **1**(1), 7–24 (1984)
17. Wang, G., et al.: Unsupervised clickstream clustering for user behavior analysis. In: Proceedings of the 2016 CHI Conference on Human Factors in Computing Systems. ACM (2016)
18. KDD Cup 2015. http://kddcup2015.com
19. Trowler, V.: Student engagement literature review. High. Educ. Acad. **11**, 1–15 (2010)

Software-Defined Network Based Bidirectional Data Exchange Scheme for Heterogeneous Internet of Things Environment

Chao-Hsien Lee[✉] and Yu-Wei Chang

Department of Electronic Engineering, National Taipei University
of Technology, Taipei City, Taiwan (R.O.C.)
chlee@ntut.edu.tw

Abstract. The popularity of Internet of Things (IoT) triggers the rapid development of various IoT platforms, standards, and protocols. However, the differences and heterogeneity among IoT platforms, standards and protocols have become the critical difficulty to interconnect a mass of IoT devices. According to our observation, most IoT standards and platforms utilize similar light-weighted network protocols, i.e., CoAP and MQTT. Thus, in this paper, one bidirectional data exchange scheme based on software-defined network (SDN) is proposed to let two IoT devices with different IoT protocols, i.e., CoAP and MQTT, be able to communicate each other more straightforward. The experimental results show that the proposed scheme costs more time to establish one traffic flow. Once the traffic flow is established, the transmission performance is a little lower than the existing mechanism, i.e., Ponte.

Keywords: Interoperability · Internet of Things (IoT)
Software-defined network (SDN) · CoAP · MQTT

1 Introduction

Internet of Things (IoT) is composed of (1) high-capacity and full-functional devices, e.g., laptops and phones, and (2) small-size and low-complexity constrained devices, e.g., sensors and wearable devices. In order to satisfy all possible machine-to-machine (M2M) communication scenarios, not only academic researchers but also industrial engineers join to define various IoT standards, platforms and protocols. For example, European Telecommunications Standards Institute (ETSI) has proposed one M2M communication standard called smartM2M which is extended from oneM2M. Open Mobile Alliance (OMA) has also defined another M2M communication standard called Light-Weight M2M (LWM2M). AllSeen Alliance and Open Connectivity Foundation (OCF) have also announced their own platforms, i.e., AllJoyn and IoTivity respectively. Furthermore, regarding the network protocol stack, Institute of Electrical and Electronics Engineers (IEEE) has proposed IEEE 802.15.1 (Bluetooth), IEEE 802.15.4 (Zigbee), and IEEE 802.11ah (Wi-Fi HaLow) to interconnect large groups of sensors and devices. Internet Engineering Task Force (IETF) has established IPv6 over Low Power WPAN (6lowpan), IPv6 over Networks of Resource-constrained Nodes (6lo), IPv6 Maintenance

© Springer Nature Singapore Pte Ltd. 2018
J. C. Hung et al. (Eds.): FC 2017, LNEE 464, pp. 414–422, 2018.
https://doi.org/10.1007/978-981-10-7398-4_42

(6man), Routing over Low Power and Lossy networks (roll), and Constrained RESTful Environments (core) working groups (WGs). Therefore, the current IoT environment becomes highly heterogeneous, keeps changing and evolves rapidly.

How to enhance interoperability among IoT devices has become one critical issue based on the aforementioned observation. Recently, one technique called software-defined network (SDN) is able to separate the control plane and the data plane of traditional network equipment. Initially, SDN is created to handle the changing business requirements inside the cloud networking environment quickly. That is, SDN is designed to make more flexible and agile to support the virtualized server and storage infrastructure in the data center. Since the IoT environment has similar characteristics, one SDN-based Bidirectional Data Exchange Scheme (SDN-BDES) is proposed in this paper. Two common IoT application-layer protocols, i.e., Constrained Application Protocol (CoAP) and Message Queuing Telemetry Transport (MQTT), are taken into consideration. In the proposed SDN-BDES, bidirectional data exchange between two different protocols, i.e., from CoAP to MQTT and vice versa, is allowed without any further modification in the end devices.

This paper is organized as follows. Section 2 introduces the development and applications of both IoT protocols, i.e., CoAP and MQTT. Section 3 presents our proposed SDN-BDES. Section 4 shows the performance evaluation of our SDN-BDES. Finally, Sect. 5 concludes this paper.

2   Related Work

MQTT, which is one popular application-layer IoT protocol now, is built on top of TCP and adopts the subscribe/publish model. MQTT was invented early for satellite communication and then applied to home automation. Recently, MQTT has been widely utilized IoT services and applications. Kang et al. implemented one IoT service about room temperature control and fire alarm using MQTT over Amazon Web Service (AWS) [1]. Dhar and Gupta presented an intelligent parking IoT service using MQTT [2]. Grgić et al. proposed a real-time web-based solution for monitoring and tracking temperature and moisture values in the agricultural drying process [3]. The key characteristic of MQTT is light-weighted and easily implemented. Kodali and Mahesh demonstrated a low cost implantation of MQTT [4]. In addition, Luzuriaga et al. provided how to handle mobility using MQTT without any mobility protocols, e.g., Mobile IP [5].

CoAP is another application-layer IoT protocol. Different from MQTT, CoAP is built on top of UDP and can be seem as the light-weighted and binary-format Hypertext Transfer Protocol (HTTP). Thus, CoAP is operated on the client/server model. More and more services and applications adopts CoAP as the communication protocol. For example, Ugrenovic and Gardasevic selected CoAP for web-based healthcare monitoring [6]. Chen et al. discussed how to use CoAP to share resources and services between clouds of mobile devices [7]. Cho et al. improved the notification system using CoAP [8]. Based on their experimental results, the service utilization is improved and the average transmission rate is increased. Similar to MQTT, CoAP also has the characteristics of easy implementation and extension. Tanganelli et al. presented one open-source Python CoAP library called CoAPthon [9]. On the other hand, Chun and Park proposed CoAP based mobility management protocol called CoMP [10].

Some works has investigated and compared the difference among IoT protocols. Kayal and Perros implemented one smart parking application in order to compare CoAP, MQTT, XMPP, and WebSocket [11]. Some works has discussed how to integrate heterogeneous IoT networks horizontally. Al-Fuqaha et al. compared CoAP, MQTT, XMPP, AMQP, REST and DDS [12]. Then, one generic IoT protocol extended from MQTT was proposed to support rich QoS features. Žitnik et al. extended OM2M, which follows oneM2M, to support various network protocols, e.g., HTTP, CoAP, MQTT, Zigbee, and WebSocket [13]. Bellavista and Zanni extended the Kura framework to propose a flexible architecture based on the interworking between MQTT and CoAP [14].

3 Architecture and Operations

Regarding the IoT environment, sensors are responsible for gathering data from the realistic environment. Thus, sensors usually plays the role of IoT servers because they are the data source. On the other hand, nodes in the Internet take charge of requesting data for analysis and should be considered as the IoT clients. However, the main functionality of the existing IoT platforms is to provide the IoT data storage. That is, the IoT platforms themselves are neither IoT servers nor IoT clients. Thus, when the IoT platforms support more than one IoT protocols [15, 16], some kind of nature conflict exists among IoT protocols. For example, in the MQTT, the MQTT server should publish its data to the MQTT broker and the MQTT client can subscribe data from the MQTT broker. It is obvious that the IoT platform can be seem as MQTT broker. On the contrary, CoAP is defined between servers and clients directly. If the IoT platform is regarded as CoAP server, it means that the IoT platforms become the data source. Otherwise, the IoT platform requires data from the Internet if the IoT platform is regarded as CoAP client.

According to the aforementioned observation, in order to preserve the original communication scenarios, this paper is motivated to enable bidirectional data exchange between CoAP and MQTT. Based on the software-defined network (SDN), Fig. 1 depicts the proposed SDN-based Bidirectional Data Exchange Scheme (SDN-BDES). In our SDN-BDNS, the SDN network intercepts all flows from CoAP to MQTT, and vice versa. If flows belong to the same protocols, they just follow the original communication scenarios and the SDN network would ignore these flows. The flow rule to distinguish the traffic belonging to heterogeneous protocols from the one belonging to homogeneous protocols is based on the URI rule defined in this paper. Eq. (1) depicts our URI rule. For example, if one packet is transmitted from CoAP to CoAP, the URI can be represented as "coap://10.0.0.1/coap/10.0.0.1/LED". On the other hand, if the packet is transmitted from MQTT to CoAP, the URI should be represented as "mqtt://192.168.0.1/coap/10.0.0.1/LED".

$$<\text{source protocol}>://<\text{destination}>/<\text{target protocol}>/<\text{target}>/<\text{topic}> \quad (1)$$

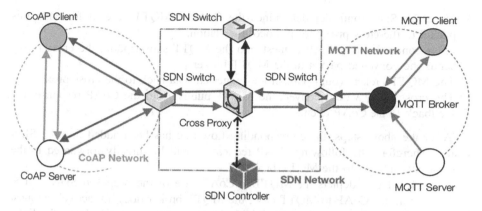

Fig. 1. The proposed SDN-based Bidirectional Data Exchange Scheme (SDN-BDES).

Once the traffic belonging to heterogeneous protocols is intercepted, the SDN-BDES divides into (1) the CoAP-to-MQTT case, in which packets are from one CoAP client to one MQTT broker, and (2) the MQTT-to-CoAP case, in which packets are from one MQTT client to one CoAP server. First, Fig. 2 depicts the CoAP-to-MQTT case of the proposed SDN-BDES. Since all MQTT data are stored in the MQTT broker, the SDN-BDES is responsible for translating the CoAP request into the MQTT subscription in the CoAP-to-MQTT case. The detailed execution steps are as follows.

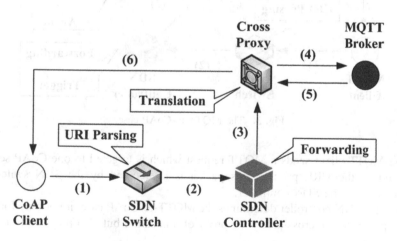

Fig. 2. The CoAP-to-MQTT case.

1. One CoAP client sends a CoAP request which is targeted to one MQTT server.
2. Based on the URL parsing, this packet is intercepted by the SDN switch and redirected to the SDN controller.

3. Once the SDN controller determines the CoAP-to-MQTT case, it forwards this packet to the cross proxy for protocol translation.
4. After converting the CoAP request into the MQTT subscription, the cross proxy sends the converted packet to the MQTT broker.
5. The MQTT broker would publish the requested data back to the cross proxy.
6. The cross proxy would convert the MQTT publish into the CoAP response and feedback to the CoAP client.

After the above steps, the corresponding flow rule has been added into the SDN switch. Therefore, the following CoAP requests would be directly forwarded to the cross proxy and then to the MQTT broker.

Secondly, Fig. 3 depicts the MQTT-to-CoAP case of the proposed SDN-BDES. Different from the CoAP-to-MQTT case, the MQTT broker does not actively request data from the MQTT server in the original communication scenario. Thus, the SDN controller must play the role of being the MQTT broker's agent to request from the CoAP server in the MQTT-to-CoAP case. The detailed execution steps are as follows.

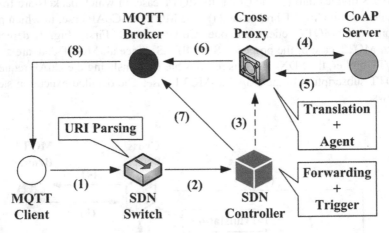

Fig. 3. The MQTT-to-CoAP case.

1. One MQTT client sends a MQTT request which is targeted to one CoAP server.
2. Based on the URL parsing, this packet is intercepted by the SDN switch and redirected to the SDN controller.
3. Once the SDN controller determines the MQTT-to-CoAP case, it not only forwards this packet to the cross proxy for protocol translation but also prepares to become the agent.
4. After converting the MQTT subscription into the CoAP request, the cross proxy sends the converted packet to the CoAP server.
5. The CoAP server would response the requested data back to the cross proxy.
6. The cross proxy would convert the CoAP response into the MQTT publish and feedback to the corresponding MQTT broker.

7. The SDN controller would trigger the MQTT broker to accept the MQTT publish from the cross proxy.
8. Once the MQTT broker has the requested data, it would publish to the requesting MQTT client.

After the above steps, the corresponding flow rule has been added into the SDN switch. Therefore, the following MQTT requests would be directly forwarded to the MQTT broker. At the same time, if the CoAP server has any update, it would send the update to the cross proxy and then forward to the MQTT broker after protocol translation.

4 Performance Evaluation

The proposed SDN-BDES preserves the original communication scenario of CoAP and MQTT. In order to evaluate the overhead, we compare the SDN-BDES with one famous IoT platform called Ponte. Ponte belonging to the Eclipse IoT project offers uniform open APIs to let developers create their applications sup-porting different IoT protocols, including CoAP, MQTT and HTTP [17]. Ponte has the HTTP module, the CoAP module, and the MQTT module. All data collected from three different net-works are all stored into one SQL or NoSQL database. Therefore, no matter which protocol clients utilize for communication, they can access the same resources. Figure 4 depicts the performance of the CoAP-to-MQTT case. The setup time means that one CoAP client sends the first request to the MQTT broker. Since the SDN-BDES needs to add the flow rule and forward packets to the cross proxy for protocol trans-lation, the SDN-BDES costs more time than Ponte. Furthermore, the arrival time means that data are transmitted from the MQTT server to the CoAP client. Since the flow rule has been added into the SDN switch, the SDN-BDES costs a little time than Ponte.

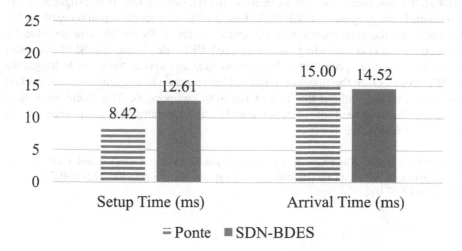

Fig. 4. The performance of the CoAP-to-MQTT case

Figure 5 depicts the performance of the MQTT-to-CoAP case. Ponte does not support that the MQTT broker would actively send requests to the CoAP server. Ponte only supports that the MQTT client can subscribe data which have been pushed by the CoAP server. The performance results can be referred to the CoAP-to-MQTT case. Hence, in the MQTT-to-CoAP, we only presents the SDN-BDES's performance results. As depicted in Fig. 5, the SDN-BDES costs much time, i.e., about 428 ms, to establish the flow. That is because the SDN-BDES needs the SDN controller to act as the agent and then actively trigger the CoAP server to push their data to the MQTT broker. However, once the flow rule has been added into the SDN switch, the SDN-BDES costs only 13.12 ms for data exchange.

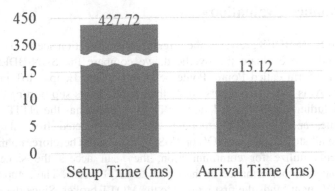

Fig. 5. The performance of the MQTT-to-CoAP case

5   Conclusion

In this paper, one bidirectional data exchange scheme based on SDN called SDN-BDES has been proposed to increase the interoperability of heterogeneous IoT protocols. In the proposed SDN-BDES, there is no further modification on existing IoT protocols and the corresponding implementation. The SDN switch is responsible for intercepting packets according to our defined URI rule. Then, the SDN controller forwards to the cross proxy for protocol translation and acts as the agent to trigger the MQTT broker. Once the flow is established between two heterogeneous ends, packets can be exchanged without the help of the SDN components. The future work is to minimize the overhead of protocol translation, especially the setup time of the MQTT-to-CoAP case.

Acknowledgements. The research is supported by the Ministry of Science and Technology of the Republic of China (Taiwan) under the grant number MOST 105-2221-E-027-087 and 105-2218-E-027-012.

References

1. Kang, D.H., et al.: Room temperature control and fire alarm/suppression IoT service using MQTT on AWS. In: Proceedings of International Conference on Platform Technology and Service (PlatCon), Busan, pp. 1–5 (2017)
2. Dhar, P., Gupta, P.: Intelligent parking Cloud services based on IoT using MQTT protocol. In: Proceedings of International Conference on Automatic Control and Dynamic Optimization Techniques (ICACDOT), Pune, pp. 30–34 (2016)
3. Grgić, K., Špeh, I., Heđi, I.: A web-based IoT solution for monitoring data using MQTT protocol. In: Proceedings of International Conference on Smart Systems and Technologies (SST), Osijek, pp. 249–253 (2016)
4. Kodali, R.K., Mahesh, K.S.: A low cost implementation of MQTT using ESP8266. In: Proceedings of the 2nd International Conference on Contemporary Computing and Informatics (IC3I), Greater Noida, India, pp. 404–408 (2016)
5. Luzuriaga, J.E., Cano, J.C., Calafate, C., Manzoni, P., Perez, M., Boronat, P.: Handling mobility in IoT applications using the MQTT protocol. In: Proceedings of Internet Technologies and Applications (ITA), Wrexham, pp. 245–250 (2015)
6. Ugrenovic, D., Gardasevic, G.: CoAP protocol for Web-based monitoring in IoT healthcare applications. In: Proceedings of the 23rd Telecommunications Forum Telfor (TELFOR), Belgrade, pp. 79–82 (2015)
7. Chen, N., Li, X., Deters, R.: Collaboration & mobile cloud-computing: using CoAP to enable resource-sharing between clouds of mobile devices. In: Proceedings of IEEE Conference on Collaboration and Internet Computing (CIC), Hangzhou, pp. 119–124 (2015)
8. Cho, C., Kim, J., Joo, Y., Shin, J.: An approach for CoAP based notification service in IoT environment. In: Proceedings of International Conference on Information and Communication Technology Convergence (ICTC), Jeju, pp. 440–445 (2016)
9. Tanganelli, G., Vallati, C., Mingozzi, E.: CoAPthon: Easy development of CoAP-based IoT applications with Python. In: Proceedings of IEEE 2nd World Forum on Internet of Things (WF-IoT), Milan, pp. 63–68 (2015)
10. Chun, S.M., Park, J.T.: Mobile CoAP for IoT mobility management. In: Proceedings of the 12th Annual IEEE Consumer Communications and Networking Conference (CCNC), Las Vegas, NV, pp. 283–289 (2015)
11. Kayal, P., Perros, H.: A comparison of IoT application layer protocols through a smart parking implementation. In: Proceedings of the 20th Conference on Innovations in Clouds, Internet and Networks (ICIN), Paris, pp. 331–336 (2017)
12. Al-Fuqaha, A., Khreishah, A., Guizani, M., Rayes, A., Mohammadi, M.: Toward better horizontal integration among IoT services. IEEE Commun. Mag. 53(9), 72–79 (2015)
13. Žitnik, S., Janković, M., Petrovčič, K., Bajec, M.: Architecture of standard-based, interoperable and extensible IoT platform. In: Proceedings of the 24th Telecommunications Forum (TELFOR), Belgrade, pp. 1–4 (2016)
14. Bellavista, P., Zanni, A.: Towards better scalability for IoT-cloud interactions via combined exploitation of MQTT and CoAP. In: Proceedings of IEEE 2nd International Forum on Research and Technologies for Society and Industry Leveraging a better tomorrow (RTSI), Bologna, pp. 1–6 (2016)

15. Huh, J.H., Kim, D.H., Deok Kim, J.: oneM2M: extension of protocol binding: reuse of binding protocol's legacy services. In: Proceedings of International Conference on Information Networking (ICOIN), Kota Kinabalu, pp. 363–365, 13–15 January 2016

16. Al-Fuqaha, A., Guizani, M., Mohammadi, M., Aledhari, M., Ayyash, M.: Internet of things: a survey on enabling technologies, protocols, and applications. IEEE Commun. Surv. Tutor. 17(4), 2347–2376 (2015)

17. Ponte. http://www.eclipse.org/ponte/

Multi-mode Halftoning Using Stochastic Clustered-Dot Screen

Yun-Fu Liu[(✉)], Jing-Ming Guo[(✉)], and Shih-Chieh Lin[(✉)]

Department of Electrical Engineering, National Taiwan University of Science
and Technology, Taipei, Taiwan
yunfuliu@gmail.com, jmguo@seed.net.tw,
ggookey123@yahoo.com.tw

Abstract. The conventional dual-mode halftoning methods achieve high quality halftone patterns by easing smooth artifact and preserving image details. However, the boundaries between low- and high-frequency image regions still present undesired texture in some particular cases, which significantly degrades the visual quality. In this work, a multi-mode halftoning method is proposed to deal with the object map artifact and boundary artifact simultaneously. In addition, the new absorptance-frequency stacking constraint is also employed to solve the noisy textures of the halftone outputs. As documented in the experimental results, high quality halftone outputs can be obtained, proving that the proposed method can be a competitive candidate for electrophotographic printers.

Keywords: Digital printing · Digital halftoning · Screening
Clustered-dot dithering · Direct binary search

1 Introduction

Digital halftoning is a technique of converting a continuous-tone image into a binary image, and it has been widely used in devices which render a limited number of colors, e.g., inkjet printers which generate a dot or not at each location. When dots are spatially close enough to each other, the human visual system (HVS) considers more intensities because of its lowpass natural.

Although the dual-mode halftoning algorithms improve the quality on smooth and detailed regions, some researchers [9–12] indicate that the object boundary in between smooth and detailed regions still presents some undesired texture. The latest research about boundary artifact is the seamless halftoning proposed in Allebach et al. [9] work. They generated object map by page description language (PDL) and divided a document into smooth and detailed areas as per the object map. However, the object map also introduces some new issues when it renders documents by the dual-mode halftoning algorithm. First, if the adjacent smooth and detailed areas of a document are with similar colors, the different halftone textures may introduce the visible "object map artifact" which creates artificial boundary between different areas. Second, one object in the object map may contain both smooth and detailed areas. Figure 1 shows an example of this artifact. Consequently, with the object map, the screens used for both detailed and smooth areas are the same and it makes the dual-mode halftoning meaningless.

© Springer Nature Singapore Pte Ltd. 2018
J. C. Hung et al. (Eds.): FC 2017, LNEE 464, pp. 423–431, 2018.
https://doi.org/10.1007/978-981-10-7398-4_43

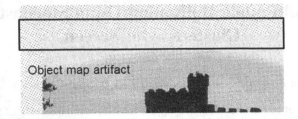

Fig. 1. The object map artifact appears when using dual-mode halftoning algorithm and object map

The multi-mode halftoning algorithm is proposed to address this object map artifact in this paper. The fundamental idea is to check the local spatial frequency and switch among various halftone patterns as per the local variance. Consequently, more than two screens can be utilized to reduce the boundary artifact. To overcome the problem of inhomogeneity among different areas, the screen generation algorithm inspired by the inter-iterative clustered-dot direct multi-bit search (CLU-DMS) [13] is proposed in this paper. This algorithm not only generates one mask with the stacking constraint, but multiple masks associates to various spatial frequencies. This new method is termed the "absorptance-frequency stacking constraint" (AF stacking constraint) in this paper. This method generates the screens with a similar texture, and the rendered image is with very good quality.

The remaining of this paper is organized as follows. Section 2 introduces the proposed inter-iterative CLU-DBS halftone algorithm, and it is applied in Sect. 3 to generate screens of different spatial frequencies. Section 3 presents the proposed screen design method with the AF stacking constraint. Section 4 details the proposed multi-mode halftoning algorithm, and it presents the way of render an image with screen set. Sections 4 and 5 draw the experimental results and conclusions, respectively.

2   Inter-iterative CLU-DBS

To design a high-quality screen set, the inter-iterative clustered-dot direct multi-bit search (CLU-DMS) [13] which generates both blue- and green-noise multitone patterns. As the focus of this study is on the printers of two tones, the multitone capability of CLU-DMS is constrained to binary form, as thus termed the inter-iterative clustered-dot direct binary search (ICLU-DBS) in this work. In contrast to the conventional CLU-DBS [14], the ICLU-DBS utilized the inter-iterative CLU-DMS to achieve significantly better texture homogenous.

To generate halftone patterns with blue- and green-noise spectrums, a two-component cost function formulated with the initial filter and the update filter is used. These two filters are modeled as Gaussian kernels with standard deviations σ_{init} and σ_{update}, where $\sigma_{init} \leq \sigma_{update}$. If $\sigma_{init} = \sigma_{update}$, a halftone pattern with blue-noise spectrum is rendered; $\sigma_{init} < \sigma_{update}$, a halftone pattern with green-noise spectrum is rendered.

3 Screen Design

The multi-mode halftoning algorithm adopts more than two screens to render an image. Consequently, the screen set $T_{\Delta\sigma}$ is generated first. The construction of a screen $T_{\Delta\sigma}$ can be decomposed into L masks, and each mask is produced by the ICLU-DBS.

To generate the masks for constructing a screen, the traditional stacking constraint can generate screens of different spatial frequencies, but it cannot enforce the screens with similar texture. To address this problem, the new screen generation algorithm which is named absorptance-frequency (AF) stacking constraint is proposed in this paper, it generates not only a mask with the stacking constraint over absorptance levels, but also over all spatial frequencies to avoid the above issue.

4 Multi-mode Halftoning

To avoid the object map artifact, the variance map is utilized instead of the object map. Switching the screens according to the variance map can adopt more than two screens to reduce the boundary artifact because the spatial frequency can be changed gradually by using more screens of different spatial frequencies.

5 Experimental Result

The HP LJ300-400 color MFP M375-M475 Color LaserJet printer is utilized in the following simulation printed at 300 dpi. The printed results are scanned at 600 dpi for analysis and discussion. In addition, the Uncompressed Colour Image [15] dataset is employed in the experiments.

5.1 Parameters

Figure 2 shows the influences under various numbers of screens (N). When the number of screens is increasing, the boundary between areas of different dot densities becomes seamlessly because the clustered-dot sizes can be changed smoothly. Finally, this boundary issue is significantly eased when sufficient screens are utilized.

5.2 Stacking Constraint

This subsection shows the comparison results between the traditional stacking constraint and the proposed AF stacking constraint. As for the traditional way, these midtone masks are utilized to render each of the mask, and the traditional stacking constraint is used to construct the mask set for constructing screens. Since all the midtone masks are independent to each other, the textures of the yielded screens from different spatial frequencies are also varied.

Figure 3 shows the influence when screens of two different spatial frequencies are utilized to render the top and bottom parts of a constant tone image. As it can be seen in Fig. 3(b), a clear horizontal line, termed the "discontinuous texture problem" appears in

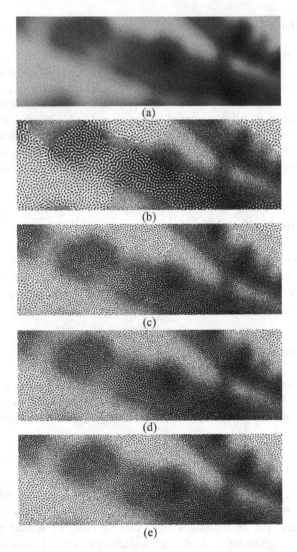

Fig. 2. Multi-halftoning result with various numbers of screens. (a) Original image. (b)–(e) The number of screens $N = [2, 4, 6, 7]$

the middle of the rendered result generated by the traditional method. However, this issue is significantly eased as shown in Fig. 3(c) by using the screens, of which the textures are spatially similar to each other.

Figure 4 shows the multi-mode halftoning results with the traditional stacking constraint and the proposed AF stacking constraint. As it can be seen, the result generated by the conventional method is noisy because it suffers from the discontinuous texture problem, in particular at the background areas which are rendered with large clustered-dots. In addition, it introduces more severe discontinuous texture

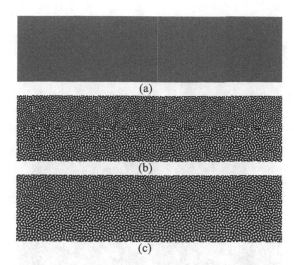

Fig. 3. Screened results with different sets of screens on the (a) constant image of absorptance level 0.3. (b) Rendered with the typical stacking constrained screens. (c) Rendered with the proposed AF constrained screens

problem. Conversely, Fig. 4(c) shows the one generated with the proposed AF stacking constraint. Apparently, the above artifact is reduced, which demonstrates the superiority of the proposed method.

Figure 5 shows a rendered ramp result with the proposed AF stacking constraint. Apparently, a homogenous result is generated over various spatial frequencies and intensities.

5.3 Comparison

Although the former dual-mode halftoning improves the image quality on smooth and detailed regions, it still suffers from the boundary artifact until some former schemes [9–11] solve this problem. Among these, Allebach et al. work [9] has been proved to achieve superior performance than that of the works [10, 11]. Consequently, in this study we simply include Allebach et al. [9] method for comparison.

As shown in Fig. 6(b), when using the seamless algorithm [9], the boundary between the areas rendered by different screens is apparent because of the object map artifact. In addition, the screens used for rendering the entire detailed region are identical without considering whether dramatic different properties on the spatial frequencies. Conversely, as it can be seen in Fig. 6(c), the screens can be switched across various regions. Thus, it can successfully remove the object map artifact.

Figure 7 shows the transition regions of the screened outputs for comparison. Figure 7(b) shows the result of the seamless algorithm [9]. Its transition region is blended by different screens. Conversely, the result of the proposed algorithm as shown in Fig. 7(c) is more natural compared to that of the seamless method regarding the visibility of the artifact in the middle of the result.

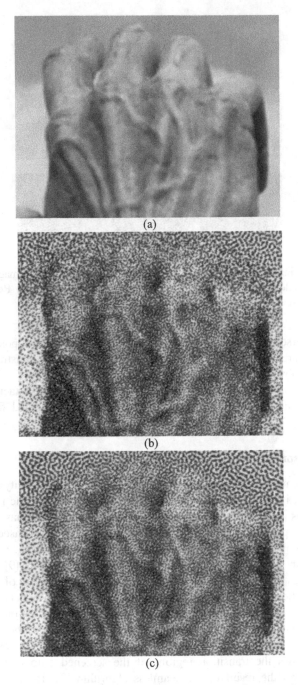

Fig. 4. Multi-halftoning result with similar or different textures of screens. (a) Original image. (b) without similar textures. (c) with similar textures

Intensity

Spatial frequency

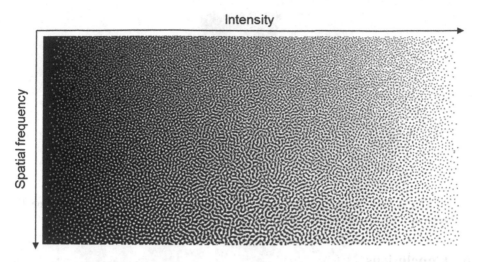

Fig. 5. Ramp image with different spatial frequency. AF stacking constraint generates the masks not only with the stacking constraint at various absorptance levels but also at various spatial frequencies

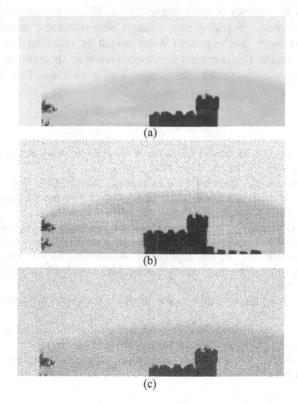

(a)

(b)

(c)

Fig. 6. Comparison of seamless and multi-mode halftones. (a) Original image. (b) Seamless halftone [9]. (c) Proposed multi-mode halftone.

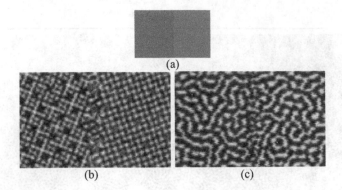

Fig. 7. Comparison on the transition area. (a) Original image. (b) Seamless halftone [9]. (c) Proposed multi-mode halftone.

6 Conclusions

In this paper, the multi-mode halftoning algorithm is proposed for laser printers to improve the smooth artifact and the boundary artifact, as well as preserve the rendered details. In this algorithm, the use of variance map can adaptively switch the screens to adapt the spatial frequencies of each image region. Moreover, the ICLU-DBS is proposed to generate the high-quality screen set. A new stacking constraint is also proposed to remove with the unwanted noise which is introduced when using the conventional approach. Experimental results demonstrate apparent superiority of the proposed method on artifact reduction compared with the state-of-the-art methods.

References

1. Lin, G., Allebach, J.P.: Multilevel screen design using direct binary search. J. Opt. Soc. Am. A: Opt. Image Sci. Vis. **19**(10), 1969–1982 (2002)
2. Lau, D.L., Gonzalo, R.: Modern Digital Halftoning, 2nd edn. CRC Press, Arce (2008)
3. Kacker, D., Camis, T., Allebach, J.P.: Electrophotographic process embedded in direct binary search. IEEE Trans. Image Proces. **11**, 234–257 (2002)
4. Ostromoukhov, V., Nehab, S.: Halftoning with gradient-based selection of dither matrices, U.S. Patent 5,701,366, assigned to Canon Information Systems, Inc. (1997)
5. Huang, J., Bhattacharjya, A.: An adaptive halftone algorithm for composite documents. In: Proceedings of SPIE, Color Imaging: Processing, Hardcopy and Applications IX, vol. 5293, pp. 425–433 (2004)
6. Floyd, R.W., Steinberg, L.: An adaptive algorithm for spatial greyscale. Proc. Soc. Inf. Disp. **17**, 75–77 (1976)
7. Daly, S., Feng, X.: Methods and systems for adaptive dither structures, U.S. Patent 7,098,927, assigned to Sharp Laboratories of America, Inc. (2004)
8. Gupta, M.R., Bowen, J.J.: Ranked dither for high-quality robust printing. J. Opt. Soc. Am. A **25**, 1454–1458 (2008)

9. Park, S.J., Shaw, M.Q., Kerby, G., Nelson, T., Tzeng, D.-Y., Bengtson, K.R., Allebach, J.P.: Halftone blending between smooth and detail screens to improve print quality with electrophotographic printers. IEEE Trans. Image Proces. **25**(2), 601–614 (2016)

10. Lin, Q.: Adaptive halftoning based on image content, U.S Patent 5,970,178, assigned to Hewlett-Packard Development Company (1999)

11. Kritayakirana, K., Tretter, D., Lin, Q.: Adaptive halftoning method and apparatus, U.S Patent 6,760,126, assigned to Hewlett-Packard Development Company

12. Hel-Or, H.Z., Zhang, X., Wandell, B.A.: Adaptive cluster dot dithering. J. Electron. Imag. **8**, 133–144 (1999)

13. Liu, Y.F., Guo, J.M.: Clustered-dot screen design for digital multitoning. IEEE Trans. Image Proces. **25**(7), 2971–2982 (2016)

14. Goyal, P., Gupta, M., Staelin, C., Fischer, M., Shacham, O., Allebach, J.P.: Clustered-dot halftoning with direct binary search. IEEE Trans. Image Proces. **22**(2), 483–487 (2013)

15. Database of Uncompressed Colour Image. http://homepages.lboro.ac.uk/~cogs/datasets/ucid/ucid.html

Performance Assessment Under Different Impulsive Noise Models for Narrowband Powerline Communications

Yu-Xain Chen[1], Rong-Sian Lai[1], Shao-Hang Lu[2],
and Ying-Ren Chien[1(⊠)]

[1] Department of Electric Engineering, National Ilan University, Yilan, Taiwan
yrchien@niu.edu.tw
[2] Graduate Institute of Automation and Control,
National Taiwan University of Science and Technology, Taipei, Taiwan

Abstract. For narrow-band powerline communications (NB-PLC), the dominant impulsive noise (IN) is cyclostationary noise and periodic noise. However, some existing works still apply the Middle class-A (MCA) IN model in an NB-PLC system. By applying the cyclic spectral analysis technique, we show that the MCA IN model cannot capture the nature of cyclostationary noise and periodic noise.

Keywords: Narrow-band powerline communication (NB-PLC)
Cyclic spectral analysis

1 Introduction

Due to low cost of deployment and ability to communicate across transformers, a narrowband powerline communication (NB-PLC) technology is attractive as a communication solution for enabling smart grid systems. The operating band for NB-PLC is about 3 kHz to 500 kHz, which is distinct from the broadband PLC (BB-PLC). The channel impairments, noise characteristics, and standards are reported in [1]. It has been stated that the dominated noise is cyclostationary noise and periodic impulsive noise (IN). The periodic noise component is wide-sense periodic noise and is compelling below 10 kHz [2]. Moreover, due to the shared nature of the powerline network, the uncoordinated interference from either NB-PLC or BB-PLC results in the random and impulsive noise. These uncoordinated interferences are similar to asynchronous IN and follow a Middleton Class-A (MCA) or Gaussian mixture distribution [3]. However, in [4], the authors have been reported that the MCA model is not suitable for modeling the impulsive noise in NB-PLC environment. This claim was verified by comparing the empirical PDF of the measured noise with the best fitted PDF of MCA model. However, some existing works adopted the MCA model for NB-PLC applications [5–8].

In this paper, we numerically analyze the cyclostationary IN in frequency domain to show that MCA model exhibits quite different behavior from the measured

© Springer Nature Singapore Pte Ltd. 2018
J. C. Hung et al. (Eds.): FC 2017, LNEE 464, pp. 432–437, 2018.
https://doi.org/10.1007/978-981-10-7398-4_44

impulsive; moreover, we compare the resulting bit error rate (BER) performance with respect to different IN models via computer simulation.

2 Cyclostationary Impulsive Noise

A cyclostationary process is a signal having statistical properties that vary cyclically with time and has a periodic instantaneous auto-correlation function [9]. An important special case of cyclostationary signals is one that exhibits cyclostationary in second-order statistics, such as the autocorrelation function. These are called wide-sense cyclostationary signals, and are analogous to wide-sense stationary processes. The exact definition differs depending on whether the signal is treated as a stochastic process or as a deterministic time series. A stochastic process $x(t)$ of mean function $E[x(t)]$ and auto-correlation function:

$$R_x(t, \tau) = E[x(t + \tau)x^*(t)] \tag{1}$$

where the symbol $*$ denotes complex conjugation operation, is said to be wide-sense cyclostationary with period T_0 if both $E[x(t)]$ and $R_x(t, \tau)$ are cyclic in time with period T_0. This periodic function can be expressed using Fourier series as follows:

$$R_x(t, \tau) = \sum_{n=-\infty}^{\infty} R_x^{n/T_0}(\tau)e^{j2\pi\frac{n}{T_0}t} \tag{2}$$

where n/T_0 is called the cyclic frequency; $R_x^{n/T_0}(\tau)$ is the cyclic autocorrelation function and is obtained with inverse Fourier series as follows:

$$R_x^{n/T_0}(\tau) = \frac{1}{T_0} \int_{-T_0/2}^{T_0/2} R_x(t, \tau)e^{-j2\pi\frac{n}{T_0}t} dt \tag{3}$$

From the frequency domain point of view, we can define the "cyclic spectrum" by taking Fourier transform of the cyclic autocorrelation function at cyclic frequency α as follows:

$$S_x^\alpha(f) = \int_{-\infty}^{\infty} R_x^\alpha(\tau)e^{-j2\pi f\tau} d\tau \tag{4}$$

with

$$R_x^\alpha(\tau) = \int_{-\infty}^{\infty} x(t - \tau/2)x^*(t + \tau/2)e^{-j2\pi\alpha t} dt \tag{5}$$

Note that all non-zero α describe the cyclostationary characteristics of the signal.

In addition, akin to the concept of spectral coherence, the cyclic spectral coherence function for a single process $x(t)$ can be defined as follows:

$$C_x^\alpha(f) = \frac{S_x^\alpha(f)}{\sqrt{S_x^0(f + \alpha/2)S_x^0(f - \alpha/2)}} \tag{6}$$

This cyclic coherence function is normalized between 0 and 1, which indicates the strength of the correlation between spectral components spaced apart by an amount of α.

3 Numerical Analysis Results

Figure 1 illustrates a time domain trace for an MCA IN model with impulsive index $A = 0.01$ and the background-to-impulsive noise power ratio $\Gamma = 0.01$. A numerical analysis results for this MCA IN signal is shown in Fig. 2. Notice that neither cyclic spectral or cyclic coherence results exhibit significant components along with the cyclic frequency. However, the measured IN for the NB-PLC environment showed that the IN is predominantly cyclostationary with dominant period equal to half the AC cycle [10].

To assess the resulting BER performance under MCA IN model and real measured IN traces for NB-PLC, we refer to the specification of IEEE P1901.2 [11] to setup the transmitters of the OFDM-based PLC system. We set the number of subcarrier $N = 256$, the number of samples for cyclic prefix (CP) is 30, the modulation scheme is differential binary phase shift keying (DPSK); the number of symbols per frame is 3. The magnitude response of the PLC channel model is illustrated in Fig. 3 and we assume that perfect one-tap equalizers are available. For the real measured IN, we adopt the data set from office space [10].

For fairness consideration, we multiply the measured IN data with a constant gain according to the background-to-impulsive noise power ratio. The resulting BER curves are shown in Fig. 4. Note that although the water-falling trends of the BER curves are similar to each other; however, the MCA IN model does not capture the nature of cyclostationary IN and periodic IN.

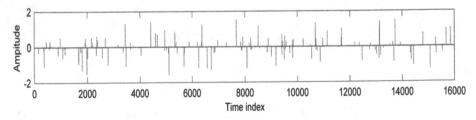

Fig. 1. Time trace for an MCA IN model ($E_b/N_0 = 10$ dB).

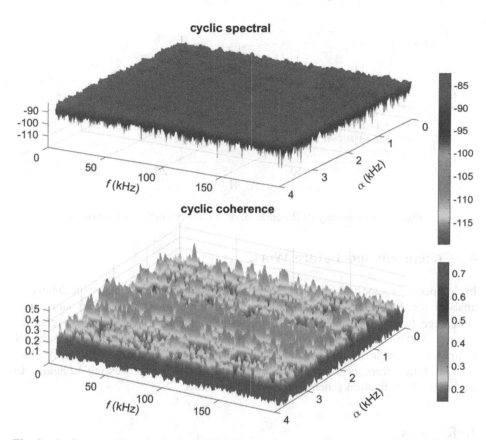

Fig. 2. Cyclic spectral analysis for MCA IN model (top: cyclic spectrum density function; bottom: cyclic coherence function).

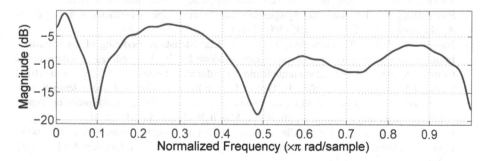

Fig. 3. The magnitude response of the PLC channel with the sampling frequency 1.2 MHz.

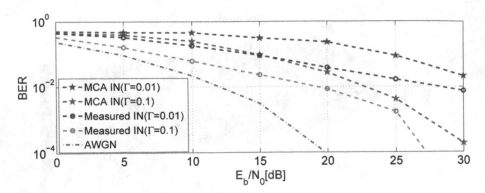

Fig. 4. The resulting BER performance curves with different IN models.

4   Conclusion and Future Work

In this paper, we apply the framework of cyclostationary signal analysis technique to analyze the MCA IN. The resulting cyclic spectrum density function and cyclic coherence function do not exhibit significant components along with the cyclic frequency. Some existing papers have been applied MCA IN model to design the IN mitigation algorithm or analyze the system performance might lead incorrect results.

Our future work is to apply the cyclostationary signal processing technique to design an IN mitigation scheme for NB-PLC systems.

References

1. Nassar, M., Lin, J., Mortazavi, Y., Dabak, A., Kim, I.H., Evans, B.L.: Local utility power line communications in the 3–500 kHz band: channel impairments, noise, and standards. IEEE Signal Process. Mag. **29**(5), 116–127 (2012)
2. Rieken, D.W.: Periodic noise in very low frequency power-line communications. In: Proceedings of International Symposium on Power Line Communications and its Applications, April 2011, pp. 295–300 (2011)
3. Nassar, M., Gulati, K., Mortazavi, Y., Evans, B.L.: Statistical modeling of asynchronous impulsive noise in powerline communication networks. In: Proceedings of International Conference IEEE Global Telecommunications Conference, December 2011, pp. 1–6 (2011)
4. Cortés, J.A., Sanz, A., Estopiñán, P., García, J.I.: On the suitability of the Middleton class A noise model for narrowband PLC. In: Proceedings of International Symposium on Power Line Communications and its Applications, March 2016, pp. 58–63 (2016)
5. Kim, Y., Bae, J.N., Kim, J.Y.: Performance of power line communication systems with noise reduction scheme for smart grid applications. IEEE Trans. Consum. Electron. **57**(1), 46–52 (2011)
6. Korki, M., Hosseinzadeh, N., Vu, H.L., Moazzeni, T., Foh, C.H.: Impulsive noise reduction of a narrowband power line communication using optimal nonlinearity technique. In: Proceedings of International Conference Australasian Telecommunication Networks and Applications, November 2011, pp. 1–4 (2011)

7. Matanza, J., Alexandres, S., Rodriguez-Morcillo, C.: Difference setsbased compressive sensing as denoising method for narrow-band power line communications. IET Commun. 7(15), 1580–1586 (2013)
8. Matanza, J., Alexandres, S., Rodriguez-Morcillo, C.: Compressive sensing techniques applied to narrowband power line communications. In: Proceedings of International Conference on Signal Processing, Computing and Control (ISPCC), September 2013, pp. 1–6 (2013)
9. Antoni, J.: Cyclic spectral analysis in practice. Mech. Syst. Signal Process. 21, 597–630 (2007)
10. Nieman, K.F., Lin, J., Nassar, M., Waheed, K., Evans, B.L.: Cyclic spectral analysis of power line noise in the 3–200 khz band. In: Proceedings of International Symposium on Power Line Communications and its Applications, March 2013, pp. 315–320 (2013)
11. IEEE standard for low-frequency (less than 500 kHz) narrowband power line communications for smart grid applications, IEEE Std 1901.2-2013, pp. 1–269, December 2013

Cerebral Apoplexy Image Segmentation Based on Gray Level Gradient FCM Algorithm

Wenai Song[1], Xiaoliang Du[1], Qing Wang[2,3(✉)], Yi Lei[1],
WuBin Cai[1], and Xiaolu Fei[4]

[1] School of Software, North University of China, Taiyuan 030051, China
[2] Guangdong Province Key Laboratory of Popular High Performance
Computers of Shenzhen University, Shenzhen, China
[3] Research Institute of Information Technology, Tsinghua University,
Beijing 100084, China
qing.wang@tsinghua.edu.cn
[4] Information Center, Xuanwu Hospital of Capital Medical University,
Beijing, China

Abstract. Fuzzy clustering algorithm as a more successful segmentation algorithm has been successfully applied in the medical field. However, the traditional Fuzzy C-means clustering (FCM) algorithm has the disadvantages of time-consuming, noise-sensitive and non-consideration of neighborhood information in the segmented brain MRI (MRI), and proposes a corresponding solution to these problems. Firstly, Canny operator and morphological processing method is employed to extract the brain MRI of image contour information, reducing the image background brings a series of calculation problem. Secondly, before the FCM image segmentation, the adaptive adjustment of the weight coefficient in the neighborhood is realized by introducing the gradient information to achieve the purpose of eliminating the noise and reducing the initial value of the image objective function. With the experiment proved above, the robustness of the algorithm is improved and effectively shorten the calculation time in the case of constant accuracy.

Keywords: Image segmentation · Stroke · FCM · Gray gradient

1 Introduction

The name of the stroke is cerebral apoplexy, mainly using MR images for disease diagnosis and treatment in medicine. But in clinical practice, a sequence of brain MR images is about 200 or so, to find and determine the lesion from a large number of unclear magnetic images, it is a time-consuming and laborious work for doctors. At present, there are mainly for brain MR image segmentation algorithm: threshold segmentation, fuzzy clustering, random field model, active contour model and so on [1, 2]. Although there are abundant research results in the field of medical image segmentation, the brain tissue image segmentation is ineffective due to the low contrast in the magnetic field image, the variability of tissue features, the unclear boundary between

© Springer Nature Singapore Pte Ltd. 2018
J. C. Hung et al. (Eds.): FC 2017, LNEE 464, pp. 438–447, 2018.
https://doi.org/10.1007/978-981-10-7398-4_45

different soft tissue and lesion. Therefore, for different clinical diseases often need different ways to segment.

In view of the above problems, this paper uses FCM algorithm to segment the brain magnetic images, prepare for the extraction of the lesion. Fuzzy C-means, referred to as FCM, The concept of fuzzy is introduced, which is the generalization of K-means algorithm [3]. FCM algorithm uses iterative optimization objective function to obtain the fuzzy classification of the data set, has a good convergence. Therefore, the successful application of the algorithm in medical image segmentation [4]. However, there are many defects in traditional FCM image segmentation: (1) the amount of calculate, clustering is a nonlinear optimization process, and image segmentation is a large sample classification problem, the iterative algorithm is computationally intensive, (2) the use of spatial information, takes into account the gray features, ignoring the inherent rich spatial information in the image, so that the segmentation of the region is discontinuous, effective use of the spatial information can improve the quality of the division, but the amount of calculate is increased.

To solve the problem above, the first step is to extract the edge of the magnetic image of the brain by using the Canny operator, for the kernel image, is obtained by morphological hole filling and mask operation. In the second step, the noise in the image is often the extreme value of the gradient in the neighborhood, and the gradient value of each pixel in the image is calculated. The gradient value is used to give different weights in the neighborhood of the neighborhood. The other gray values give the new gray value to the noise point, and the non-extreme value is not handled, so that most of the details of the image are preserved while ensuring the elimination of the noise, so as to reduce the number of iterations.

2 Extraction Edge Based on Canny Operator

Edge detection is based on the edge of the pixel gray value changes violently, solve the problem of image segmentation by detecting the edge between different uniform areas [5]. The basic idea of Canny edge detection is to first choose a certain Gauss filter to smooth the filter, and then use non-extreme suppression technology to deal with the final edge image. Specific steps are as follows:

- Smooth the image with a Gaussian filter;
- The magnitude and direction of the gradient are calculated using the finite difference of the first order derivative;
- The magnitude of the gradient is non-maximal;
- Using a double threshold algorithm and connecting edges.

As show in Fig. 1(a), after the Canny operator, there are often some breakpoints in the image, Split the target image and background by the hole filling technique (show in Fig. 1b) and the mask operation (show in Fig. 1c). And then set the background gray value t=0, which eliminates the noise in the background of the image and reduces the computational complexity of the algorithm.

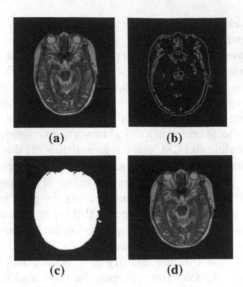

(a)   (b)

(c)   (d)

Fig. 1. (a) Original drawing (b) Edge extraction, (c) Hole filling, (d) The final segmentation results

3 FCM Image Segmentation Algorithm

Fuzzy clustering is first proposed by Dunn [6] and improved by Bezdek et al., as a general fuzzy C-means clustering algorithm, FCM algorithm based on least squares is proposed in the related literature. And Bezdek [7] proved its convergence, says that the algorithm converges to an extreme value.

The definition $\{x_i, i = 1, 2, \ldots, n\}$ is the sample set of n samples, C is the set number of classes, m_j is the center of each cluster, $\mu_j (x_i)$ is the i_{th} sample For the j_{th} class membership function. The objective function defined by the membership degree can be written as:

$$J_{FCM} = \sum_{j=1}^{C} \sum_{i=1}^{n} [\mu_j(x_i)]^P \|x_i - m_j\|^2 \tag{1}$$

In Eq. (1), the constant p > 1 controls the degree of blurring of the clustering results. The membership function requires the following conditions:

- For any of the j and i, $\mu j(xi) \in [0, 1]$;
- For any of the i, $\sum_{j=1}^{C} \mu_j(x_i) = 1$
- For any of the j, $0 < \sum_{j=1}^{n} \mu_j(x_i) < n$

Under the constraints of the above conditions, the minimum value of the objective function is obtained, and the partial derivatives of the JFCM to the cluster center m_j and the membership function μ_j are zero, respectively, and the formulas of (2) and (3) are obtained:

$$m_j = \frac{\sum_{i=1}^{n} \left[\mu_j(x_i)\right]^P x_i}{\sum_{i=1}^{n} \left[\mu_j(x_i)\right]^P} j = 1, 2, \ldots, C \tag{2}$$

$$\mu_j(x_i) = \frac{1/\left(\|x_i - m_j\|^2\right)^{1/(P-1)}}{\sum_{i=1}^{n} 1/\left(\|x_i - m_j\|^2\right)^{1/(P-1)}} \tag{3}$$

(2) and (3) are obtained by iterative method. When the algorithm converges, the clustering centers and the samples belong to different categories of membership values, and the fuzzy clustering is completed.

4   Adaptive Filtering Based on Gradient

The traditional fuzzy C-means algorithm does not consider the pixels in the neighborhood, so it is sensitive to the noise. For these defect, Proposed FCM_S algorithm [8], FCM_S1 and FCM_S2 [9] respectively, although these algorithms have taken the neighborhood into account, it fixed in the field coefficients, for different neighborhoods within the different pixels should have a different weight can achieve a better get rid of the noise effect.

4.1   Neighborhood Gradient

In the improvement of neighborhood problems, someone have adopted the distance way to give different weights to different pixels in the neighborhood, but in application there are problems show in Fig. 2. Figure 2(a) center pixel Gray value of 68, A pixel gray value of 50 its membership is different from the center point, according to the distance principle A point is given a higher weight, B pixel and center point membership of the same, but was given a smaller weight, The final weight assignment results shown in Fig. 2(b). So the use of distance in practical applications often lead to false assignments.

50	55	75
	A	B
72	75	70
60	68	80

(a)

0.1	0.4	0.1
	A	B
0.4	1	0.4
0.1	0.4	0.1

(b)

Fig. 2. (a) Gray value, (b) Based on distance value

A new neighborhood solution is proposed to solve the above problem. The gray value transformation in the brain magnetic image corresponds to the transformation of the gradient value. Therefore, according to the transformation of the gradient value, it is more scientific to assign different weights to the pixels in the neighborhood than the distance-based assignment method. The larger the gradient value is, the smaller the weight given, and the smaller the gradient value is, the greater the weight given. So the use of domain-wide gradient values to normalize the neighborhood points to give different weights. The formula is as follows:

$$X = \frac{1}{N-1}\sum_{j=1}^{N} \alpha_j x_j \quad N = 1, 2, \ldots, 8 \tag{4}$$

$$\alpha_j = 1 - \frac{\mu_j - \mu}{\sum_{j=1}^{N}\|\mu_j - \mu\|} \quad j = 1, 2, \ldots 8 \tag{5}$$

Where μ_j is the gradient value corresponding to the pixel j in the neighborhood, μ is the average of the gradient values in the neighborhood, α_j is the weight of the original image corresponding to the pixel, and the pixel in the neighborhood is the gray value of the pixel i of the generated new image.

4.2   Define Replacement Rules

The gradient value of the noise in the image is often the extreme value in the neighborhood, and the gradient value of the ordinary pixel is often not the extreme value in the neighborhood. Therefore, in order to achieve the purpose of removing the extreme value, before the replacement of gray scale to determine whether the neighborhood is the extreme value, if this is the case, replace it, otherwise do not deal with. This eliminates the noise while preserving the details of the image as much as possible (Fig. 3).

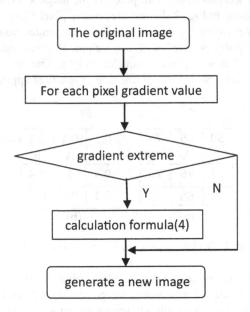

Fig. 3. Flow chart

Compared with the original image (Fig. 4(a)) and the post-processing (Fig. 4(b)), it was found that the details of the image were sufficiently retained while the extreme points were processed.

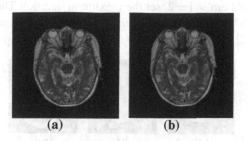

(a)   **(b)**

Fig. 4. (a) The original image, (b) Removal of extreme image

5   Analysis of Results

5.1   Experimental Evaluation Criteria

In this paper, from the point of view of clustering, two criteria are selected in many evaluation criteria, and the three algorithms involved are compared. The relevant criteria for selection are as follows:

1. The first quantitative evaluation criterion is called the Bezdek division coefficient [10], which is defined as follows:

$$V_{PC} = \frac{1}{n} \sum_{i=1}^{C} \sum_{j=1}^{n} \mu_{ij}^2$$

As can be seen from the V_{PC}, a good cluster should make the image pixels belong to a class as much as possible, and other class membership should be as small as possible. So a good clustering V_{PC} value should be as large as possible.

2. The second quantitative evaluation standard is called the Xie-Beni [11], as defined below:

$$V_{XB} = \frac{\sum_{i=1}^{C} \sum_{j=1}^{n} \left(\mu_{ij}^2 \right)^m \|y_k - v_j\|^2}{N \min_{\forall j \neq k} \|v_j - v_k\|}$$

As can be seen from the definition of V_{XB}, the value looks for a balance between intraclass compactness and interclass class separation. In the image segmentation, where the molecular part/sample number N, said the class compactness, the smaller the value the better. The denominator indicates the distance between classes, and the larger the value is, the better. Thus, a good clustering segmentation, its V_{XB} value should be as small as possible.

5.2   Experimental Results

In this experiment, we use the true 100 real brain magnetic image, the size is 512 * 512, the selected clustering number k = 5, select the fuzzy factor value m = 2, select the iteration stop threshold em = 1e−2, set the maximum iteration The number of times is 500. Figures 5 and 6 below selected two brain magnetic images as a description.

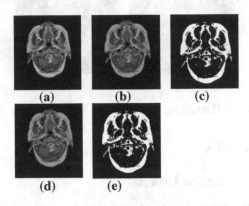

Fig. 5. (a) The original image, (b) FCM, (c) FCM segmentation, (d) This paper clustering, (e) This paper segmentation

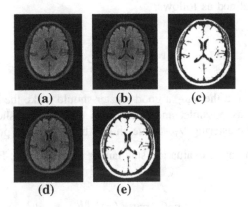

Fig. 6. (a) The original image, (b) FCM, (c) FCM segmentation, (d) This paper clustering, (e) This paper segmentation

It is found that the initial value of the image function is significantly reduced by the algorithm, and the number of iterations is obviously reduced. Figure 5 as an example to illustrate the initial value of its function and the number of iterations reduced, as shown in Fig. 7.

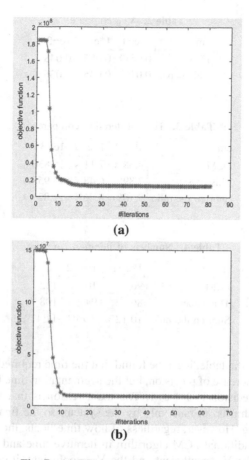

Fig. 7. (a) FCM, (b) Improved algorithm

By simple comparison, it can be found that the number of image iterations processed in the general magnetic image is about 10% lower than that of the traditional FCM algorithm with the adaptive gradient value. The specific parameters are as follows.

5.3 Split Quality Comparison

The effects of the improved algorithm and the traditional FCM on the 100 real nuclear magnetic images are compared with the following Tables 1, 2, 3 and 4:

Table 1. V_{PC} compare

em	1*e−1	1*e−2	1e−3
FCM	0.898	0.890	0.892
This paper	0.906	0.897	0.913

Table 2. V_{XB} compare

em	1*e−1	1*e−2	1e−3
FCM	0.0579	0.055	0.048
This paper	0.046	0.045	0.036

Table 3. Time of iteration compare

em	1*e−1	1*e−2	1e−3
FCM	8.68 s	7.11 s	15.830 s
This paper	7.89 s	6.48 s	13.958 s
Shorten the ratio	9.10%	8.86%	11.82%

Table 4. Numbers of iteration compare

em	1*e−1	1*e−2	1e−3
FCM	96	103	122
This paper	86	94	108
Shorten the ratio	10.12%	8.73%	11.47%

From the data in the table, it can be found that the time required for segmentation is increased with the increase of precision, but the segmentation time is longer than that of 1*e−2 when the precision is 1 * e−1, The main reason is that the initial clustering center of the algorithm is set randomly by the system, so the final segmentation time will be some floating. However, regardless of how time floats, the improved algorithm is superior to the traditional FCM algorithm in iterative time and iteration times.

By comparing the V_{pc} coefficient and the V_{xb} coefficient, it can be found that the improved algorithm has the same V_{pc} coefficient as that of the traditional FCM, and the V_{xb} coefficient of the inter-class distance coefficient is greatly improved. The improved algorithm is finally more representative of the classification center. In general, the improved algorithm achieves the purpose of improving the operation efficiency of FCM, shortening the iteration time and less the number of iterations, while improving the image segmentation effect compared with the traditional FCM.

6   Conclusion

The FCM algorithm is widely used in medical image processing, because of its low efficiency; the improved algorithm does not achieve a good balance between efficiency and accuracy. Therefore, aiming at this problem, proposes an adaptive FCM algorithm based on gradient value. The core is to use the gradient value as the pixel weight in the neighborhood, which reduces the sensitivity of the algorithm to the extreme point and reduces the initial objective function and reduce the number of iterations. However, the initial clustering center of this method is randomly generated, which leads to a series of problems such as useless iteration and clustering time with some instability. In order to reduce the iterative time and improve the efficiency of the algorithm, select the iterative center and add the neighborhood factor in the function expression for further study, further shorten the algorithm running time and improving the image segmentation accuracy.

Acknowledgements. This work is supported by China National Key Technology Research and Development Program project with no. 2015BAH13F01 and Guangdong Province Key Laboratory of Popular High Performance Computers Research Program.

References

1. Wang, A.-M., Shen, L.-S.: Study survey on image segmentation. Measur. Control Technol. **19**(5), 1–6 (2000)
2. Bezdek, J.C., Clarke, L.P.H.: Review of MR image segmentation techniques using pattern recognition.Med. Phys. **20**(4), 1033–1048 (1993)
3. Pal, N.R., Bezdek, J.C.: On cluster validity for the fuzzy c-means model. IEEE Trans. Fuzzy Syst. **3**(3), 370–379 (1995)
4. Cai, J.-Xi., Yang, F., Feng, G.-C.: Degeneracy improved semi-supervised fuzzy clustering with application in MR image segmentation. J. Image Graph. **16**(5), 784–791 (2011)
5. Duan, H.-Y., Shao, H., Zhang, S.-Z.: An improved algorithm for image edge detection based on canny operator. J. Shanghai Jiaotong Univ. **50**(12), 1861–1865 (2016)
6. Pham, D.L.: Spatial models for fuzzy clustering. Comput. Vis. Image Underst. **84**(2), 285–297 (2001)
7. Chen, S.C., Zhang, D.Q.: Robust image segmentation using FCM with spatial constraints based on new kernel-induced distance measure. IEEE Trans. Syst. Man Cybern. Part B Cybern. **34**(4), 1907–1916 (2004)
8. Ahmed, M.N., Yamany, S.M., Mohamed, N., et al.: A modified fuzzy c-means algorithm for bias field estimation and segmentation of MRI data. IEEE Trans. Med. Imaging **21**(3), 193–199 (2002)
9. Szilagyi, L., Benyo, Z., Szilagyi, S., et al.:MR brain image segmentation using an enhanced fuzzy c-means algorithm. In: 25th Annual International Conference of IEEE Engineering in Medicine and Biology Society, pp. 724–726. IEEE Computer Society, Washington, DC (2003)
10. Bezdek, J.C.: Cluster validity with fuzzy sets. Cybernertics **3**(3), 58–73 (1974)
11. Xie, X.L., Benii, G.: A validity measure for fuzzy clustering. IEEE Trans. Pattern Anal. Mach. Intell. **13**(8), 841–847 (1991)

Data Explosion Model for Public Safety Data Processing and Its Application in a Unified Security System

Pan Gao[1], Zheng Xu[1(⊠)], and Shuhong Gao[2]

[1] The Third Research Institute of the Ministry of Public Security,
Shanghai 201142, China
13817917970@163.com
[2] Shenzhen Senior High School, Shenzhen, China

Abstract. In recent years, with safe city and the construction and development of the intelligent City project, has become a public security authority security control video surveillance systems, combat crime, and effective means to prevent emergency incidents. With the rapid development of network communication technology and mobile intelligent terminals (such as smart phones, tablets, etc.) the rapid proliferation of smart terminals have been carrying video surveillance, audio, speed sensors and sensing devices. Video equipment parts in high-end smart terminal can carry over parts of the lower end of video surveillance equipment. Intelligent terminal mass popularity makes building a people-centric sensing and computing networks possible in order to achieve the perfect fusion of the physical world and the digital world. Effective integration of different information spaces of information can enhance public safety and effective sensing and detection. According to public safety incidents of multi-source information fusion is proposed surge of data model, and the model is defined. And witness in one system by the model was verified. The unity of witness systems have been developed in several Beijing bus station and train station.

Keywords: Public safety · Data fusion · Surge models

1 Introduction

In recent years, with the construction and development of safe city and smart city project, video surveillance system has become an effective means of public security organs of public security management and control, to combat crime and prevent emergency emergencies. As before the end of 2012, the country has built more than 20 million video surveillance equipment, including direct access to the public security organs in more than 3 million, more than 95% criminal cases are involved to video surveillance equipment. However, the cost of dedicated video surveillance equipment is high, the average video surveillance equipment average 60 thousand yuan/per point, HD equipment 100 thousand yuan/per point. The further development of video surveillance system is facing a huge cost bottleneck, can not achieve a wider range of popularity, the need for a fundamental breakthrough in technology to significantly reduce the cost of construction and maintenance. With the rapid development of

J. C. Hung et al. (Eds.): FC 2017, LNEE 464, pp. 448–452, 2018.
https://doi.org/10.1007/978-981-10-7398-4_46

network communication technology and intelligent mobile terminal (such as intelligent mobile phone, tablet computer) the rapid popularization of intelligent terminals, has been carrying audio, video monitoring, acceleration sensor sensing equipment. Some high-end smart terminal can carry video equipment has been part of the low-end video surveillance equipment. With the popularity of intelligent terminals, it is possible to construct a human centered perception and computing network, in order to realize the perfect integration of the physical world and the digital world. The effective integration of information in different spatial information can enhance the effective detection and detection of public security incidents.

In recent years, the Internet of things has aroused widespread concern around the world, known as the next trillion level of the industry, the market outlook will far exceed the computer, the Internet, mobile communications, etc. Following the computer, the Internet, the Internet of things led to another wave of information industry revolution. The world's science and technology powers will put the Internet of things in the future development of an important position in the strategy, invested heavily in in-depth research and exploration, and actively seize the commanding heights of the development of things, cultivate new economic growth points. Urban perception is an important form of Internet of things in the city. With the rapid development of network communication technology and intelligent mobile terminal (such as intelligent mobile phone, tablet computer) the rapid popularization of intelligent terminals, has been carrying audio, video monitoring, acceleration sensor sensing equipment. With the popularity of intelligent terminals in the city, it is possible to construct a human centered urban sensing and computing network. In the sensing node calculation of large-scale city perception (including fixed camera coverage of the mobile intelligent terminal in the city, etc.) the information is generally huge and diverse, modal volume generated fast and huge value, there are very obvious characteristics of big data.

In this paper, we propose a data surge model for solving the problem of multi-source information fusion of public security events. And the model is verified to witness one system. The development of witness one system has been in the city of Beijing bus station and train station was carried out.

2 Related Works

The development of the Internet of things today, demand for thorough perception is more and more intense, with advances in wireless communication and sensor technology, the market for smart mobile phone, tablet computer, vehicle sensing device and mobile terminal integrated sensor has more and more, powerful perception, storage, communication and computation ability. With the explosion of the wireless mobile terminal devices, after more than ten years to develop the awareness of how to use the specific sensor deployment provides service, networking will provide a larger and more complex, thorough and comprehensive service through the use of these ubiquitous mobile devices, which develops into an era the new [1]. In recent years, with the idea of mobile sensing scholars "Crowdsourcing" combination of mobile devices will ordinary users as the basic sensing unit, collaborate consciously or unconsciously through the mobile Internet, the formation of "crowdsourcing networks" [2–9], realize the sensing task distribution and sensing data collection, the city completed the task of perception large, complex.

3 Data Surge Model

According to the characteristics of public security incidents, such as multiple sources, various types of data and wide range of data, we believe that there are several challenges.

Multi source heterogeneous data awareness ability of public security incidents. Based on the lack of the traditional model of conscious fixed deployment, it brings a new opportunity for the development of public security, but also brings many challenges. On the one hand, the number of nodes in sensor network, more and more types of a wider range of transmission sensing data more diverse, with great potential for the formation of the node's extensive participation, perceived content rich, scalable, low cost of full coverage of the perception of the network; on the other hand, perception between nodes is usually loose the coupling, the specific performance due to resource constrained nodes leads to lower willingness to participate because the node is unconscious and non professional perception of low quality, lead to data redundancy due to the diverse ways of network communication and dynamic result data transmission efficiency is not high, because the nodes of distributed deployment leads to a lack of effective collaborative resource optimization mechanism. Therefore, a problem of public security big data computing is facing is: how to solve the data redundancy, low node participation perceived low quality, the transmission efficiency is not high, the lack of collaborative resource optimization mechanism and the contradiction between the extensive data acquisition effectively.

The lack of cognitive ability of massive multi modal public safety data. On the one hand, the public security data with multi modal features, which contains both structured data and unstructured data contained more and semi-structured data; on the other hand, the public security data has many associated features, including time correlation, spatial relevancy, logical relationship. These characteristics increase the difficulty of representing and managing the large amount of multi-modal data, affect the effectiveness and accuracy of the understanding of media data, and reduce the efficiency of the use of public security data. The traditional data management and analysis technology is mainly aimed at a single structured data, it is difficult to express multi modal data and describe the relationship between them, facing the challenge of cognitive depth and analysis efficiency. Therefore, a problem of public security big data computing faces is: how the massive multimodal data representation and management, how to analyze and find different correlation between the data, comprehensive value for later use ready.

Insufficient capacity of public security data aggregation. In the complex urban environment, the public security data show significant fragmentation characteristics, which are independent of fragmented information on the micro scale, the user does not have much value. However, the information on the macro is interrelated, and it is the fusion and structural description of a particular environment or thing. Needs to establish the fragmentation information deep fusion and aggregation, to achieve the conversion of information from the debris to swarm intelligent information, realize the city public safety incident fast, comprehensive, complete and correct cognition and grasp. Need to study the public safety incident fragmentation information scene reconstruction method based on existing methods, overcome technical bottlenecks, make full use of the cross sensing information of swarm intelligence sensing network acquisition, and ultimately achieve the public safety incident complete scene reconstruction. Therefore, a problem of public security big data computing faces is: how to put the value through the unified expression of aggregation and fusion, efficient integration method, the formation of fragmented information value of the polymerization mechanism, value reconstruction problem solve fragmentation information.

Data surge: a single data surge can be seen as an individual, a group, a data set of events. These data are described in terms of a particular individual, group, event, etc. Data can come from different data sources, such as cell phone, video monitor, etc.

Surge timeliness: Data surge is not always exist, with timeliness. In some cases, there is a data surge, most of the time, the data is stable, there is no surge. Therefore, the timeliness of data surge is an important time threshold for public security event detection.

Unicom: data from the surge wave to surge on bottom is connected, can be regarded as the internal data set on China unicom. Data within a single surge is associated.

Surge height: below above sea level height depends on the depth of plane waves Yu Hai Chung bottom, can data structure data representation depends on events inside as. The surge height can be seen as the amount of data, and the wave height can be seen as an abnormal amount of data, for the detection of waves can be seen as the detection of public safety incidents of abnormal data.

Surge separation line: sea level and data acquisition bottleneck line with timeliness, is the representation of the data and the separation of deep data lincs.

Surge Island: in the sea level and above the bottleneck line, data sets and data sets between the island. These data are not connected to each other.

Surge correlation: there is a connection between the bottom of the surge and the underlying data.

Surge line: bottlenecks are inversely proportional to the height dimension of the data collection and the bottle neckline. Beyond the bottleneck, the dimension of the data collection becomes less, since there is no correlation between the data. Only multi dimensional data acquisition and data mining can break through the bottleneck of data acquisition to achieve the multi-dimensional correlation between the data.

The surge of sparsity: an update of the higher density, the greater the amount of data share, data association is stronger, directly visible to a small amount of data acquisition can support data collection bottle neckline breakthrough. The ideal data acquisition mode of the information system should be the connection between the data acquisition bottleneck and the data link.

4 Conclusions

In recent years, with the construction and development of safe city and smart city project, video surveillance system has become an effective means of public security organs of public security management and control, to combat crime and prevent emergency emergencies. With the rapid development of network communication technology and intelligent mobile terminal (such as intelligent mobile phone, tablet computer) the rapid popularization of intelligent terminals, has been carrying audio, video monitoring, acceleration sensor sensing equipment. Some high-end smart terminal can carry video equipment has been part of the low-end video surveillance equipment. With the popularity of intelligent terminals, it is possible to construct a human centered perception and computing network, in order to realize the perfect integration of the physical world and the digital world. The effective integration of information in different spatial information can enhance the effective detection and detection of public security incidents. In this paper, we propose a data surge model for solving the problem of multi-source information fusion of public security events. And the model is verified to witness one system. The development of witness one system has been in the city of Beijing bus station and train station was carried out.

Acknowledgment. This work is supported by National Key R&D Program of China (No. 2017YFC0803700). The authors of this paper are members of Shanghai Engineering Research Center of Intelligent Video Surveillance. This work was supported in part the National Natural Science Foundation of China under Grant 61300202, 61332018, 61403084. Our research was sponsored by Program of Science and Technology Commission of Shanghai Municipality (No. 15530701300, 15XD15202000, 16511101700), in part by the technical research program of Chinese ministry of public security (2015JSYJB26, 2015QZX002), was sponsored by CCF-Venustech Open Research Fund (Grant No. CCF-VenustechRP2017006).

References

1. Ganti, R.K., Ye, F., Lei, H.: Mobile crowdsensing: current state and future challenges. IEEE Commun. Mag. **49**(11), 32–39 (2011)
2. Rai, A., Chintalapudi, K.K., Padmanabhan, V.N., Sen, R.: Zee: zero-effort crowdsourcing for indoor localization. In: Proceedings of ACM MobiCom, pp. 293–304 (2012)
3. Eriksson, J., Girod, L., Hull, B., Newton, R., Madden, S., Balakrishnan, H.: The pothole patrol: using a mobile sensor network for road surface monitoring. In: Proceedings of ACM MobiSys, pp. 29–39 (2008)
4. Koukoumidis, E., Peh, L.S., Martonosi, M.R.: SignalGuru: leveraging mobile phones for collaborative traffic signal schedule advisory. In: Proceedings of ACM MobiSys, pp. 127–140 (2011)
5. Ra, M.R., Liu, B., La Porta, T.F., Govindan, R.: Medusa: a programming framework for crowd-sensing applications. In: Proceedings of ACM MobiSys, pp. 337–350 (2012)
6. Yan, T., Kumar, V., Ganesan, D.: Crowdsearch: exploiting crowds for accurate real-time image search on mobile phones. In: Proceedings of ACM MobiSys, pp. 77–90 (2010)
7. Yang, D., Xue, G., Fang, X., Tang, J.: Crowdsourcing to smartphones: incentive mechanism design for mobile phone sensing. In: Proceedings of ACM MobiCom, pp. 173–184 (2012)
8. Rachuri, K.K., Mascolo, C., Musolesi, M., Rentfrow, P.J.: Sociablesense: exploring the trade-offs of adaptive sampling and computation offloading for social sensing. In: Proceedings of ACM MobiCom, pp. 73–84 (2011)
9. Packer, H.S., Samangooei, S., Hare, J.S.: Event detection using Twitter and structured semantic query expansion. In: Proceedings of ACM Workshop on Multimodal Crowd Sensing, pp. 7–14 (2012)

Intelligent Video Analysis Technology of Public Security Standard Sets of Data and Measurements

Huan Du[1], Zheng Xu[1(✉)], Zhiguo Yan[1], and Shuhong Gao[2]

[1] The Third Research Institute of the Ministry of Public Security,
Shanghai, China
13817917970@163.com
[2] Shenzhen Senior High School, Shenzhen, China

Abstract. Video surveillance technology has become an indispensable method of public security work. As the number of monitored devices is increasing, resulting in a large amount of video data, resulting in great strength brought by police officers. From the early days of license plate recognition to the nearest human face than, vehicles feature recognition technology, are typical applications of intelligent video analysis, has produced positive results in public security work. With the progress of artificial intelligence techniques, especially the depth of learning in the field of video analysis to continuously refresh the recognition task, can see into the future will have more intelligent video analysis technology penetration into the field of public safety. However, intelligent analysis techniques in the field of public security still faces great challenges and bottlenecks.

Keywords: Intelligent video analysis · Public security · Standard sets

1 Introduction

Analysis of existing video technology is often in the context of experimental testing and analysis, generally only implemented in a specific environment. But the actual monitor environment is often more complex and diverse, intelligent analysis of this has led to most of the existing products and public safety needs there is a gap.

Existing video analysis technology features a single, unable to meet the variety of needs of police. For example, can search similar faces do not search people with moles, but "some moles on the face" of public security detection may be an important clue. Video analysis technologies capable of processing the data of the existing smaller increases computing device space and calculation of cost, lack of effective top design to meet public security video real time analysis of data [1].

Police actual combat personnel and high communication cost of video analysis technician, led video analysis technology process was too cumbersome. Led above one of the main reasons is that traditional intelligent video analysis technology usually in the academic study of standard data sets and related tasks (such as object detection and recognition based on ImageNet data [2–4]). Video contains scenes of academic research was simple and high quality task definition is also simple; scenes in public security

© Springer Nature Singapore Pte Ltd. 2018
J. C. Hung et al. (Eds.): FC 2017, LNEE 464, pp. 453–456, 2018.
https://doi.org/10.1007/978-981-10-7398-4_47

surveillance data more complex and complicated task and needs more. Recently, has set the maximum data exposed on the ImageNet's performance has exceeded 90%, whether academic or application requirements have been met is a bottleneck. In such a context, the proposed video analysis for public safety data set specification and related tasks and assessment criteria are more important. Not only have significant research value, and reduces the communication costs of security personnel and researchers, promoting video technology for public safety health and efficient growth, fully functional video analysis technology in public security safety [5, 6].

2   Alert Analysis

Alert district analysis alarm is intelligent video analysis system of a most important typical application, it similar Yu traditional of movement detection, but and than original of movement detection has more high reliability and trust degrees, traditional of movement detection just in need special of concern regional delineation a (or some) rectangle sensitive regional, dang the regional pixel occurred changes, and and reached must degree Shi triggered alarm, this detection way exists some obviously of defects, (1) sensitive regional in light or color suddenly changes Shi, very easy occurred errors alarm; (2) By set of sensitive regional within cannot contains any can movement or at any time changes of scene, and objects, as flashing of lights, and water of fluctuations, and tree shadow of shake and so on, intelligent video analysis of alert district alarm not simple to according to pixel brightness, and color of changes and judge whether exception, but through on may appeared of target size, and form, and movement law recognition Hou only triggered alarm, intelligent video analysis system can accurate to judge, and difference out people, and car, and ship, target of different, Only for need concern of target appeared only to out tips, and excluded tree shadow, and waves of interference, as: in road Shang excluded vehicles of movement only recognition out road in the was through of exception, so, alert district analysis alarm can should complex of environment, to out real of violation security rules of alert, also, due to alert district of number, and shape, and size, and location set are can free set, is not is single of rectangle square, so, more for prevention target regional of set more detailed, and precise.

Cordon analysis is in video image in the human set a article "cordon", main for on beyond the cordon of behavior to out tips, this tips is can according to different target to for of, usually, cordon of through direction is can set of, it can is one-way through alarm or two-way through alarm, so-called one-way alarm is from a direction through alarm and addition a direction through is not alarm, it can for on various only allows one-way through of regional for fortification, and two-way way is for those shall not into, and The fortification of the area or the door.

3   Video Enrichment

Video surveillance systems in recent years has been widely applied in intelligent transportation, sensitive public and State security departments. These video surveillance systems can provide people in security and forensic evidence provided strong protection.

Widely using of video monitoring system brings has mass of monitoring video data, and mass of video data caused has two a significantly of problem: (1) due to storage space of limit, most history video in storage several time Hou on will was delete, to led to has video in the of useful information lost; (2) due to data volume of huge, to artificial browse of way on history video data for query and retrieved not only is took, and also will for data more, time long and produced many people for omissions.

To solve these two problems, efficient video storage, viewing and analysis technology is particularly important. In the video store and browsing, video technology can concentrate under the premise of not moving target information is lost, in terms of both time and space to concentrate the original video, thus shortening the length of the original video, and ease the pressure on storage. In addition, due to the complete record of the moving objects spatio-temporal information in the original video, video enrichment can also support video fast facts back and moving target. Therefore, the video enrichment for mass-efficient use of video has a very large value.

4 Vehicle License Plate Recognition

With the development of pattern recognition, intelligent traffic system for vehicle license plate character recognition has become an important part of, from complex backgrounds, it can accurately extract, recognize car number plates, vehicle information such as type, occupies a very important position in the traffic control and surveillance, with a wide range of applications. Therefore, the identification of license plates have become a modern traffic engineering in the field of research focuses on and one of the hot issues.

Due to the environment, to identify the vehicle models complex and the effects of vehicle license plate location is not fixed, bring some difficulties to the choice of license plate location method. Pollution, defects of the license plate itself can also affect the rate of recognition. Some vehicles because of the bad weather or road conditions, making license plates being contaminated by dust, dirt, and other vehicles for a long time, characters on the license plate has some defect, when severe, is difficult to distinguish between characters on the license plate of the human eye. These conditions will affect the system identification led to mistaken identification. Obviously, to improve positioning system accuracy and character segmentation system will encounter a lot of difficulties.

There is no doubt that if vehicle license plate character recognition system with high adaptability and robustness, that is, to a certain noise or distortion of the character is able to correctly identify the image, it will greatly relieve the pressure on vehicle license plate location and character segmentation system. Therefore, design a good license plate character recognition system for anti-jamming performance of automatic vehicle license plate recognition system as a whole is beneficial. Meanwhile, character in some ways reflects the General characteristics of the graph, is a special kind of graph, its automatic recognition has been the subject of much attention because it solutions for production and life is of great practical significance.

Because the vehicle is a motor vehicle only manage identity symbols, plays an irreplaceable role in traffic management, vehicle license plate recognition system should therefore have a high accuracy rate, the environmental light conditions, shooting position and vehicle speed and other factors have greater safety threshold, and asked to meet real-time requirements.

5 Conclusions

With the widespread use of intelligent technology and markets are no longer satisfied with existing technology types, but look for newer algorithms, more business, more integrated systems, and strive to achieve breakthroughs in the breadth and depth of the application. To this end, the security industry has begun to conduct research on next-generation intelligent video analysis technology, put forward a number of new products, new application model, the new system architecture. These new technologies, new products and market integration are developed, to seek new development.

Acknowledgment. This work is supported by National Key R&D Program of China (No. 2017YFC0803700). The authors of this paper are members of Shanghai Engineering Research Center of Intelligent Video Surveillance. This work was supported in part the National Natu-ral Science Foundation of China under Grant 61300202, 61332018, 61403084. Our research was sponsored by Program of Science and Technology Commission of Shanghai Municipality (No. 15530701300, 15XD15202000, 16511101700), in part by the technical research program of Chinese ministry of public security (2015JSYJB26, 2015QZX002), was sponsored by CCF-Venustech Open Research Fund (Grant No. CCF-VenustechRP2017006).

References

1. Zheng, X., Mei, L., Chuanping, H., Liu, Y.: The big data analytics and applications of the surveillance system using video structured description technology. Cluster Comput. **19**(3), 1283–1292 (2016)
2. Zheng, X., Mei, L., Liu, Y., Chuanping, H., Chen, L.: Semantic enhanced cloud environment for surveillance data management using video structural description. Computing **98**(1–2), 35–54 (2016)
3. Zheng, X., Chuanping, H., Mei, L.: Video structured description technology based intelligence analysis of surveillance videos for public security applications. Multimedia Tools Appl. **75**(19), 12155–12172 (2016)
4. Chuanping, H., Zheng, X., Liu, Y., Mei, L.: Video structural description technology for the new generation video surveillance systems. Front. Comput. Sci. **9**(6), 980–989 (2015)
5. Zheng, X., Liu, Y., Mei, L., Chuanping, H., Chen, L.: Semantic based representing and organizing surveillance big data using video structural description technology. J. Syst. Softw. **102**, 217–225 (2015)
6. Zheng, X., Zhi, F., Liang, C., Mei, L., Luo, X.: Generating semantic annotation of video for organizing and searching traffic resources. IJCINI **8**(1), 51–66 (2014)

The Analysis of the Cyberspace Security Using Immune Factors Network Algorithm

Zheng Xu[1(✉)], Yuan Tao[1], and Shuhong Gao[2]

[1] The Third Research Institute of the Ministry of Public Security, Shanghai, China
13817917970@163.com
[2] Shenzhen Senior High School, Shenzhen, China

Abstract. This paper proposed a new classification model of information systems. Immune Network algorithm based on classification model is proposed. It can be positive from a large number of security information to access and extract useful information. It also can analyze the consequences of the threat information and effective measures in a timely manner. Classification protection threat information can be shared in a timely manner. Emergency response, announcements and alerts can be completed in a timely manner.

Keywords: Threat intelligence · Classified protection model
Immune factors network algorithm

1 Introduction

Cyberspace security has risen to national strategies for cyberspace security strategy and action plan. Therefore, classification and protection of critical information infrastructure and important information systems process has been released by the Government. With the development and promotion of new technologies such as cloud computing, Internet of things, the existing security system is challenged. Based on the technical level protection model for large data analysis and threat information is proposed to make this defense, which can improve to a dynamic defense, passive defense, and the proactive defense.

According to the traditional network security threat defense and detect agency information system based on feature detection. APT attack signatures and 0 day attacks are rarely. Therefore, the existing APT attack and 0 day attacks are often detected unusual network behavior, and traditional prevention and detection mechanisms is used to effectively prevent. By APT and 0 days attacks continues to grow, from traditional defense in response to these new threats into active defense, it offers a new way to monitor and manage networks.

Threat intelligence [1–3] faced with new threats, is the inevitable result of the evolution of law based on the threat Center. This article is based on study of the third-party threat intelligence data, further excavations and evaluation of improved protection of classified data. So you can analyze the consequences of the threat information and effective measures in a timely manner, classification protection threats and the timely

© Springer Nature Singapore Pte Ltd. 2018
J. C. Hung et al. (Eds.): FC 2017, LNEE 464, pp. 457–462, 2018.
https://doi.org/10.1007/978-981-10-7398-4_48

sharing of information. If tactical threat intelligence is analysis, APT and 0 day attack can be detected, emergency response also can be effectively done. If strategic threat intelligence is analysis, bulletins and early warning can be timely done.

This paper proposed a new classification model of information systems. Immune Network algorithm based on classification model is proposed. It can be positive from a large number of security information to access and extract useful information. It also can analyze the consequences of the threat information and effective measures in a timely manner. Classification protection threat information can be shared in a timely manner. Emergency response, announcements and alerts can be completed in a timely manner.

The organization of this paper is as follow. The Sect. 2 gives the related work. The Sects. 3 and 4 introduce the proposed method. The Sect. 5 gives the analysis of the proposed method. Conclusions are summarized in the Sect. 6.

2   Related Work

OSI security architecture [4] can ensure security of information exchange among heterogeneous computers. The model defines five categories of security services and eight categories of security mechanisms. Non-repudiation security services, data integrity, data confidentiality, access control, and authentication. Security includes encryption, digital signature, access control, data integrity, authentication, communication services, routing, control and notarization.

PDR [5] model can ensure protection, detection, and response. In this model, is a continuous cycle of prevention, detection and response, so that the defense can continue to strengthen, and the risk of leaks can be reduced. PDR model is based on the theory of time. The basic idea is that all activities related to information security are time-consuming, so information security capabilities can be measured by time scales.

With the development of information security technology, many model. These models include the OSI (open System Interconnection) security architecture, PDR (protection, detection, response) security model SSE-CMM (systems security engineering capability maturity model), IATF (information assurance technical framework) and the hierarchical model.

IATF [7] information protection technology is divided into four key areas: networks and infrastructures, regional borders, computing environment and supporting facilities. In each of these focus areas, introduces unique security requirements and technical measures, appropriate selection, so security needs to conduct a comprehensive analysis of information systems, and security mechanisms can be duly taken into account. SSE-CMM [6] is the standard implementation of information security engineering evaluation; the main application is the information security engineering capability. The model covers the entire life of the project cycle, including management activities, activities and project activities.

Classification for information security model [8] is the multi-level protection of information systems. Multiple levels from the first to the fifth level. The higher the level of information more important, more stringent security requirements.

3   Intelligent Classification Based on Threat Model

Classification for information security model includes five requirements into the module. These modules include the taxonomy, the system records, construction and reform, evaluation, supervision and inspection. These five modules are in circulation. Classification model of information system as shown in Fig. 1.

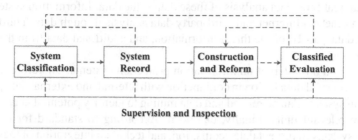

Fig. 1. The classified protection model of information system.

Taxonomy module is to card the internal information systems and information resources, the target of protection can clearly focus on. Information systems of the Executive Agent for this unit. This step is the authority of the competent departments. Report of the results of this step is the classification system. implementation is responsibility for security of information systems, involving the liability and responsibility can clear. This step is implementation of the information system of the operating unit. This step is the police unit of the competent departments. Outcome of the work of this step is information system security records.

Module is a balance of construction and reform enterprise business continuity and the national security needs of the Government. This step is the implementation of a security service provider. This step of information systems is a business unit of the competent departments. This step's work is a modification of the information system security building. In Fig. 2 classified protection of data fusion process.

Classification and evaluation of classification and evaluation modules to ensure objectivity, impartiality and security should be implemented by major rating agencies and evaluators. This step is the implementation of a testing and assessment Department. This step is the police unit of the competent departments. Results of this step are information systems test and evaluation report. Supervision and inspection of modularity is the key to implementation of hierarchical protection, supervise and inspect it every year. This step is the implementation of the public safety unit. Results of this step is to check the feedback of information security.

4   Threat Intelligence Analysis Factor Chain of Classified Protection

Static data is information system assets, evaluating data evaluation report, inspection and classification of data protection, which is the main data of the data analysis. Seepage test of dynamic data includes data, surveillance data, net monitor data, log data and site survey data, real-time data analysis of these data is the data. Information system threats from the Internet, so Internet and third-party data analysis through data. Third part data is Internet data, part III of the threat information, and additional data from these data is the data analysis.

Threats from the Internet on information protection systems. Therefore, the classification of protected data fusion in conjunction with internal and external data protection information system. Can be related and data mining to identify potential network threats and inform relevant units. Finally, can do it. According to standard for information protection systems to protect data acquisition, and technical integration of these data for analysis of data. Using regular expression method and protocol reduction, normalizing the logs and traffic data for various types of equipment. These data are normalized data elements, for analysis and storage.

These data elements associated with a database and knowledge base to complement the information, provide database for subsequent analysis. Information stored in distributed search engine and knowledge base of the platform, to provide data that is associated with fast search function. The data of classified protection are include of static, dynamic and third party data. These data should be fusion so as to effectively analysis. The data fusion process of classified protection is shown in Fig. 2.

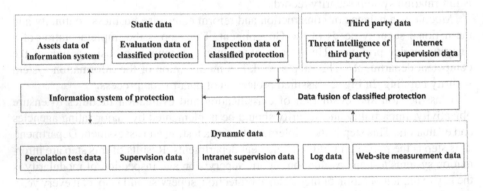

Fig. 2. The data fusion process of classified protection.

5 Threat Intelligence Situational Awareness of Classified Protection

Import collected through offline static data, dynamic data by switching the information collection, uses Web crawlers collect third party data. These data attributes, efficient retrieval and storage services. Fusion data Association analysis and clustering analysis. After analyzing all these results are stored to the database platform. In this article, Elasticsearch as the storage and retrieval of large data analysis platform, custom modifications and analysis procedures complete platforms for data analysis.

Index analysis results are stored in Elasticsearch as multiple copies of multiple blocks in a distributed system. Therefore, you can improve the performance and reliability of the data, you can ensure that. Elasticsearch memory index can be used to ensure that the last entered data can achieve near-real-time query.

The information exchange through the information exchange interface servers. Analysis Services and data mining by Hadoop and storm. All data are stored in the Elasticsearch. Finally, the results page displays the dimensional rendering of the page. Thus, situational awareness system architecture consists of 4 modules: data collection, data analysis, data integration, and data representation.

6 Conclusions

Based on data analysis and threat intelligence technology classification model of information systems, threat intelligence analysis factors chain is building intelligent situational classification and threat protection. Useful information about security threats to take the initiative from a large number of security information in the access and extraction. Consequences of threat information and analysis of effective measures in time. Tactical and strategic information systems threat intelligence available, and APT and 0 day attack can be detected, appropriate defensive measures you can select first aid response, announcements and alerts can be classified to protect, and the timely sharing of threat information. Therefore, APT and 0 day attack can effectively detect, first aid, bulletins and alerts can be completed in a timely manner.

Acknowledgment. This work is supported by National Key R&D Program of China (No. 2017YFC0803700). The authors of this paper are members of Shanghai Engineering Research Center of Intelligent Video Surveillance. This work was supported in part the National Natural Science Foundation of China under Grant 61300202, 61332018, 61403084. Our research was sponsored by Program of Science and Technology Commission of Shanghai Municipality (No. 15530701300, 15XD15202000, 16511101700), in part by the technical research program of Chinese ministry of public security (2015JSYJB26, 2015QZX002), was sponsored by CCF-Venustech Open Research Fund (Grant No. CCF-VenustechRP2017006).

References

1. Ye, J., Xu, Z., et al.: Secure outsourcing of modular exponentiations under single untrused programme model. J. Comput. Syst. Sci. https://doi.org/10.1016/j.jcss.2016.11.005
2. Yang, Y., Xu, Z., et al.: A security carving approach for AVI video based on frame size and index. Multimed. Tools Appl. https://doi.org/10.1007/s11042-016-3716-4
3. Osanaiye, O., Cai, H., Choo, R., Dehghantanha, A., Xu, Z., Dlodlo, M.: Ensemble-based multi-filter feature selection method for DDoS detection in cloud computing. EURASIP J. Wirel. Commun. Netw. **2016**, 130 (2016)
4. Information Assurance Technical Framework (IATF), V3.1, NSA (2003)
5. GB/T 22239-2008 Baseline for classified protection of information system (2008)
6. GB/T 22240-2008 Information security technology- Classification guide for classified protection of information system (2008)
7. GB/T 25058-2010 Implementation guide for classified protection of information system (2010)
8. GB/T 28448-2012 Testing and evaluation requirement for classified protection of information system (2012)

The Police Application and Developing Direction of UAV

Qianjin Tang and Zheng Xu[(✉)]

The Third Research Institute of the Ministry of Public Security, Shanghai, China
13817917970@163.com

Abstract. To improve the actual application ability of UAVs for public security, and promote the development of the UAV industry in the police market, the paper analyses the demand and application status of multi axis rotor UAV (unmanned aerial vehicle) for different job categories of public security. It is pointed out that the multi rotor UAV exists the problems in application and the development direction of police multi axis rotor UAV.

Keywords: Police unmanned aerial vehicle (UAV) · Police application
Task load · Video detection · Communications relay

1 Introduction

In recent years, developed from military applications of UAV technology spillovers, mature hardware chain, reduction in costs, drones are agriculture, electricity oil and disaster detection, forestry, meteorology, land and resources, police, marine hydraulics, surveying and mapping, urban planning and other civil industry application, its market size of explosive growth trends [1]. Application of UAV in policing activities dating back to the 2008 Beijing Shandong public security systems and the acquisition of 4 sets of UAV system, Hebei, Liaoning and other places has procured various types of police UAVs, as currently more than 150 police buy drones. Police with no machine is oriented police industry video investigation, and counter-terrorism dimension stability, and traffic monitoring, and drug smuggling, and fire rescue, and large security, and command scheduling, police works activities of mobile type professional equipment, can and video investigation car, and communications command car, platform for tie, formed space integration of reconnaissance and command scheduling system, loaded of professional equipment can meet variety specific police business of combat need [2, 3]. Police force in the country at the present stage with lack of police equipment technology content is not too high, many police status such as cooperative engagement capability is weak, and UAV applications can greatly improve the effectiveness of police.

UAV unmanned helicopters, fixed-wing UAVs and multiple-spindle rotary-wing unmanned aerial vehicles and many other types, early in this century in the public security sector to explore applications, and there are many successful cases [4, 5]. Due to more axis suspense no machine operation convenient, and can vertical landing, and hover effect good, and volume small, and price relative cheap, currently in police with

© Springer Nature Singapore Pte Ltd. 2018
J. C. Hung et al. (Eds.): FC 2017, LNEE 464, pp. 463–467, 2018.
https://doi.org/10.1007/978-981-10-7398-4_49

aviation has is will not missing of models, no helicopter and fixed wing no machine despite has its range time long, and anti-wind capacity good, and load volume big, advantages, but its cost high, and operation complex, and took off conditions requirements relative high of limitations also obviously, currently main in Xinjiang, to configuration the class models. Based only on multiple-axis rotary wing unmanned police for analysis.

2   Multi-axis Rotary Wing Unmanned Alert Application

Police axis rotor UAV system, mainly composed of UAV platforms, payload, ground stations and wireless transmission link. A model is used within Visual distance, their wireless link only UAV platforms and station data transmission between; another pattern nonvisual use, you need to transfer information to the command center, which require long distance wireless transmission link.

Payload capabilities determine the police drone of the completion of the mission and police drones the most important feature, payload of pluralism, combat, modular is the ideal objective of police drones load.

In practical applications, wireless data links are also important factors in determining applications, particularly wireless broadband transmission effect of drone war has a key role, which is related to the current application with video detection as the main mode, even some combat, material cast can get with video images, such as on-site judging.

Currently police with no machine also in to airlines took mainly of primary application stage, application form and application scene very simple, in counter-terrorism at bursting, and highway traffic monitoring, and forest fire, and emergency air defense, and ban species shovel HIV, and large activities security, and group event disposal, variety combat task in the, airlines took image can from General Shang see site situation of panorama, this on command decision has important role.

The application is a general model of UAV, but to meet the police get, or need some special requirements, vehicle parking surveillance to obtain evidence, need to drone camera with a zoom lens, clearly read vehicles' license plates.

Task loading function determines the application of police drones scenes and effects, but is limited by the load of the shaft-rotor UAV capabilities, time and other factors, in addition to the basic application model, are exploring the application of mainly in the following categories:

1. carrying non-visible light Imaging payload. Relative to the visible light, not visible light images with visible light does not have the information, there are thermal and hyperspectral camera payload. Airborne thermal imaging can search in a wide range of heat source, can be used for night-time runaways, monitoring of forest fire danger, fire fighting monitoring police activities. For multiple axes rotary wing unmanned aerial hyperspectral cameras is relatively small, mainly import-oriented. Because of the hyperspectral imager on the continuous spectrum of the same object at the same time imaging, directly reflects the spectral characteristics of the object to be observed, even the surface composition of substances and making detection capabilities significantly improve and target detection from qualitative to quantitative

analysis possible. Using airborne hyperspectral imager for drug cultivation at the investigation, resolution from similar plants of marijuana, opium plants, enhance investigation efficiency. Focuses on core collection of the tasks currently load device localization and low cost.

2. carry payload based on image analysis. Intelligent video analysis technology and processing module development, application of artificial view image mode transitions toward the intelligent, even the industry has proposed the concept of drone data [6]. Suspected based on intelligent image analysis tasks can load vehicles and personnel tracking, number of passenger flow and vehicular traffic automatic statistical analysis, highway emergency lanes occupied in violation of traffic law enforcement and forensics, these applications can greatly reduce the intensity of police activities, improve the efficiency of law enforcement.

3. equipped with grab body payload. Catch body according to the ground control officer's instructions, to complete the Agency's closing and opening, complete the pre-designed policing tasks. Now part of the public security organs are exploring applications include: toss life jackets, throw tear gas, firing tear gas, grab samples of suspicious evidence, PA warning, auxiliary hit and so on. But the limited carrying capacity of the shaft-rotor UAV, these applications often have significant limitations, its mandate is quite limited. Foreign countries and there were reports police application of UAV, such as America's next generation of police armed drones, to plant a variety of lethal or non-lethal weapons, deadly weapons are mainly used for the eradication of harmful, facts of the crime are clear of terrorists and non-lethal weapons are mainly used for Suppression of crowd disturbances. Application mode of using drones equipped with lethal weapons there may be legal issues, is unable to determine the mode of legal responsibility.

4. carrying communications relay payload. Special terrain such as mountains, often require communications erection on high relay base station, and miniaturization of base stations to load on shaft-rotor UAV, which will create a commanding height in the shortest possible time to ensure smooth communication. In view of the General multiple-spindle rotary-wing unmanned aerial vehicle battery life shorter, Taiwan have tethered rotor UAV communication relay, to ensure 24-h uninterrupted.

3   Development Direction

Enhance the security of UAV can avoid the drone crash causing property damage and injuries on the ground. Organization somewhere in the UAV field test, 12 different types of multi-spindle rotary-wing drone had half the crash, and a drone aircraft crash burn, this fully illustrates the drone of security needed strengthening. Should technically protect the drones in an abnormal state of automatic termination, automatic return of technology, enhance the UAV's automatic obstacle-avoidance, nobody has the use does not function in security incidents. Therefore, in order to enhance UAV under strong electric field and magnetic field environment reliability test.

Meanwhile, public security is a special Department, perform tasks a lot is hidden, confidentiality requirements, tasks not at liberty to reveal. In the use of unmanned aircraft carrying out their duties, must be data-link of transmission data, communications, can guarantee the security of the content.

A prominent feature of public security work is emergency, emergency, this feature requires the use of police equipment must quickly pull out quickly at the time. Many police drone of combat units and unmanned aerial vehicle accessories to achieve modular, with pluggable functions can be realized in minutes with quick installation and dismantling. And fast way to goofy is the trend. If 4 drones similar to the structure of the rotor shaft only need simple two actions can be retracted, particularly suited to outdoor temporary jobs expanded, for contracted State, which converted just push a button on the shaft.

Enhance the UAV's battery life is an eternal topic, battery life and short range is restricted worldwide problem of rotor UAV applications. To increase the capacity of the lithium-ion battery tends to lead to increased weight and shorten the time only balance battery capacity and life time. Therefore, the new development direction of energy supply is currently developing hydrogen fuel cells, solar cells and other new ways of energy supply.

Analysis of UAV mission load, you should first operational mission analysis of drone, combined with a specific policing activities, development meet the actual requirements of the payload, payload of intelligent, integrated, information system and the Exchange.

Intelligent application of UAV is based on police activities, enhance the value of payload data, technical realization and application requirements from the current perspective, with special emphasis on depth of UAV image application, includes the image of track and field measurement and reconstruction, site information, image recognition.

4   Conclusions

Police use UAV is not simply an unmanned aerial vehicle hardware, it is a whole system, is an important part of the whole business system. Police with no machine of application, from surface see is no machine platform, and task load in works, but this behind, actually covers has at police programme developed, and no machine and task load control, and police case thought with of, and police command and scheduling, and emergency plans prepared and implementation,, these related system of convergence and tie, formed has a perfect of, and Qian joint Hou pass of communications command control system, research no alert works application must and police business system height fusion. No machine in technology increasingly mature of today, how more in-depth application to police works work of all field, except no machine equipment itself, more key is combined the police itself business, depth research, and custom no alert works combat application solution programme, development supporting software, and application system, and not visited with looks, has a "showy", only down research technology, to will police with no machine of role play of better.

Acknowledgment. This work is supported by National Key R&D Program of China (No. 2017YFC0803700). The authors of this paper are members of Shanghai Engineering Research Center of Intelligent Video Surveillance. This work was supported in part the National Natural Science Foundation of China under Grant 61300202, 61332018, 61403084. Our research was sponsored by Program of Science and Technology Commission of Shanghai Municipality (No. 15530701300, 15XD15202000, 16511101700), in part by the technical research program of Chinese ministry of public security (2015JSYJB26, 2015QZX002), was sponsored by CCF-Venustech Open Research Fund (Grant No. CCF-VenustechRP2017006).

References

1. Xu, Z., Mei, L., Hu, C., Liu, Y.: The big data analytics and applications of the surveillance system using video structured description technology. Cluster Comput. **19**(3), 1283–1292 (2016)
2. Xu, Z., Mei, L., Liu, Y., Hu, C., Chen, L.: Semantic enhanced cloud environment for surveillance data management using video structural description. Computing **98**(1–2), 35–54 (2016)
3. Xu, Z., Hu, C., Mei, L.: Video structured description technology based intelligence analysis of surveillance videos for public security applications. Multimed. Tools Appl. **75**(19), 12155–12172 (2016)
4. Hu, C., Xu, Z., Liu, Y., Mei, L.: Video structural description technology for the new generation video surveillance systems. Front. Comput. Sci. **9**(6), 980–989 (2015)
5. Xu, Z., Liu, Y., Mei, L., Hu, C., Chen, L.: Semantic based representing and organizing surveillance big data using video structural description technology. J. Syst. Softw. **102**, 217–225 (2015)
6. Xu, Z., Zhi, F., Liang, C., Mei, L., Luo, X.: Generating semantic annotation of video for organizing and searching traffic resources. IJCINI **8**(1), 51–66 (2014)

The Study on Verification Systems of Face and ID Card for Public Security

Zhiguo Yan and Zheng Xu^(✉)

The Third Research Institute of the Ministry of Public Security,
Shanghai 201024, China
13817917970@163.com

Abstract. The integration system of face and ID card is used to detect person in the important sections such as government place and road. In order to do the effective security of the important region, in this paper, we propose the integration system of face and ID card. The basic framework, modules, and case system are given.

Keywords: Integration system of face and ID card · Public security
Face detection

1 Introduction

Witnesses in one system for core district, the key government departments, transportation distribution of real time video streaming sites, such as key personnel have been supervised. According to the actual demand dispatched to key personnel in accordance with the specific requirements for site security, the selective libraries, and storage operations to improve performs real-time response time, reduce the rate of false alarms, improve system efficiency, effective early warning services practice. Mainly consists of the following subsystems. Institute of Ministry of public security, the third image application based on video surveillance has made a lot of work [1–7].

(1) Channel witness verification, videos have been supervised system

The system is deployed in an important place in channel-style surroundings, can be integrated with baggage security screeners. System has a witness who saw the combination of authentication and video performs two functions, you can switch function according to the actual situation:

Passing when staff carry identity cards, buffet swiping at the right-hand side of ID card readers, card e-photos and field surveillance photos meet the identity requirements are released, otherwise the system alarm;

Passing people without identity documents when passage by moving the right hand side button dynamically dispatched a function, system monitoring current staff photos and backstage photos of key personnel, library staff, warning, or release.

(2) Identity image information collection system

The systems deployed in channel-scene, scene layout gate, arrangement of gates in front of multiple cameras and turnstiles integrated ID card reader. Personnel passage, buffet swiping pass card signal triggers the camera to capture, can be captured from

J. C. Hung et al. (Eds.): FC 2017, LNEE 464, pp. 468–473, 2018.
https://doi.org/10.1007/978-981-10-7398-4_50

different angles of the portrait, and identity formation binding electronic photos and other identifying information. The multi-angle scene photograph can dramatically restore the actual passage of personnel under the surveillance photos, with lighting, angles, partial occlusion, variability, can provide high quality training for deep learning systems, and helped to establish the actual science fair of face recognition technology evaluation standards.

2 Basic Description

Based on the above requirements, the project consists of the following equipment and application systems (platform):

Traffic distribution center is an important place for ticket information, equipment and system application retention module interface, in a subsequent project through a secure gateway and the Ministry of public security information system for key library as well as ticketing, field photos, status information, ticketing information and associated intelligent real-time collision.

Independent you want to test devices are placed in the hall outside of customs clearance import channels, camera and card reader is installed on the device, device management service system platform with key staff and ticket information, identity information and ticketing information real time identification can be made. For autonomous customs clearance (first gate), which links-card validation. According to the waiting Hall at an outdoor space.

Layout portrait collection at the luggage security identification system as a second pass, in strict compliance with these standards after the implementation is complete, portrait than the system can capture clear facial features.

3 Technology Implementation Plan

From the functional point of view, face authentication system is an authentication technology based on Terminal cameras, its facial recognition system, as shown in the flowchart. It through a Terminal currently used by the camera video interception of facial image, and uploaded to the server through the HTTP framework Protocol and storage photos were compared, confirms the user's identification. When the user enters the login screen, the terminal will automatically incoming server users through the camera image of face recognition systems, at the same time, user login ID will be returned back, face recognition system based on a user ID called server-side user storage pictures and facial feature analysis and identification of the current picture, determine whether the user is legitimate, so as to release or deny to the user logon request.

Enter two face photos of face authentication system, given the similarity of the two men face, such as the similarity is greater than a certain threshold, the system considers the two men belong to the same person, or think it's two people belonging to different people. Witness verification of specific applications, categorized under face authentication technologies, than on photo card store is the distribution of electronic photos and field monitoring portraits.

To witness the validation system, traffic officers when independent credit card, electronic photos and ID number in the ID card will be read automatically and site surveillance camera captures facial images compared. Due to the credit card you can get the ID number of pedestrians (ID) and identification of electronic photos offline, system without logged on to the remote server to get the current Portrait Gallery photos.

Use face recognition technology and second-generation card, we can technically solve two major problems: using second-generation card chip reading technologies solve the proliferation of fake ID, registration unit is unable to verify the authenticity of identity problems using face recognition technology to solve an impostor, witnesses, and registration units difficult problems.

In actual application in the, consider to ID electronic photos of resolution only 102*126, to improve II generation card chip read photos and site camera collection photos of validation performance, we specifically optimization has people face validation algorithm, through used people face Super resolution reconstruction technology, on II generation card chip photos for Super resolution reconstruction, using reconstruction Hou of people face image for people face recognition, experiment showed that validation performance more existing algorithm has larger upgrade.

Will automatically import the registration module to the management platform, after the approval of the administrator to verify a user's identity, and users clear positive portrait photo into database and electronic ID card, when used in important places in and out of the portrait and access (white list). Similarly, portrait collection to key personnel, and dispatched the emergence of key personnel, namely for blacklisting mechanism. System operator via Terminal front face camera capture card access personnel clear photos to upload to the server, and can also store local staff ID photos to upload (requires photo pixel is greater than 400*400, hat front view). In order to improve the success rate of logging on to the system, like validation, automatic face recognition module using multi-features fusion algorithm, the system supports many of the same people positive clear photo upload.

Registration can be reported as finished portrait face feature extraction of data modeling, and build facial template (facial features file) Save to the database. In the face when searching (search), adds the specified model a portrait, and then compared it with all templates in the database to identify and ultimately will be based on the list of people most similar to similar values listed.

We use a deep learning architecture, scenes are in the large amount of data under the training support through DL model scene, not the same face image mapping to the same image, will feature maps to the class differences between larger, class differences within a smaller space. In real-world scenarios on the surveillance video testing algorithm for variable lighting and angles, with occlusion, fuzzy, age range and other complications is very robust.

Human face identification verification (authentication) and two kinds of searches than the mode. Verification is to specify the capture portrait or portrait and registered in a database than to check to determine whether they are the same person, also known as the "1:1 portrait". Search the comparison refers to, from the database of registered owner search as looking for a specified figures exist, also known as the "1:N man", where n is the number of all inbound pictures.

To 1:1 figures involved in the application of the system validation and 1:N portrait, Terminal cameras capture portrait than an identity card electrical and photo portrait of belonging to 1:1... Borrowed, stolen identity cases, the cameras capture the portraits and background like 1:N than to key personnel to determine whether the current Internet users belong to a focus on people.

General schema for face authentication service into the access layer and the service layer. Among them, the access layer provides services based on HTTP protocol interface, main access user requests, login, assigning features such as application servers, returns the result; services provide specific identification services, complete simple computing tasks, and return the results to the access layer, the current pass user release or rejection.

More portraits than support from the core algorithms of races, different age groups, different environmental conditions of image data, support the identification of images in multiple formats rather than on. Face detection based on manifold, local Gabor binary pattern (LGBP, Local Gabor Binary Pattern), Ada Gabor features and LGBP characteristics pose estimation techniques, supporting facial chunking of attitude adjustment, automatic/manual, hand-set eyes, mouth position, support single image composite template than a peer.

In order to map the two photos to compare in the same space, different depths based on neural network, we propose a double layer depth of heterogeneous neural network model. In this model each layer is a deep Web (respectively two pictures as inputs), while training in the use of dichotomous loss functions and the corresponding weight differences in the two network regularization can realize different spatial mapping to the same features. In the feature space and identity face images of class differences within the same small differences between different face image class larger, enhances the characteristics of discrimination.

Using sophisticated multi-tier architecture, make the whole platform has the advantages of stable, safe and expandable. Platform using distributed architecture, making full use of hardware resources, and good scalability.

Using multiple machines, multiple threads in parallel, partitioning the bulk model, fast and efficient using advanced facial feature extraction technology, adapt to a variety of photo source, storage should not less than 99.95%; supports automatic models, handmade model, batch input, than it is when the input in different ways of modeling.

Use a 64-bit platform technology, efficient use of memory than all the physical memory load on the template, reducing the number of access to the database or template files, improve the speed and shorten the response time. Meanwhile, the distributed architecture, which enables the system to an increase in unit, close to increase linearly platform than on performance.

Optimizing strategy of dynamic templates loaded, ensure normal operation of the platform under the premise, new photo dynamic storage, and can immediately participate in comparison.

Automatic data import and automatic modeling function, automatic data updates and update time can be set in accordance with the needs and conditions of updates, updates to multiple data sources, and data import, the remote model.

Platform to achieve synchronization than the recognition, than alarm, asynchronous, asynchronous recognition than different than on models such as alarms, and provide means of technical realization and protection of related patterns. Manager, system support 1:N and 1:1 different than the way additional filter criteria of classification than the right, improve speed and efficiency. Can increase the age, gender, region multiple constraints, you can set the threshold value, the combined ratio compared to the library.

4 Structure and Function Modules

Using b/s structure in the framework of the platform, multi-tier, distributed mechanism for better load balancing. The whole platform has the advantages of stable, secure, extensible.

Larger than the storage capacity, compute-intensive features, platform using distributed alignment, full use of hardware resources to calculate than the speed boost, and the architecture is easier to extend, according to actual needs, by adding hardware to finish the upgrade of the performance.

The platform can be used independently realized portrait collection online business process based on networks, portrait, like validation, business statistics, business requirements, implementing online faces data validation in real time core functionality, also available through Web Services methods for existing e-learning systems and other business systems to provide real-time or asynchronous services.

The first database design capacity of the platform can support up to 100,000 people, and supports the mega-like extensions, in the case of hardware supports the capability to meet scale to tens. Support for trainees, believers, monks, Shi Ku amounts to more than 30 business library, library capacity designed to meet the basic needs in the next five years.

Performs the actual need for real-time video stream, up to 100 access to front-facing camera is designed to support, supports 2000 users online at the same time.

5 Conclusions

Witness systems is an important supporting technology to ensure public safety. This paper describes the basic framework of unity of witness systems, modules, schemas, and so on. And the practical case.

Acknowledgment. This work is supported by National Key R&D Program of China (No. 2017YFC0803700). The authors of this paper are members of Shanghai Engineering Research Center of Intelligent Video Surveillance. This work was supported in part the National Natural Science Foundation of China under Grant 61300202, 61332018, 61403084. Our research was sponsored by Program of Science and Technology Commission of Shanghai Municipality (No. 15530701300, 15XD15202000, 16511101700), in part by the technical research program of Chinese ministry of public security (2015JSYJB26, 2015QZX002), was sponsored by CCF-Venustech Open Research Fund (Grant No. CCF-VenustechRP2017006).

References

1. Xu, Z., Mei, L., Liu, Y., Hu, C., Chen, L.: Semantic enhanced cloud environment for surveillance data management using video structural description. Computing **98**(1–2), 35–54 (2016)
2. Hu, C., Xu, Z., Liu, Y., Mei, L.: Video structural description technology for the new generation video surveillance systems. Front. Comput. Sci. **9**(6), 980–989 (2015)
3. Xu, Z., Liu, Y., Mei, L., Hu, C., Chen, L.: Semantic based representing and organizing surveillance big data using video structural description technology. J. Syst. Softw. **102**, 217–225 (2015)
4. Hu, C., Bai, X., Qi, L., Chen, P., Xue, G., Mei, L.: Vehicle color recognition with spatial pyramid deep learning. IEEE Trans. Intell. Transp. Syst. **16**(5), 2925–2934 (2015)
5. Hu, C., Bai, X., Qi, L., Wang, X., Xue, G., Mei, L.: Learning discriminative pattern for real-time car brand recognition. IEEE Trans. Intell. Transp. Syst. **16**(6), 3170–3181 (2015)
6. Hu, C., Xu, Z., Liu, Y., Mei, L., Chen, L., Luo, X.: Semantic link network-based model for organizing multimedia big data. IEEE Trans. Emerg. Topics Comput. **2**(3), 376–387 (2014)
7. Dai, J., Zhao, Y., Liu, Y., Qi, L., Hu, C.: Cloud-assisted analysis for energy efficiency in intelligent video systems. J. Supercomputing **70**(3), 1345–1364 (2014)

Erratum to: Frontier Computing

Jason C. Hung, Neil Y. Yen, and Lin Hui

Erratum to:
J. C. Hung et al. (Eds.): Frontier Computing, LNEE 464,
https://doi.org/10.1007/978-981-10-7398-4

The original version of the book was inadvertently published without the affiliation "School of Computer Science and Engineering, University of Aizu, Aizu-Wakamatsu, Japan" of second editor "Neil Y. Yen" in frontmatter, which has to be now included. The erratum book has been updated with the change.

The updated online version of this book can be found at
https://doi.org/10.1007/978-981-10-7398-4

© Springer Nature Singapore Pte Ltd. 2018
J. C. Hung et al. (Eds.): FC 2017, LNEE 464, p. E1, 2018.
https://doi.org/10.1007/978-981-10-7398-4_51

Author Index

© Springer Nature Singapore Pte Ltd. 2018
J. C. Hung et al. (Eds.): FC 2017, LNEE 464, pp. 475–477, 2018.
https://doi.org/10.1007/978-981-10-7398-4

Printed in the United States
By Bookmasters